# Springer Tracts in Mechanical Engineering

Springer Tracts in Mechanical Engineering (STME) publishes the latest developments in Mechanical Engineering - quickly, informally and with high quality. The intent is to cover all the main branches of mechanical engineering, both theoretical and applied, including:

- Engineering Design
- Machinery and Machine Elements
- Mechanical Structures and Stress Analysis
- Automotive Engineering
- Engine Technology
- Aerospace Technology and Astronautics
- Nanotechnology and Microengineering
- Control, Robotics, Mechatronics
- MEMS
- Theoretical and Applied Mechanics
- Dynamical Systems, Control
- Fluids Mechanics
- Engineering Thermodynamics, Heat and Mass Transfer
- Manufacturing
- Precision Engineering, Instrumentation, Measurement
- Materials Engineering
- Tribology and Surface Technology

Within the scope of the series are monographs, professional books or graduate textbooks, edited volumes as well as outstanding PhD theses and books purposely devoted to support education in mechanical engineering at graduate and post-graduate levels.

**Indexed by SCOPUS. The books of the series are submitted for indexing to Web of Science.**

Please check our Lecture Notes in Mechanical Engineering at http://www.springer.com/series/11236 if you are interested in conference proceedings.

To submit a proposal or for further inquiries, please contact the Springer Editor **in your country**:

Dr. Mengchu Huang (China)
Email: mengchu.Huang@springer.com
Priya Vyas (India)
Email: priya.vyas@springer.com
Dr. Leontina Di Cecco (All other countries)
Email: leontina.dicecco@springer.com

More information about this series at http://www.springer.com/series/11693

Chao Jiang · Xu Han · Huichao Xie

# Nonlinear Interval Optimization for Uncertain Problems

Springer

Chao Jiang
College of Mechanical and Vehicle Engineering
Hunan University
Changsha, China

Xu Han
Hebei University of Technology
Hongqiao District
Tianjin, China

Huichao Xie
College of Material Science and Engineering
Central South University of Forestry and Technology
Changsha, China

ISSN 2195-9862 ISSN 2195-9870 (electronic)
Springer Tracts in Mechanical Engineering
ISBN 978-981-15-8545-6 ISBN 978-981-15-8546-3 (eBook)
https://doi.org/10.1007/978-981-15-8546-3

Translation from the Chinese language edition: 区间不确定性优化设计理论与方法, © Science Press 2017. Published by Science Press. All Rights Reserved



This Springer imprint is published by the registered company Springer Nature Singapore Pte Ltd.
The registered company address is: 152 Beach Road, #21-01/04 Gateway East, Singapore 189721, Singapore

# Contents

# Abbreviations

| | |
|---|---|
| AAO | All-at-once approach |
| ANN | Artificial neutral network |
| AR | Augmented reality |
| BLISS | Bi-level integrated system synthesis approach |
| BP | Back propagation |
| CBF | Constant binder force |
| CO | Collaborative optimization approach |
| CPU | Central processing unit |
| CSSO | Concurrent subspace optimization approach |
| DE | Differential evolution |
| DOE | Design of experiments |
| FEM | Finite element method |
| FFC | Femur force criteria |
| FORM | First-order reliability method |
| GA | Genetic algorithm |
| HBF | High binder force |
| HIC | Head injury criterion |
| IDE | Interval differential evolution |
| IDF | Individual disciplinary feasible approach |
| IEEE | Institute of Electrical and Electronics Engineers |
| IP-GA | Intergeneration projection genetic algorithm |
| KTT | Karush-Kuhn-Tucker |
| LBF | Low binder force |
| LHD | Latin hypercube design |
| MDF | Multidisciplinary feasible approach |
| MDO | Multidisciplinary design optimizaiton |
| NSGA-II | Improved non-dominated sorting genetic algorithm |
| P-ICR | Preference-based interval comparison relation |
| RBF | Radial basis functions |
| RPDI | Reliability-based possibility degree of interval number |
| SLP | Sequential linear programming |
| SQP | Sequential quadratic programming |

| | |
|---|---|
| V-ICR | Value-based interval comparison relation |
| WIC | Weighted injury criterion |
| μGA | Micro-genetic algorithm |
| μMOGA | Micro multi-objective genetic algorithm |

# Chapter 1
# Introduction

**Abstract** This chapter introduces the engineering background and research significance of uncertain optimization and analyzes the research status of several mainstream uncertain optimization methods, in which the research status and main technical problems of interval optimization method are emphasized. Finally, this chapter gives the framework for this book.

## 1.1 The Research Significance of Uncertain Optimization

Optimization is the method used to select the best one out of a variety of decisions (including finite or infinite), which can bring remarkable improvement for structural or system performance. Therefore, optimization is widely applied in many fields such as industry, agriculture, national defense, transportation, etc. Traditional analysis and design optimization are generally based on deterministic systematic parameters and optimization models, and solved by traditional deterministic optimization methods [1–6]. However, errors or uncertainties related to material properties, geometric characteristics, boundary conditions, initial conditions, measuring deviation, assembly deviation, etc., are inevitable in many practical engineering problems. Although such errors or uncertainties tend to be small in most situations, the coupling effect may cause a significant deviation on the systematic response, which will consequently influence the system performance, even lead to failure. Here are some practical engineering problems involving uncertainties [7–9]:

(1) When carrying out dynamic analysis for gears, the meshing error, including pitch deviation, tooth shape deviation, and transmission error, etc., which are mainly caused by the structural complexity of the gear, tooth deformation, manufacturing error, and assembly error, is one of the major dynamic excitations in gear meshing [10]. Besides, the gear stiffness will also vary in different mesh positions. Therefore, it is difficult to precisely describe these parameters using deterministic values [11].
(2) The integrated dynamic model of a complex machine tool assembly or system is hard to establish in physical coordinate due to the difficulty to accurately establish the dynamic model of machine tool joint surface. When the machine



C. Jiang et al., *Nonlinear Interval Optimization for Uncertain Problems*, Springer Tracts in Mechanical Engineering, https://doi.org/10.1007/978-981-15-8546-3_1

tool is operating, the contact state, gap state, attachment state, and the slide state between different components will dramatically influence the stiffness and damping characteristics of the structure, leading to high uncertainties in the system. Besides, the external disturbance will also influence even entirely change the contact state and slide state of the joint, thus change the stiffness and damping characteristics of the structure [12–14].

(3) For the nuclear power plant structure, high uncertainties exist in the structural parameters of the prestressed concrete containment and the reinforced concrete shear wall, such as material strength, structural stiffness, damping, and spectral value, peak acceleration, lasting time of the seismic force [15].

(4) The characteristics of the vehicle vertical side impact can be summarized as follows: after the collision, the motions of two vehicles are in the same quadrant. The vehicles will rotate by different directions and then collide at least one more time [16]. However, the identification of speed and the accident analysis is challenging due to the uncertain parameters in the accident [17], such as the vehicle load, centroidal displacement after collision, and friction coefficient of the pavement.

(5) When carrying out reliability analysis for a liquid rocket engine, the internal uncertainties such as the varying pipeline pressure and turbine efficiency must be taken into consideration. These uncertainties are often risen by the parts installation errors, the measuring errors in liquid flow experiments, and the use of statistical data [18].

The major causes of uncertainties in practical structures and systems can generally be summarized as follows: the structural manufacturing errors and installation errors; the calculation and measuring errors on parameters; the varying parameters in different working conditions such as external loads that cannot be accurately measured; the mismatching between the theoretical model and physical model. Uncertainties, more or less, exist in most engineering problems. However, the simplification has to be adopted in most situations due to mathematical difficulty and inconvenience. For example, the multiple uncertainties can be simplified into single uncertainty and the uncertain problem into a deterministic one. Dialectically, certainty is relative, while uncertainty is absolute. Classical optimization theories and methods are incompetent for the design optimization of the structures or systems that involve uncertainties. Therefore, we should resort to the uncertain optimization or uncertain programming methods that can fully consider the effect of uncertainties in the solution procedure to model and solve the problems. Uncertain optimization is an extension of the classical optimization theory, in which too many simplifications and assumptions are not required, and thus it enables us to construct a more objective and accurate optimization model and consequently obtain a more reliable design.

The research on uncertain optimization has great significance in both theory and engineering. In the past, more than half a century, the uncertain optimization theory and methods have been widely studied, and have attracted more and more attention. Currently, the uncertain optimization has become an important research direction of advanced design theory and related fields. Many uncertain optimization methods have been successfully applied in different actual engineering fields, such as production planning [19–22], network optimization [23], vehicle scheduling [24–26], energy [27, 28], facility location [29, 30], and structural optimization [31–35], etc. The researches on these problems have demonstrated the effectiveness of the uncertain optimization theory in practical application and also have revealed a lot of engineering application prospect of uncertain optimization methods.

## 1.2 Stochastic Programming and Fuzzy Programming

The research on uncertain optimization problems originated from the studies of Bellman and Zadeh [36, 37] and Charnes and Cooper [38], in the middle of the last century. In traditional mathematical programming, the parameters in the optimization model are generally assumed as deterministic. However, such assumptions may bring large modeling errors when uncertainties are actually involved. Current mathematical programming theory often uses stochastic analysis method or fuzziness analysis method to describe the uncertain or imprecise parameters in practical problems. Two uncertain optimization theories, stochastic programming and fuzzy programming, therefore, are developed according to the way to describe the uncertainty. In stochastic programming, uncertain parameters are viewed as random variables whose probability distributions are assumed to be known exactly. In fuzzy programming, uncertain parameters are viewed as fuzzy sets whose fuzzy membership functions are assumed to be known.

### *1.2.1 Stochastic Programming*

Stochastic programming originated in the 1950s and has gradually developed with the development and application of the linear programming theory and nonlinear programming theory. Dantzig [39, 40] and Beale [41], the founders of linear programming theory, first raised the issue of stochastic programming. They applied linear programming to the flight schedule and proposed the compensated two-stage optimization problem considering the randomness of the passenger flow. Then, Wets et al. [42, 43], systematically studied this kind of problems. Charnes et al. [44] firstly proposed the probabilistic constrained programming model, which is also called the chance-constrained programming, and applied it to the refinery production and storage problem. Afterwards, Borell [45], Prekopa and Dempster [46], found the important relation between the convexity of the feasible solution set and the quasiconcavity of the probability measure in the stochastic programming problem. During the 1960s and 1970s, the theoretical research and applications of the stochastic programming achieved a remarkable progress, such as Markowitz mean-variance analysis approach [47, 48], Dupacova's penalization model [49], and Neumann-Morgenstern utility model [50–52]. Besides, Garstka [53] and Ziemba [54], applied the stochastic programming to the economic equilibrium analysis, finance risk measurement, and obtained a lot of significant conclusions. In recent years, the stochastic programming has been further developed in both theory and application. There arise many research results, such as stochastic linear programming [55–57], stochastic integral linear programming [58–60], stochastic nonlinear programming [61–64], robust stochastic programming [65–67], reliability-based design optimization [68–76]. And these stochastic programming methods can be classified into two categories according to where the random variables appear:

(1) Random variables exist in the objective function
There are two major models for this case: E-model and P-model. In E-model, the uncertain optimization is transformed into a deterministic optimization problem through optimizing the expectation of the objective function. In P-model, the uncertain optimization is transformed into a deterministic optimization problem through maximizing the probability that the objective function is higher or lower than a given value.
(2) Random variables exist in the constraints
In practical problems, there are two ways to deal with the randomness in the programming problems. The first is to solve the programming problem after the realization of the random variables is observed. The second is to make decisions according to the previous experiences before the realization of the random variables. However, if the second strategy is adopted, the decision could be verified to be infeasible after the realization of the random variables. In such a situation, how to deal with the problem will lead to different programming models. As shown in Fig. 1.1, when the random variables exist in the constraints, the stochastic programming problems can be classified into three categories according to the strategies to deal with the random variables: distribution problem, two stages (multi-stage) stochastic programming with compensation, and chance-constrained programming. For the distribution problem, after

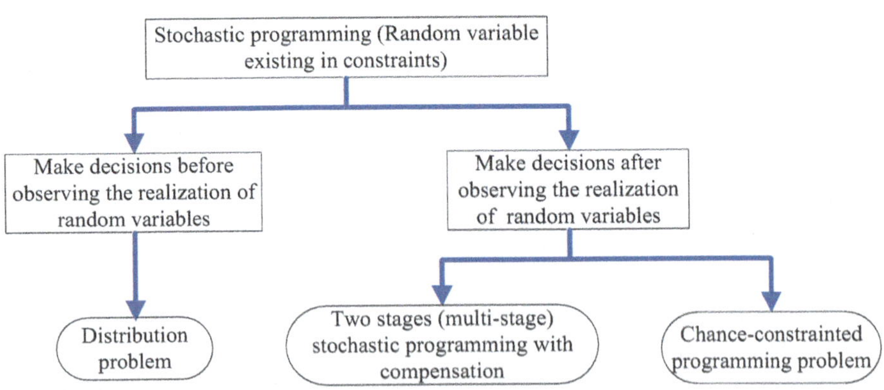


**Fig. 1.1** Three stochastic programming methods with random variables existing in constraints [77]

the realization of random variables, these random variables will become deterministic values, and consequently, the original problem is converted into a deterministic programming problem. Naturally, different observed values will lead to different deterministic programming problems and different optimal values. Therefore, for this type of problems, we should not only solve the deterministic programming problems but also obtain the probability distribution of the optimal values. Chance-constrained programming allows the violation of the constraints to some extent, while the probability that the design is feasible should be higher than a given confidence level. Similar to the chance-constrained programming, the two-stage stochastic programming with compensation makes the decision before the realization of the random variables. However, when the constraints are violated, the penalization will be adopted (introduce the compensation to satisfy the constraints). There are some relationships between these three types of stochastic programming problems, and they could be converted into each other [77].

### *1.2.2 Fuzzy Programming*

There is a significant difference between the fuzzy programming and stochastic programming in their ways of uncertain parameters modeling. In stochastic programming, uncertain parameters are described by discrete or continuous probability distribution functions. In fuzzy programming, the uncertain parameters are treated as fuzzy numbers, and the constraints are treated as fuzzy sets. The satisfaction degree of the constraints is described by a fuzzy membership function, and dissatisfaction of the constraints to a certain extent is allowed.

At present, the fuzzy programming has undergone considerable development in both theory and application. Since Bellman and Zadeh [37], proposed the fuzzy decision, many scholars have put forward a series of different methods for practical problems. Some frequently-used methods were introduced in [37, 78]. According to Inuiguchi and Ramik [79], the fuzzy programming methods can be classified into three categories as follows:

(1) Fuzzy programming with allowance
This type of method was firstly proposed in [37], and was used to deal with those decision problems with a fuzzy objective and fuzzy constraints. It was then improved by Zimmermann et al. [80].
(2) Fuzzy programming with uncertainties in objective and constraints
This type of method focuses on the uncertain coefficients in objective function and constraints, rather than the fuzzy objective and constraints. Dubois and Prade [81] firstly provided the solution method for systems of linear fuzzy constraints. Over the years, many different methods [82–85], were proposed to solve the linear programming problems with fuzzy coefficients.

(3) Fuzzy programming with allowance and uncertainties
Luhandjula et al. [86] introduced the target value into objective function and constraints with fuzzy coefficients. Inuiguchi et al. [87] further developed this kind of method based on the possibility theory.

According to the characteristics of the objective function and constraints, the fuzzy programming can also be divided into linear programming and nonlinear programming [88]. Since the fuzzy programming was proposed, most researches are confined to linear programming problems. Through constructing different equivalent models, the fuzzy linear programming model can be converted into the deterministic model, and thus can be solved by traditional mathematical programming methods. The research on this filed has been reaching maturity. However, it is still hard to find an effective solution method for fuzzy nonlinear programming problems due to the complexity of the objective function and constraints, as well as the irregularity of the feasible region. Although a series of research achievements have been published [89–95], this research direction is still at its primary stage.

### *1.2.3 Troubles and Difficulties in Stochastic Programming and Fuzzy Programming*

Up to now, the research on stochastic programming has achieved abundant results, which are widely applied to practical engineering problems. Nevertheless, there still exist inadequacies and difficulties in this field as follows [96]:

(1) Precise probability distributions of the uncertain parameters, which generally require a large number of samples to construct, are necessary for stochastic programming. However, limited by measuring technique, economy or other reasons, enough sample information is hard to obtain in many practical engineering problems. Therefore, when solving the practical stochastic programming problems, engineers often make assumptions or approximations about the distribution types or distribution parameters of random variables. However, some researches indicate that a slight error in probability distribution may lead to large deviation in uncertainty analysis results [97].
(2) Not all the nonlinear programming algorithms can be used to efficiently solve the stochastic programming problems (applicable algorithms include barrier function method and supporting hyperplane method, etc.). Besides, most of the solvable problems are limited to the linear constrained programming problems, or the problems in which each random variable takes only a finite number of discrete values. The main reason for the difficulty of solving is the huge computational requirements. When calculating the function values or gradients of the constraints at the iteration point, multidimensional integral with respect to the decision variables is needed, which generally brings vast computational amounts. Therefore, it is necessary to develop some more efficient

algorithms. Monte Carlo method is often used to calculate this multidimensional integral, i.e., the probability that a multidimensional vector falls into a given region. However, the conventional Monte Carlo method is extremely computationally expensive. To make this method feasible, many special techniques and approaches that can reduce the variance of estimations should be employed [97]. For some more complex constraints, we can only resort to some approximate methods [98, 99].

In fuzzy programming, we often use fuzzy membership function to describe the satisfaction degree of the constraints, the expectation level of the objective function, and the variation of the imprecise coefficients. When making decisions on the fuzzy problems, we often treat the fuzzy objective and constraints equally by obtaining the intersection of their fuzzy sets, and then take the decision with maximum membership as the optimal decision. The whole solution procedure is based on the precise fuzzy membership functions of uncertain parameters. However, in practical application, the fuzzy membership function is often determined by limited data or the decision maker's experience, which may bring in significant error. Essentially, stochastic programming and fuzzy programming are both probabilistically based, where the stochastic programming adopts objective probability, while the fuzzy programming adopts subjective probability [100]. Therefore, enough information is generally needed for both stochastic programming and fuzzy programming. Unfortunately, obtaining enough information on uncertainties is often difficult or expensive in practical engineering problems, which thus limits the application of the two kinds of methods.

## 1.3 Uncertain Optimization Based on Non-probabilistic Modeling

Obtaining the accurate probability distributions or fuzzy membership functions of the uncertain parameters are often difficult in practical engineering problems. However, it is relatively easy to obtain the possible variation range or bounds of the uncertain parameters, and the sample information needed is much less. Therefore, many scholars are committed to developing the non-probabilistic uncertainty modeling methods based on boundary characterization [97, 101], and correspondingly a class of non-probabilistic uncertain optimization methods are proposed, which can help solve the problems that the traditional stochastic programming and fuzzy programming are incompetent to solve. At present, there are two major types of such methods. The first is the optimization method based on convex model considering only the worst case, which is called convex model optimization in this paper for convenience. The second one is the so-called interval optimization method based on interval analysis theory. Despite some similarities between these two kinds of methods, the solution strategies of them are actually based on different frameworks:

(1) The convex model optimization generally uses a convex set such as ellipsoid to describe the boundary of the uncertainty domain of the involved multidimensional uncertain parameters. However, interval optimization only uses an interval to describe the uncertainty of each uncertain parameter, and hence the uncertainty domain of parameters is a multidimensional cuboid. Therefore, in terms of the way to model uncertainty, the interval optimization is a special case of the convex model optimization since the multidimensional interval set is a kind of convex set.

(2) For the treatment of uncertain optimization functions, the convex model optimization is generally based on the worst-case strategy, i.e., it only considers the worst case of objective function and constraints under uncertainty, which thus leaves less space for decision makers to participate and control the optimization. On the contrary, the interval optimization method generally uses a mathematical model to quantitatively describe the possibility that the constraints will be satisfied under uncertainties and adopts multiple criteria to ensure the performance of the objective function, which makes it more flexible than the convex model optimization. The interval optimization allows the decision makers to control the optimization model more flexibly based on their experiences and preferences. Therefore, there is a bigger decision space by using interval optimization. According to the way to deal with the optimization model, the convex model optimization can be regarded as a special case of the interval optimization, and this is the biggest difference between these two kinds of methods.

(3) Generally, both convex model optimization and interval optimization need to convert the uncertain optimization problem into a deterministic optimization problem, which is generally a double-loop nested optimization. In the convex model optimization, the nested optimization only involves one bound of the objective function and each constraint. However, for the interval optimization, both upper and lower bounds of the objective function and constraints will be involved, and the obtained nested optimization is usually a multi-objective optimization problem. Therefore, the numerical solving of the interval optimization seems harder than that of the convex model optimization.

(4) Regarding the research contents, the research on the convex model optimization mainly focuses on the structural problems, and it is usually combined with the finite element method (FEM) to construct corresponding optimization algorithms. On the contrary, presently the research on the interval optimization mainly focuses on the mathematical programming theory, and the research objects are usually the problems with explicit functions.

Hereafter, the basic conceptions, solution methods, and research status of the convex model optimization and the interval optimization will be summarized.

### 1.3.1 Convex Model Optimization

In the 1990s, there had been many important studies on the theory and applications of convex model uncertainty analysis. By using the convex model approach, we first need to construct the uncertainty domain of the uncertain parameters using a convex set model. At present, the dominating convex set models that are applied in engineering include [97]: ① uniform bound convex model; ② ellipsoidal-bound convex model; ③ envelope-bound convex model; ④ instantaneous energy-bound convex model; ⑤ cumulative energy-bound convex model. References [102] and [103], proposed other expressions of the convex sets to describe the geometric imperfection and dynamic loading. In recent years, there have also been a series of new convex models to deal with more complex uncertainties, such as multi-ellipsoid convex model [104], multidimensional parallelepiped model [105], super ellipsoid model [106], convex model process [107], etc.

The first application of convex model theory to uncertain optimization can date back to 1994. Elishakoff et al. [31] applied the worst-case strategy to the design of a structure under uncertain loading, thus proposed an uncertain optimization model, which can be expressed as follows:

$$\begin{cases} \min\limits_{\mathbf{x}} f(\mathbf{x}, \mathbf{p}) \\ \text{s.t. } g_j(\mathbf{x}, \mathbf{p}) \geq 0,\ j = 1, 2, \ldots, l,\ \mathbf{p} \in C_p \end{cases} \tag{1.1}$$

where $f$ and $g$ represent the objective function and the constraint, respectively; $\mathbf{x}$ represents the design vector; $\mathbf{p}$ represents the uncertain vector whose uncertainty domain belongs to a convex set $C_p$; $l$ represents the number of the constraints. Using Elishakoff's method, Eq. (1.1), can be transformed into a deterministic optimization problem:

$$\begin{cases} \min\limits_{\mathbf{x}} \max\limits_{\mathbf{p} \in C_p} f(\mathbf{x}, \mathbf{p}) \\ \text{s.t. } \min\limits_{\mathbf{p} \in C_p} g_j(\mathbf{x}, \mathbf{p}) \geq 0,\ \ j = 1, 2, \ldots, l \end{cases} \tag{1.2}$$

Obviously, Eq. (1.2), is a double-loop nested optimization problem, which consists of the outer loop for updating the design vector and the inner loop for calculating the worst response of the objective function and constraints in $C_p$.

Based on the Elishakoff's work, more efforts were then put into the research on convex model optimization. Lombardi [108] improved Elishakoff's method and proposed a two-steps method to solve the nested optimization problem, which promoted the computational efficiency. Pantelides and Ganzeli [109–111] proposed the modeling approach of the ellipsoidal convex model, and applied it to the design of the truss structure and multiple-span beam. Pantelides and Ganerli [112] systematically compared the fuzzy programming and the convex model optimization methods. Qiu [113–115] proposed several methods to efficiently solve the response boundary of the structure with an uncertainty domain described by convex set, providing potential numerical tools for convex model optimization. Au et al. [116] proposed a robust optimization method based on the convex model. Gurav et al. [117] proposed an enhanced anti-optimization method for efficiently carrying out uncertain optimization, and applied it to the design of a microelectromechanical system. Guo et al. [118] discussed the global optimum of the inner uncertainty analysis and obtained the confidence upper bound of the extreme structural response by solving the Lagrange duality problem of the inner uncertainty analysis, thus ensured the safety of the uncertain optimization results.

Over the past decade, the classical first-order reliability method (FORM) in the field of probability-based structural reliability analysis was introduced into convex model analysis, based on which a series of reliability-based convex model design optimization methods were proposed. Based on the convex set method, Guo and Lv [119, 120], proposed the non-probabilistic reliability index to measure the safety of a structure with uncertainty and applied it to convex model optimization. For the situation where the ellipsoidal model and the interval model exist simultaneously, Cao and Duan [121], presented the non-probabilistic reliability index and further proposed a sequential linearization method [122], to solve the reliability-based convex model optimization problem. Kang and Luo [123] proposed an optimization method based on the target performance to solve the reliability-based optimization problems based on the convex model. Afterwards, Kang and Luo [124] further applied the convex model optimization to structural topology optimization under uncertainties. In the multiobjective optimization field, Li et al. [125] proposed a design optimization method using the convex model for solving multi-objective problems.

Despite the great progress that has been achieved in convex model optimization, the research is still not complete theoretically and algorithmically, and there are some technical difficulties in this area. The first difficulty lies in the modeling of the convex model, especially for the problems with multisource uncertainties. How to construct the accurate multidimensional uncertainty domain through samples is still a difficulty to be broken through. The second difficulty is the vast computational cost caused by the double-loop nested optimization. How to develop an efficient decoupling algorithm with good convergence for the problems with relatively high dimensional design variables is a key factor of the application of the convex model optimization to engineering problems.

### 1.3.2 Interval Optimization

Interval optimization is also called interval programming or interval number programming. In interval optimization, the varying range of an uncertain parameter is described by an interval. To be specific, we only need to obtain the upper and lower bounds of each uncertain parameter, rather than the precise probability distribution or fuzzy membership function. In the interval mathematics [101], the interval is defined as a new type of number, i.e., the interval number. In this book, the words "interval number" and "interval" represent the same mathematical meaning. Over a relatively long period of time, the research on interval optimization was mainly focused on linear programming problems, while in recent years the nonlinear interval optimization has been obtaining more attention. For linear problems, to solve the interval optimization problem, we generally transform the uncertain optimization problem into a deterministic problem by the introduction of order relation of interval number or minimax regret criterion. Hereinafter, the research progress on the interval optimization will be reviewed from the following three aspects:

(1) Linear interval optimization based on the order relation of interval number

A general linear interval optimization problem can be expressed as

$$\begin{cases} \min\limits_{\mathbf{x}} \sum\limits_{j=1}^{n} \left[c_j^L, c_j^R\right] x_j \\ \text{s.t.} \sum\limits_{j=1}^{n} \left[a_{ij}^L, a_{ij}^R\right] x_j \le \left[b_i^L, b_i^R\right],\ i = 1, 2, \ldots, l \end{cases} \tag{1.3}$$

where [*] represents the uncertain coefficients described by the interval.

Tanaka et al. [82], Rommelfanger [126], Ishibuchi and Tanaka [127], transformed the interval optimization into a deterministic optimization problem by using order relation of interval number. For the case that the constraints in Eq. (1.3), are deterministic, the interval objective can be converted into multiple deterministic objective functions based on the order relation of interval number. For interval constraints, they can be measured by the satisfaction degree of the order relation [128–130], and whereby transformed into deterministic constraints. For the problems where the coefficients in objective function and constraints are interval numbers, Tong [131], proposed a method to solve the possible objective function interval, which represents two extreme situations of the objective function and constraints. For the same problem, Liu and Da [132], presented a solution method based on the fuzzy possibility degree of interval constraint. Based on the probability method, Zhang et al. [133], developed a new type of possibility degree of interval number to solve the multi-attribute decision-making problems. Xu and Da [134] studied the relationship between different possibility degrees of interval number and developed a new possibility degree method for interval number ranking to solve the alternative ranking problem involving uncertainties. A more general method for handling the uncertain objective function with interval coefficients was proposed by Chanas and Kuchta [135], to transform the uncertain problem into a deterministic problem. Sengupta and Pal [136] reviewed and summarized the current interval number ranking models, and then proposed two new ranking methods. Sengupta et al. [137] defined a linear interval programming problem as a generalization of the traditional linear programming under uncertainties. In Sengupta' work, the inequality constraints containing interval coefficients are simplified based on their interval ranking methods. Lai et al. [138] transformed the interval optimization problem with interval coefficients existing in both objective function and constraints into a conventional linear optimization problem based on two types of order relations of interval number. A new order relation of interval number was proposed by Chen et al. [139], to deal with the interval inequality constraints.

(2) Linear interval optimization based on the minimax regret criterion

Consider a linear programming problem involving interval coefficients in the objective function.

$$\begin{cases} \min_{\mathbf{x}} \mathbf{c}^{\mathrm{T}}\mathbf{x} \\ \text{s.t.}\, \mathbf{x} \in \Omega \\ \mathbf{c} \in \Gamma = \left\{ \mathbf{c} \middle| c_i^L \le c_i \le c_i^R, i = 1, 2, \ldots, n \right\} \end{cases} \tag{1.4}$$

where $\Omega$ is a non-empty bounded polyhedron set; $\mathbf{c}$ represents the interval coefficient vector of the objective function.

Inuiguchi et al. [140, 141] proposed the definitions of the necessary optimal solution set and the possible optimal solution set for Eq. (1.4), and gave a programming method with interval objective function based on the minimax regret criterion. Whatever the interval coefficient vector $\mathbf{c}$ is, any element in the necessary optimal solution set is the optimal solution of Eq. (1.4). However, the elements in the possible optimal solution set are the optimal solution only when $\mathbf{c}$ is chosen on some special values. For any given $\mathbf{x}$, the regret $R$ of the detaermination and the worst (maximum) regret $R_{\max}$ of the determination can be, respectively, defined as.

$$R(\mathbf{c}, \mathbf{x}) = \max_{\mathbf{y}\in\Omega} \left( \mathbf{c}^{\mathrm{T}}\mathbf{y} - \mathbf{c}^{\mathrm{T}}\mathbf{x} \right), \qquad R_{\max}(\mathbf{x}) = \max_{\mathbf{c}\in\Gamma} R(\mathbf{c}, \mathbf{x}) \tag{1.5}$$

The purpose of the minimax regret criterion is to calculate the minimum value of the maximum regret value by constructing the following optimization problem:

$$\begin{cases} \min_{\mathbf{x}} \left( \max_{\mathbf{c},\mathbf{y}} \left( \mathbf{c}^{\mathrm{T}}\mathbf{y} - \mathbf{c}^{\mathrm{T}}\mathbf{x} \right) \right) \\ \text{s.t.}\, \mathbf{x}, \mathbf{y} \in \Omega, \ \ \mathbf{c} \in \Gamma \end{cases} \tag{1.6}$$

Inuiguchi and Sakawa [141] studied the minimax regret criterion-based interval optimization problems with interval coefficients existing in the objective function and proposed a solution method based on the relaxation procedure [142]. Afterwards, Inuiguchi and Sakawa [143], Mausser and Laguna [144], improved this algorithm. Kouvelis and Yu [145] applied the minimax regret criterion to a site selection problem and obtained a polynomial algorithm. For the linear programming problems with interval coefficients in the objective function, Mausser and Laguna [146], developed a minimax regret based heuristic optimization method. Averbakh and Lebedev [147] proved that the minimax regret based optimization with interval coefficients in the objective function is NP-hard. Dong et al. [148] applied the linear interval optimization method based on the minimax regret criterion to the design of a power management system. Rivaz and Yaghoobi [149] proposed a multi-objective linear interval programming method based on the minimax regret criterion.

(3) Nonlinear interval optimization

The above-mentioned researches aim at the linear interval optimization problems, in which the objective function and constraints are all linear functions with respect to the design variables or the interval variables. Most practical engineering problems, however, are nonlinear. Further research on the nonlinear interval optimization is, therefore, necessary for the application of interval optimization to engineering practice. However, it is much more complex and challenging to solve a nonlinear interval optimization problem than its linear counterpart, which causes the underdevelopment of this field. Ma [150] began to study the nonlinear interval optimization relatively early and proposed a triple-objective robust optimization method combining the expectation, uncertainty degree, and regret degree of the objective function. In Ma's method, the interval of the uncertain objective function is obtained through two optimizations with respect to the uncertain variables in every iteration. According to the characteristics of process based industrial system, Cheng et al. [151], presented a general form of the multi-objective optimization problem for an uncertain system with interval parameters and proposed a hierarchical optimization strategy combining the genetic algorithm (GA) and conventional nonlinear optimization method to solve the multi-objective nonlinear interval optimization problems. Jiang [152] introduced possibility degree of interval number to transform the nonlinear interval constraints into deterministic constraints. Wu [153] presented the Karush–Kuhn–Tucker (KTT) condition of an optimization problem with the interval-valued objective function, and proposed two types of interval optimization methods based on the partial orderings of interval number. To solve the problems with nonlinear objective function and linear constraints, Wu et al. [154] developed a nonlinear interval optimization method based on the satisfactory performance and applied it to a waste treatment facility planning project. Gong and Sun [155] systematically studied the evolutionary algorithm for interval-based multi-objective optimization problems. The conception of the degree of interval constraint violation was proposed by Cheng et al. [156]. Based on this, they further developed a direct solution method for nonlinear interval optimization, which avoids transforming the original problem into deterministic optimization. At present, the nonlinear interval optimization methods have been applied in some engineering fields, such as the vehicle suspension system design [157], structural design [158–161], vibration control system design [162], product portfolio planning [163], power scheduling [164, 165], and optimal water allocation [166, 167], etc.

## 1.4 Current Problems in Interval Optimization

Compared with the probability method and fuzzy method, interval analysis method shows some prominent advantages in terms of uncertainty modeling because of its less dependency on the sample size and higher comprehensibility, which makes it easy for engineers to understand and use. With the introduction of interval analysis for uncertainty modeling, the uncertain optimization is expected to expand its research object and application area, thus effectively promoting the design level of future engineering systems or structures under uncertainties. The interval optimization has been growing into a mainstream uncertain optimization method like stochastic programming and fuzzy programming. So far, although interval optimization has been studied for more than 30 years, and has gained significant achievements, it is still at its primary stage and has not formed a mature and impeccable theoretical system, which largely limits its expansion and application in engineering. The main technical difficulties in interval optimization can be summarized as follows:

(1) Current researches on interval optimization mainly focus on the linear programming problems, while most engineering problems are nonlinear and even strongly nonlinear. Therefore, the research on nonlinear interval optimization will be pretty significant and it concerns the practicality and vitality of the whole interval optimization theory. Just like stochastic programming and fuzzy programming, in interval optimization, we generally need to transform the uncertain optimization into a deterministic optimization, which then can be solved by conventional optimization methods. However, such a transformation requires an effective mathematical transformation model, which serves as the foundation of nonlinear interval optimization. In this aspect, how to ensure the equivalence between the original problem and the transformed problem, and how to ensure the versatility of the transformation model are still challenging tasks. Besides, the current transformation model lacks diversity, thus it could be incompetent for diversified practical engineering problems that have different requirements and emphases in the optimization process. Therefore, how to develop different transformation models for different problems will be of great significance and deserve further researches in the future.

(2) For nonlinear interval optimization, the converted deterministic optimization is generally a complicated double-loop nested optimization problem, which is the biggest difference between nonlinear and linear interval optimizations. Although the original uncertain optimization problem is continuous and differentiable with respect to the design variables and uncertain variables, it is hard to ensure continuity and differentiability of the converted deterministic optimization problem. Therefore, the traditional gradient-based optimization methods are usually invalid for solving it. How to develop a series of non-gradient-based methods or how to improve and expand current gradient-based methods to nonlinear interval optimization, therefore, becomes significant research topics in this field.

(3) The low-efficiency problem caused by the double-loop nested optimization is the bottleneck of nonlinear interval optimization. Most current researches on nonlinear interval optimization are aimed at the analytical problems that are relatively simpler than practical engineering problems, and hence the low-efficiency problem is not obvious. However, for practical engineering problems, the optimization model is usually implicitly obtained by some numerical analysis techniques, such as finite element method (FEM) and multi-body dynamic model. A single calculation of such numerical analysis models is generally time-consuming. Therefore, the numerical analysis model based double-loop nested optimization will lead to extremely low computational efficiency. Actually, the low-efficiency problem has been a major technical difficulty existing in nonlinear interval optimization, and it also has been an important issue concerned by scholars.

(4) With the development of modern industry, the optimization problem in practical engineering is becoming more and more complex. How to extend interval optimization to some important optimization problems, such as multi-objective optimization or multidisciplinary optimization, is an issue we are concerned about. The complexity of these problems will further increase with the involvement of nonlinearity, and thus corresponding nonlinear interval optimization methods are needed to develop to solve these problems.

## 1.5 The Research Target and Framework of This Book

In conclusion, there still exist several important technical difficulties required to be solved in current interval optimization, especially in the nonlinear interval optimization. Focusing on the above-mentioned issues, this book systematically discusses the theory and methods of nonlinear interval optimization, expecting to explore the nonlinear interval optimization in the aspects of theoretical models, practical algorithms, and engineering applications. The research idea and content of this book can be summarized as follows: First, two mathematical transformation models are developed to deal with general nonlinear interval optimization problems. Second, several efficient nonlinear interval optimization algorithms with engineering practicability are presented based on the proposed mathematical transformation models, emphasizing on solving the low-efficiency problem caused by the double-loop nested optimization. Third, the nonlinear interval optimization is extended to several important problems such as multi-objective optimization, multidisciplinary optimization, etc., and the corresponding interval optimization models and solution algorithms are established. Finally, the proposed interval optimization models and methods are utilized to deal with practical problems in mechanical engineering and related fields to demonstrate their effectiveness. The chapter arrangement of this book is given as follows:

Chapter 1 *Introduction.* This chapter introduces the engineering background and research significance of uncertain optimization and analyzes the research status of several mainstream uncertain optimization methods, in which the research status and main technical problems of interval optimization method are emphasized. Finally, this chapter gives the framework of this book.

Chapter 2 *The basic principles of interval analysis*. This chapter briefly introduces the origin of interval number, the basic conceptions of interval mathematics, the calculation rules of interval number, and the important interval overestimation problem, providing some necessary theoretical foundations for interval optimization.

Chapter 3 *Mathematical transformation models of nonlinear interval optimization*. This chapter proposes two types of mathematical transformation models for general nonlinear interval optimization problems, i.e., the transformation model based on order relation of interval number and the transformation model based on possibility degree of interval number. Therefore, the uncertain optimization problem is transformed into a deterministic optimization. Furthermore, a two-layer optimization algorithm is established to solve the transformed deterministic optimization problem.

Chapter 4 *Interval optimization based on hybrid optimization algorithm*. By combining the genetic algorithm (GA) with the artificial neural network (ANN), this chapter establishes two hybrid optimization algorithms to solve the nested optimization problem after transformation. Based on these two algorithms, two efficient nonlinear interval optimization methods are developed correspondingly.

Chapter 5 *Interval optimization based on interval structural analysis*. This chapter first introduces the conventional interval structural analysis method with small uncertainties. Based on the interval set theory and the subinterval technology, it is then extended to problems with large uncertainties. Consequently, an efficient interval optimization method based on the interval structural analysis is established.

Chapter 6 *Interval optimization based on sequential linear programming*. By introducing the sequential linear programming technique, this chapter develops an efficient nonlinear interval optimization algorithm. An iterative mechanism is also provided to ensure the convergence of the proposed algorithm.

Chapter 7 *Interval optimization based on approximation models*. This chapter creates two nonlinear interval optimization methods based on the approximation model management strategy and the local-densifying approximation technique, respectively. Both two methods can greatly improve the computational efficiency of interval optimization by using approximation models, but different strategies are selected to guarantee accuracy during the iteration.

Chapter 8 *Interval multidisciplinary design optimization*. This chapter introduces the interval model into the multidisciplinary design optimization (MDO) problem, and whereby constructs an interval MDO model for complex MDO problems with uncertainties. A solution strategy is also given to solve the interval MDO model.

Chapter 9 *A new type of possibility degree of interval number and its application in interval optimization*. This chapter first proposes a new possibility degree model of interval number to realize quantitative comparison for not only overlapping intervals, but also separate intervals and then applies it to the nonlinear interval optimization.

Chapter 10 *Interval optimization considering the correlation of parameters*. This chapter first introduces a new type of interval model, i.e., the multidimensional parallelepiped interval model, based on which an interval optimization model for multisource uncertain problems is then established. A solution algorithm is also given for this interval optimization model.

Chapter 11 *Interval multi-objective optimization.* By employing interval approach to describe the uncertainty of parameters in multi-objective optimization, this chapter proposes an interval multi-objective optimization model and also an efficient solution algorithm for this optimization model.

Chapter 12 *Interval optimization considering tolerance design.* This chapter proposes an interval optimization method considering tolerance design, which provides not only the optimal design, but also the optimal tolerances of the design variables.

Chapter 13 *Interval differential evolution algorithm.* By introducing the interval model into the existing differential evolution, this chapter proposes a novel interval differential evolution algorithm, which can directly solve the original interval optimization problem rather than transforming it to a deterministic optimization problem first.

## References


1. Himmelblau DM (1972) Applied nonlinear programming. McGraw-Hill, New York
2. Haftka RT, Gürdal Z (1992) Elements of structural optimization. Kluwer Academic Publishers, Netherlands
3. Liu WX (2002) Mechanical optimization design. Tsinghua University Press, Beijing
4. Zhu J, Zhang W, Beckers P (2009) Integrated layout design of multi-component system. Int J Numer Meth Eng 78(6):631–651
5. Cheng GD (2012) Introduction to optimum design of engineering structures. Dalian University of Technology Press, Dalian
6. Nocedal J, Wright SJ (1999) Numerical optimization. Springer, New York
7. Qiu ZP (1994) Interval analysis method of static response of structures with uncertain parameters and the eigenvalue problem. PhD thesis, Jilin Industrial University, Changchun
8. Yang XW (2000) Dynamic and static analysis for structures with interval parameters. PhD thesis, Jilin University of Technology, Changchun
9. Lian HD (2002) The interval FEM for structural analysis. PhD thesis, Jilin University, Changchun
10. Li RF, Wang JJ (1997) Geared system dynamic-vibration, shock and noise. Science Press, Beijing
11. Zhang JB, Chen ZY (1990) A novel method of stiffness identification for gears. In Proceedings of the 4th national conference on vibration theory and application, Zhengzhou, China, May 1990
12. Tong ZF, Zhang J (1992) Research on the dynamic characteristic and its identification of the joint between column and bed of a machining center. J Vib Shock 2(3):13–19
13. Zhu DP (1993) The influence of vibration environment on structural characteristics. In Proceedings of the 5th national conference on vibration theory and application, Tunxi, Huangshan, China, Oct 1993
14. Zhang J (1996) Research on dynamic modelling and parameter identification of joint between complex mechanical structures. J Mech Strength 18(2):1–5
15. Sun AR (1995) Uncertainty analysis of structural and systematic parameters of nuclear power plant. J Northeast For Univ 23(1):108–115

16. Chang CC, Lo JG, Wang JL (2001) Assessment of reducing ozone forming potential for vehicles using liquefied petroleum gas as an alternative fuel. Atmos Environ 35(35):6201–6211
17. Yuan Q, Li YB (2005) Influence of parametrical uncertainties on traffic accident reappearance of vehicle side impact. Trans Chin Soc Agric Mach 36(5):16–19
18. Wang HY, Liu HJ (2006) Stochastic simulation method of performance reliability estimation on liquid propellant rocket engine. J Rocket Propuls 32(4):26–30
19. Kazemi Zanjani M, Nourelfath M, Ait-Kadi D (2010) A multi-stage stochastic programming approach for production planning with uncertainty in the quality of raw materials and demand. Int J Prod Res 48(16):4701–4723
20. Liu BD, Zhao RQ, Wang G (2003) Uncertain programming with applications. Tsinghua University Press, Beijing
21. Alfieri A, Tolio T, Urgo M (2012) A two-stage stochastic programming project scheduling approach to production planning. Int J Adv Manuf Technol 62(1):279–290
22. Goli A, Tirkolaee EB, Malmir B, Bian GB, Sangaiah AK (2019) A multi-objective invasive weed optimization algorithm for robust aggregate production planning under uncertain seasonal demand. Computing 101(6):499–529
23. Liu BD (2001) Uncertain programming: a unifying optimization theory in various uncertain environments. Appl Math Comput 120(1–3):227–234
24. Kall P (2001) Stochastic programming: achievements and open problems. In: Kischka P, Möhring RH, Leopold-Wildburger U, Radermacher FJ (eds) Models, methods and decision support for management: essays in Honor of Paul Stähly. Physica-Verlag HD, Heidelberg, pp 285–302
25. Mousavi SM, Vahdani B, Tavakkoli-Moghaddam R, Hashemi H (2014) Location of cross-docking centers and vehicle routing scheduling under uncertainty: a fuzzy possibilistic–stochastic programming model. Appl Math Model 38(7–8):2249–2264
26. Sun L, Lin L, Li HJ, Gen M (2018) Hybrid cooperative co-evolution algorithm for uncertain vehicle scheduling. IEEE Access 6:71732–71742
27. Chen C, Wang F, Zhou B, Chan KW, Cao Y, Tan Y (2015) An interval optimization based day-ahead scheduling scheme for renewable energy management in smart distribution systems. Energy Convers Manag 106:584–596
28. Wei F, Wu QH, Jing ZX, Chen JJ, Zhou XX (2016) Optimal unit sizing for small-scale integrated energy systems using multi-objective interval optimization and evidential reasoning approach. Energy 111:933–946
29. Zhou J, Liu BD (2003) New stochastic models for capacitated location-allocation problem. Comput Ind Eng 45(1):111–125
30. Alizadeh M, Mahdavi I, Mahdavi-Amiri N, Shiripour S (2015) A capacitated location-allocation problem with stochastic demands using sub-sources: an empirical study. Appl Soft Comput 34:551–571
31. Elishakoff I, Haftka RT, Fang J (1994) Structural design under bounded uncertainty-optimization with anti-optimization. Comput Struct 53(6):1401–1405
32. Doltsinis I, Kang Z (2004) Robust design of structures using optimization methods. Comput Methods Appl Mech Eng 193(23–26):2221–2237
33. Mrabet E, Guedri M, Ichchou MN, Ghanmi S (2015) Stochastic structural and reliability based optimization of tuned mass damper. Mech Syst Signal Process 60–61:437–451
34. Bhattacharjya S, Chakraborty S (2018) An improved robust multi-objective optimization of structure with random parameters. Adv Struct Eng 21(11):1597–1607
35. Liu BS, Jiang C, Li GY, Huang XD (2020) Topology optimization of structures considering local material uncertainties in additive manufacturing. Comput Methods Appl Mech Eng 360 Article 112786
36. Bellman RE (1957) Dynamic programming. Princeton University Press, New Jersey
37. Bellman RE, Zadeh LA (1970) Decision-making in a fuzzy environment. Manag Sci 17(4):141–164
38. Charnes A, Cooper WW (1959) Chance-constrained programming. Manag Sci 6(1):73–79

39. Dantzig GB (1955) Linear programming under uncertainty. Manag Sci 1(3–4):197–206
40. Ferguson AR, Dantzig GB (1956) The allocation of aircraft to routes—an example of linear programming under uncertain demand. Manag Sci 3(1):45–73
41. Beale EM (1955) On minimizing a convex function subject to linear inequalities. J R Stat Soc Ser B (Methodological) 173–184
42. Walkup DW, Wets RJ-B (1967) Stochastic programs with recourse. SIAM J Appl Math 15(5):1299–1314
43. Wets RJ-B (1974) Stochastic programs with fixed recourse: the equivalent deterministic program. SIAM Rev 16(3):309–339
44. Charnes A, Cooper WW, Symonds GH (1958) Cost horizons and certainty equivalents: an approach to stochastic programming of heating oil. Manag Sci 4(3):235–263
45. Borell C (1974) Convex measures on locally convex spaces. Arkiv För Matematik 12(1):239–252
46. Prékopa A (1980) Logarithmic concave measures and related topics. In: Stochastic programming
47. Markowitz HM, Todd GP (2000) Mean-variance analysis in portfolio choice and capital markets. Wiley, New Hope, Pennsylvania
48. Levy H, Markowitz HM (1979) Approximating expected utility by a function of mean and variance. Am Econ Rev 69(3):308–317
49. Dupačová J (1976) Minimax stochastic programs with nonconvex nonseparable penalty functions. Progress Oper Res. North-Holland, Amsterdam
50. Pollak RA (1967) Additive von Neumann-Morgenstern utility functions. Econometrica 35(3/4):485–494
51. Roth AE (1977) The Shapley value as a von Neumann-Morgenstern utility. Econometrica 45(3):657–664
52. Rong XX (2005) Research on model and algorithm about uncertain optimization problems. PhD thesis, Shandong University, Jinan
53. Gartska SJ (1980) The economic equivalence of several stochastic programming models. In: Dempster MAH (ed) Stochastic programming. Academic Press, New York
54. Ziemba WT (1974) Stochastic programs with simple recourse. Mathematical programming in theory and practice. Amsterdam, North-Holland, June 1972
55. Birge JR, Louveaux F (1997) Introduction to stochastic programming. Springer, New York
56. Leövey H, Römisch W (2015) Quasi-Monte Carlo methods for linear two-stage stochastic programming problems. Math Program 151(1):315–345
57. Ivanov SV, Kibzun AI, Mladenovic N, Urosevic D (2019) Variable neighborhood search for stochastic linear programming problem with quantile criterion. J Global Optim 74(3):549–564
58. Sherali HD, Fraticelli BMP (2002) A modification of Benders' decomposition algorithm for discrete subproblems: an approach for stochastic programs with integer recourse. J Global Optim 22(1):319–342
59. Schrijver A (1998) Theory of linear and integer programming. Wiley, Amsterdam
60. Ahmed S, Sahinidis NV (2003) An approximation scheme for stochastic integer programs arising in capacity expansion. Oper Res 51(3):461–471
61. Bastin F (2004) Nonlinear stochastic programming. PhD Thesis, University of Namur, Namur
62. Dumskis V, Sakalauskas L (2015) Nonlinear stochastic programming involving CVaR in the objective and constraints. Informatica 26(4):569–591
63. Doagooei AR (2020) Generalized cutting plane method for solving nonlinear stochastic programming problems. Optimization 69(7–8):1751–1771
64. Krasko V, Rebennack S (2017) Two-stage stochastic mixed-integer nonlinear programming model for post-wildfire debris flow hazard management: mitigation and emergency evacuation. Eur J Oper Res 263(1):265–282
65. Takriti S, Ahmed S (2004) On robust optimization of two-stage systems. Math Program 99(1):109–126
66. Liu C, Lee C, Chen H, Mehrotra S (2016) Stochastic robust mathematical programming model for power system optimization. IEEE Trans Power Syst 31(1):821–822

67. Rahimian H, Bayraksan G, Homem-de-Mello T (2019) Identifying effective scenarios in distributionally robust stochastic programs with total variation distance. Math Program 173(1–2):393–430
68. Enevoldsen I, Sørensen JD (1994) Reliability-based optimization in structural engineering. Struct Saf 15(3):169–196
69. Du XP, Chen W (2004) Sequential optimization and reliability assessment method for efficient probabilistic design. ASME J Mech Des 126(2):225–233
70. Liang J, Mourelatos ZP, Tu J (2004) A single-loop method for reliability-based design optimization. In: ASME 2004 international design engineering technical conferences and computers and information in engineering conference, Salt Lake City, UT, Sept/Oct 2004
71. Cheng GD, Xu L, Jiang L (2006) A sequential approximate programming strategy for reliability-based structural optimization. Comput Struct 84(21):1353–1367
72. Zhang YM (2015a) Reliability-based robust design optimization of vehicle components, part I: theory. Front Mech Eng 10(2):138–144
73. Zhang YM (2015b) Reliability-based robust design optimization of vehicle components, part II: case studies. Front Mech Eng 10(2):145–153
74. Chen ZZ, Qiu HB, Gao L, Li P (2013) An optimal shifting vector approach for efficient probabilistic design. Struct Multidiscip Optim 47(6):905–920
75. Shan SQ, Wang GG (2008) Reliable design space and complete single-loop reliability-based design optimization. Reliab Eng Syst Saf 93(8):1218–1230
76. Zhang Z, Deng W, Jiang C (2020) Sequential approximate reliability-based design optimization for structures with multimodal random variables. Struct Multidiscip Optim 62(2):511–528
77. Cui D (2005) Some problems on stochastic programming. Master thesis, Shandong University of Science and Technology, Qingdao
78. Feng YJ, Wei QL (1982) The general form of multi-objective fuzzy programming solution. Fuzzy Math 2(2):29–35
79. Inuiguchi M, Ramík J (2000) Possibilistic linear programming: a brief review of fuzzy mathematical programming and a comparison with stochastic programming in portfolio selection problem. Fuzzy Sets Syst 111(1):3–28
80. Zimmermann HJ (1985) Applications of fuzzy set theory to mathematical programming. Inf Sci 36(1):29–58
81. Dubois D, Prade H (1980) Systems of linear fuzzy constraints. Fuzzy Sets Syst 3(1):37–48
82. Tanaka H, Okuda T, Asai K (1973) On fuzzy-mathematical programming. J Cybern 3(4):37–46
83. Rommelfanger H (1996) Fuzzy linear programming and applications. Eur J Oper Res 92(3):512–527
84. Buckley JJ, Feuring T (2000) Evolutionary algorithm solution to fuzzy problems: fuzzy linear programming. Fuzzy Sets Syst 109(1):35–53
85. Liu QM, Shi FG (2015) Stratified simplex method for solving fuzzy multi-objective linear programming problem. J Intell Fuzzy Syst 29(6):2357–2364
86. Luhandjula MK, Ichihashi H, Inuiguchi M (1992) Fuzzy and semi-infinite mathematical programming. Inf Sci 61(3):233–250
87. Inuiguchi M, Ichihashi H, Kume Y (1992) Relationships between modality constrained programming problems and various fuzzy mathematical programming problems. Fuzzy Sets Syst 49(3):243–259
88. Hong ZY (2001) The stability of fuzzy multi-objective nonlinear programming problem and a solving method based on the order structure of numbers. Master thesis, Harbin Institute of Technology, Harbin
89. Sakawa M, Yano H (1989) Interactive decision making for multiobjective nonlinear programming problems with fuzzy parameters. Fuzzy Sets Syst 29(3):315–326
90. Huang HZ (1997) Fuzzy multi-objective optimization decision-making of reliability of series system. Microelectron Reliab 37(3):447–449
91. Tang JF, Wang DW (1997) An interactive approach based on a genetic algorithm for a type of quadratic programming problems with fuzzy objective and resources. Comput Oper Res 24(5):413–422

92. Tang JF, Wang DW (1998) Model and method for a type of nonlinear programming problems with fuzzy resources. Fuzzy Syst Math 12(3):58–67
93. Liu BD, Zhao RQ (1998) Stochastic programming and fuzzy programming. Tsinghua University Press, Beijing
94. Wu CW, Liao MY (2014) Fuzzy nonlinear programming approach for evaluating and ranking process yields with imprecise data. Fuzzy Sets Syst 246:142–155
95. Mansoori A, Effati S (2019) An efficient neurodynamic model to solve nonlinear programming problems with fuzzy parameters. Neurocomputing 334:125–133
96. Jiang Z (2005) Optimization method of uncertain system with the interval parameter and its research on the applies about gasoline blending. PhD thesis, Zhejiang University, Hangzhou
97. Ben-Haim Y, Elishakoff I (1990) Convex models of uncertainty in applied mechanics. Elsevier Science Publisher, Amsterdam
98. Salinetti G (1983) Approximations for chance-constrained programming problems. Stoch Int J Probab Stoch Process 10(3–4):157–179
99. Ermoliev Y (1983) Stochastic quasigradient methods and their application to system optimization. Stoch Int J Probab Stoch Process (1–2):1–36
100. Guo SX (2002) Non-stochastic reliability and optimization of uncertain structural systems. PhD thesis, Northwestern Polytechnical University, Xi'an
101. Moore R (1979) Methods and applications of interval analysis. Prentice-Hall, London
102. Ben-Haim Y (1993) Convex models of uncertainty in radial pulse buckling of shells. ASME J Appl Mech 60(3):683–688
103. Ben-Haim Y (1995) A non-probabilistic measure of reliability of linear systems based on expansion of convex models. Struct Saf 17(2):91–109
104. Kang Z, Luo YJ (2009a) Non-probabilistic reliability-based topology optimization of geometrically nonlinear structures using convex models. Comput Methods Appl Mech Eng 198(41–44):3228–3238
105. Jiang C, Zhang QF, Han X, Liu J, Hu DA (2015) Multidimensional parallelepiped model—a new type of non-probabilistic convex model for structural uncertainty analysis. Int J Numer Meth Eng 103(1):31–59
106. Elishakoff I, Bekel Y (2013) Application of Lamé's super ellipsoids to model initial imperfections. ASME J Appl Mech 80(6):061006
107. Jiang C, Ni BY, Han X, Tao YR (2014) Non-probabilistic convex model process: A new method of time-variant uncertainty analysis and its application to structural dynamic reliability problems. Comput Methods Appl Mech Eng 268:656–676
108. Lombardi M (1998) Optimization of uncertain structures using non-probabilistic models. Comput Struct 67(1–3):99–103
109. Pantelides CP, Ganzerli S (1998) Design of trusses under uncertain loads using convex models. J Struct Eng 124(3):318–329
110. Ganzerli S, Pantelides CP (1999) Load and resistance convex models for optimum design. Struct Optim 17(4):259–268
111. Ganzerli S, Pantelides CP (2000) Optimum structural design via convex model superposition. Comput Struct 74(6):639–647
112. Pantelides CP, Ganzerli S (2001) Comparison of fuzzy set and convex model theories in structural design. Mech Syst Signal Process 15(3):499–511
113. Qiu ZP (2003) Comparison of static response of structures using convex models and interval analysis method. Int J Numer Meth Eng 56(12):1735–1753
114. Qiu ZP, Wang XJ (2003) Comparison of dynamic response of structures with uncertain-but-bounded parameters using non-probabilistic interval analysis method and probabilistic approach. Int J Solids Struct 40(20):5423–5439
115. Qiu ZP (2005) Convex method based on non-probabilistic set-theory and its application. National Defence Industry Press, Beijing
116. Au FTK, Cheng YS, Tham LG, Zeng GW (2003) Robust design of structures using convex models. Comput Struct 81(28–29):2611–2619

117. Gurav SP, Goosen JFL, vanKeulen F (2005) Bounded-But-Unknown uncertainty optimization using design sensitivities and parallel computing: application to MEMS. Comput Struct 83(14):1134–1149
118. Guo X, Bai W, Zhang WS, Gao XX (2009) Confidence structural robust design and optimization under stiffness and load uncertainties. Comput Methods Appl Mech Eng 198(41–44):3378–3399
119. Guo SX, Lv ZZ (2002) Optimization of uncertain structures based on non-probabilistic reliability model. Chin J Comput Mech 19(2):198–201
120. Guo SX, Lv ZZ (2003) Comparison of possibilistic reliability and stochastic reliability methods for uncertain structures. Chin J Appl Mech 20(3):107–110
121. Cao HJ, Duan BY (2005a) Approach to optimization of uncertain structures based on on-probabilistic reliability. Chin J Comput Mech 22(5):546–549
122. Cao HJ, Duan BY (2005b) Approach to optimization of uncertain structures based on non-probabilistic reliability. Chin J Appl Mech 22(3):381–385
123. Kang Z, Luo YJ (2006) On structural optimization for non-probabilistic reliability based on convex models. Chin J Theor Appl Mech 38(6):807–815
124. Kang Z, Luo YJ (2009b) Non-probabilistic reliability-based topology optimization of geometrically nonlinear structures using convex models. Comput Methods Appl Mech Eng 198(41):3228–3238
125. Li FY, Luo Z, Rong JH, Hu L (2013) A non-probabilistic reliability-based optimization of structures using convex models. Comput Model Eng Sci 95(6):453–482
126. Rommelfanger H, Hanuscheck R, Wolf J (1989) Linear programming with fuzzy objectives. Fuzzy Sets Syst 29(1):31–48
127. Ishibuchi H, Tanaka H (1990) Multiobjective programming in optimization of the interval objective function. Eur J Oper Res 48(2):219–225
128. Ramik J, Rommelfanger H (1993) A single- and a multi-valued order on fuzzy numbers and its use in linear programming with fuzzy coefficients. Fuzzy Sets Syst 57(2):203–208
129. Dubois D, Prade H (1983) Ranking fuzzy numbers in the setting of possibility theory. Inf Sci 30(3):183–224
130. Ohta H, Yamaguchi T (1996) Linear fractional goal programming in consideration of fuzzy solution. Eur J Oper Res 92(1):157–165
131. Tong SC (1994) Interval number and fuzzy number linear programmings. Fuzzy Sets Syst 66(3):301–306
132. Liu XW, Da QL (1999) A satisfactory solution for interval linear programming. J Syst Eng 14(2):123–128
133. Zhang Q, Fan ZP, Pan DH (1999) A ranking approach for interval numbers in uncertain multiple attribute decision making problems. Syst Eng Theory Practice 19(5):129–133
134. Xu ZS, Da QL (2003) Possibility degree method for ranking interval numbers and its application. J Syst Eng 18(1):67–70
135. Chanas S, Kuchta D (1996) Multiobjective programming in optimization of interval objective functions-a generalized approach. Eur J Oper Res 94(3):594–598
136. Sengupta A, Pal TK (2000) On comparing interval numbers. Eur J Oper Res 127(1):28–43
137. Sengupta A, Pal TK, Chakraborty D (2001) Interpretation of inequality constraints involving interval coefficients and a solution to interval linear programming. Fuzzy Sets Syst 119(1):129–138
138. Lai KK, Wang SY, Xu JP, Zhu SS, Fang Y (2002) A class of linear interval programming problems and its application to portfolio selection. IEEE Trans Fuzzy Syst 10(6):698–704
139. Chen MZ, Wang SG, Wang PP, Ye XX (2016) A new equivalent transformation for interval inequality constraints of interval linear programming. Fuzzy Optim Decis Making 15(2):155–175
140. Inuiguchi M, Kume Y (1992) Extensions of efficiency to possibilistic multiobjective linear programming problems. In: Proceeding of the tenth international conference on multiple criteria decision making

141. Inuiguchi M, Sakawa M (1995) Minimax regret solution to linear programming problems with an interval objective function. Eur J Oper Res 86(3):526–536
142. Shimizu K, Aiyoshi E (1980) Necessary conditions for min-max problems and algorithms by a relaxation procedure. IEEE Trans Autom Control 25(1):62–66
143. Inuiguchi M, Sakawa M (1996) Maximum regret analysis in linear programs with an interval objective function. In: Proceedings of IWSCI'96
144. Mausser HE, Laguna M (1998) A new mixed integer formulation for the maximum regret problem. Int Trans Oper Res 5(5):389–403
145. Kouvelis P, Yu G (1997) Robust discrete optimization and its applications. Kluwer Academic Publishers, Boston
146. Mausser HE, Laguna M (1999) A heuristic to minimax absolute regret for linear programs with interval objective function coefficients. Eur J Oper Res 117(1):157–174
147. Averbakh I, Lebedev V (2005) On the complexity of minmax regret linear programming. Eur J Oper Res 160(1):227–231
148. Dong C, Huang GH, Cai YP, Xu Y (2011) An interval-parameter minimax regret programming approach for power management systems planning under uncertainty. Appl Energy 88(8):2835–2845
149. Rivaz S, Yaghoobi MA (2013) Minimax regret solution to multiobjective linear programming problems with interval objective functions coefficients. CEJOR 21(3):625–649
150. Ma LH (2002) Research on method and application of robust optimization for uncertain system. PhD thesis, Zhejiang University, Hangzhou
151. Cheng ZQ, Dai LK, Sun YX (2004) Feasibility analysis for optimization of uncertain systems with interval parameters. Acta Autom Sin 30(3):455–459
152. Jiang C (2008) Theories and algorithms of uncertain optimization based on interval. PhD thesis, Hunan University, Changsha
153. Wu HC (2007) The Karush–Kuhn–Tucker optimality conditions in an optimization problem with interval-valued objective function. Eur J Oper Res 176(1):46–59
154. Wu XY, Huang GH, Liu L, Li JB (2006) An interval nonlinear program for the planning of waste management systems with economies-of-scale effects—a case study for the region of Hamilton, Ontario, Canada. Eur J Oper Res 171(2):349–372
155. Gong DW, Sun J (2013) Theories and applications of interval multi-objective evolutionary optimization. Science Press, Beijing
156. Cheng J, Liu ZY, Wu ZY, Tang MY, Tan JR (2016) Direct optimization of uncertain structures based on degree of interval constraint violation. Comput Struct 164:83–94
157. Wu J, Zhou SN (2012) Robustness optimization method for frequency and decoupling ratio of powertrain mounting systems. J Vib Shock 31(4):1–7
158. Wu JL, Luo Z, Zhang N, Zhang YQ (2015) A new interval uncertain optimization method for structures using Chebyshev surrogate models. Comput Struct 146:185–196
159. Xu B, Jin YJ (2014) Multiobjective dynamic topology optimization of truss with interval parameters based on interval possibility degree. J Vib Control 20(1):66–81
160. Cheng J, Liu ZY, Qian YM, Wu D, Zhou ZD, Gao W, Zhang J, Tan JR (2019) Robust optimization of uncertain structures based on interval closeness coefficients and the 3D violation vectors of interval constraints. Struct Multidiscip Optim 60(1):17–33
161. Wang LQ, Yang GL, Xiao H, Sun QQ, Ge JL (2020) Interval optimization for structural dynamic responses of an artillery system under uncertainty. Eng Optim 52(2):343–366
162. Li YL, Wang XJ, Huang R, Qiu ZP (2015) Actuator placement robust optimization for vibration control system with interval parameters. Aerosp Sci Technol 45:88–98
163. Badri SA, Ghazanfari M, Shahanaghi K (2014) A multi-criteria decision-making approach to solve the product mix problem with interval parameters based on the theory of constraints. Int J Adv Manuf Technol 70(5):1073–1080
164. Li YZ, Wu QH, Jiang L, Yang JB, Xu DL (2016) Optimal power system dispatch with wind power integrated using nonlinear interval optimization and evidential reasoning approach. IEEE Trans Power Syst 31(3):2246–2254

165. Huang CX, Yue D, Deng S, Xie J (2017) Optimal scheduling of microgrid with multiple distributed resources using interval optimization. Energies 10(3):339
166. Soltani M, Kerachian R, Nikoo MR, Noory H (2016) A Conditional value at risk-based model for planning agricultural water and return flow allocation in river systems. Water Resour Manag 30(1):427–443
167. Zarghami M, Safari N, Szidarovszky F, Islam S (2015) Nonlinear interval parameter programming combined with cooperative games: a tool for addressing uncertainty in water allocation using water diplomacy framework. Water Resour Manag 29(12):4285–4303

# Chapter 2
# The Basic Principles of Interval Analysis

**Abstract** This chapter briefly introduces the origin of interval number, the basic conceptions of interval mathematics, the calculation rules of interval number, and the important interval overestimation problem, providing some necessary theoretical foundations for interval optimization.

Interval number, or interval for short, consists of a pair of ordered real numbers and has the characteristics of both set and number. The mathematical analysis method that investigates interval number is called interval analysis or interval mathematics. The thought of interval mathematics could date back to the 1930s, when the British scholar Young [1], firstly described interval arithmetic and proposed the calculation rules of interval and real number set in 1931. A Russian paper published in 1951, firstly suggested to use interval arithmetic as a numerical computing tool and applied interval arithmetic to calculating the intervals of traditional mathematical expressions [2]. In the same year, Dwyer gave the calculation rules of interval number in his monograph "Linear Computations" [3]. In 1956, Polish scholar Warmus [4], also independently proposed a set of calculation rules of interval number. In the same year, Sunaga laid the algebraic foundation of applying interval to computer calculation in his master thesis [5]. Unfortunately, the thesis was handwritten thus it could not be widely spread and did not draw public attention. Sunaga's research was unknown to the public until 1958, when his another more important paper [6], that established the mathematical foundation for interval mathematics was published. However, the influence of Sunaga's research was still small due to a variety of reasons. The conception of interval analysis was firstly proposed by Moore and Yang in their academic report [7], in 1959. In 1962, Moore published his doctoral thesis [8], which helped the interval analysis really attract extensive attention. To check the numerical results automatically, Moore published a monograph "Interval Analysis" [9], based on his doctoral thesis in 1966, which is viewed as the theoretical foundation of the interval mathematics until now. In 1979, Moore published another monograph "Methods and Applications of Interval Analysis" [10], which preliminarily applied the interval mathematics to some practical engineering fields. Afterwards, the interval analysis theory achieved continuous growth and become an active branch


C. Jiang et al., *Nonlinear Interval Optimization for Uncertain Problems*, Springer Tracts in Mechanical Engineering, https://doi.org/10.1007/978-981-15-8546-3_2

of computational mathematics. In this field, there has been an international professional journal "Interval Computation", which was renamed "Reliable Computing" in 1995.

This chapter briefly introduces the origin of interval number, the basic conceptions of interval mathematics, the calculation rules of interval number, and interval overestimation problem, providing some necessary theoretical foundations for the construction of interval optimization methods in subsequent chapters.

## 2.1 The Origin of Interval Number

The rudiment of interval number originated in the story of Archimedes measuring the ratio of $\pi$ [11]: Archimedes made a circumscribed $m$-regular-polygon and an inscribed $n$-regular polygon on a unit circle and regarded the areas of these two polygons as the upper and lower bounds of area of this circle. Therefore, he obtained the area interval of the unit circle. As $m$ and $n$ increase, the interval width becomes narrower and narrower. Finally, a narrow enough interval number that contains $\pi$ can be obtained, and thus $\pi$ can be approximately obtained.

The invention of computers helped the interval number draw people's attention again. As we know, a decimal number must be converted to a binary number to be recognized by computers. Let's take a 32-bits computer as an example. According to the "IEEE Standard for Binary Floating-Point Arithmetic" [12], published by the Institute of Electrical and Electronics Engineers (IEEE), a floating-point number is packed into the computer with three parts: the sign bit, the exponent bit, and the mantissa bit, from left to right. Usually, the sign bit, which is denoted as the 31st bit for single precision floating-point numbers, represents the sign of the number. More specifically, 0 means a positive sign and 1 means negative sign. The exponent bit expressed with the 23rd–30th bits can range from $-2^7+1$ to $-2^7$, it is stored as an unsigned number by adding a fixed bias 127 to it in binary formats. Finally, the mantissa bit is the 0–22nd bits. Taking decimal 85.5 as an example, it is converted into a binary number as 1010101.1, which can be counted as $1.0101011 \times 2^6$ by using the scientific notation of binary. Since the IEEE standard requires that floating-point numbers must be standardized, this means that the left side of the decimal point of the converted binary decimal must be "1". In order to increase the counting range, it is usually omitted to free up one more binary bit. This is why the 22nd to 0th digits only need to represent the mantissa after the decimal point. As shown in Fig. 2.1, taking a decimal number 85.5 as an example, the sign bit is "0" and the exponent bit "6" is stored as an unsigned number 133 by adding a bias 127 to it, i.e., "1000 0101". The mantissa bit "010 1011" is appended with 0s to fit into 23 bits, which then becomes "010 1011 0000 0000 0000 0000".

Since the number of bits is finite, the ranges that can be expressed by the exponent bit and the digit are limited. For values exceeding such range, rounded or truncated approximation should be used. In IEEE754, it is stipulated that only the first 6–7 decimal points should be saved in a 32-bit single precision floating-point number.

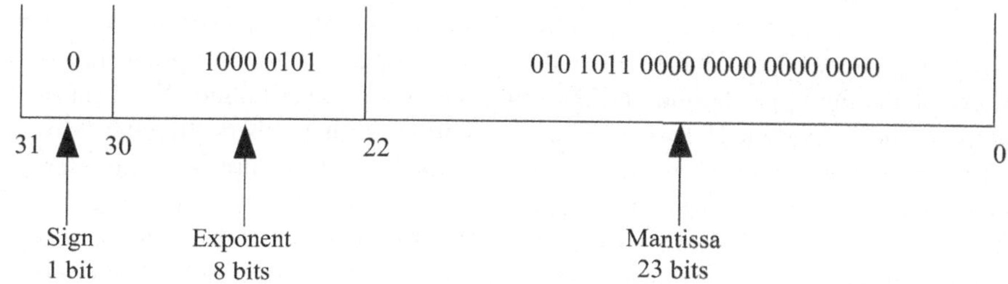


**Fig. 2.1** The binary single precision floating-point representation of a decimal number 85.5

Even if the number of digits of the computer is increased to 64 or 128 bits, it will also face a situation where the number of rational numeric digits exceeds the inherent number of digits in the system and cannot be represented. For example, the decimal number 85.49 needs to be approximated when using a 32-bit computer to store it because the fractional part 0.49 cannot be decomposed into a finite sum of powers of 2. On the contrary, it can only be converted into a binary number approximately as $1.01010101111101011100001010001111010111 \times 2^6$ (the number of digits is selected according to the accuracy requirements and in this case, 38 bits are used). The sign bit "0" and exponent bit "6" can be accurately expressed. But the mantissa bit of the number contains 38 digits, which are far beyond the 23-digit representation range. According to the provisions of IEEE754 for the treatment of overflow values, it can only be expressed as shown in Fig. 2.2.

The actual value of the above binary number is 85.48999786376953125, which is very close to the original one 85.49. Usually, it is assumed that such an approximation is negligible and would not result in a too big effect on computation. But this is not always true. Considering an infinite sequence [13]: $\{x_n\}$, $x_0 = 1$, $x_1 = \frac{1}{3}$: $i \geq 1$, $x_{i+1} = \frac{13}{3}x_i - \frac{4}{3}x_{i-1}$. If using floating-point numbers to represent this sequence, it is easy to reach the wrong conclusion that the sequence diverges. However, according to mathematic induction, this sequence equals to $\left(\frac{1}{3}\right)^n$ and will converge to 0 as $n$ increases. This problem will be even more serious for irrational numbers which must be approximated in order to be stored in computer. Moore incisively realized that such unreliability in representing irrational number is totally caused by the

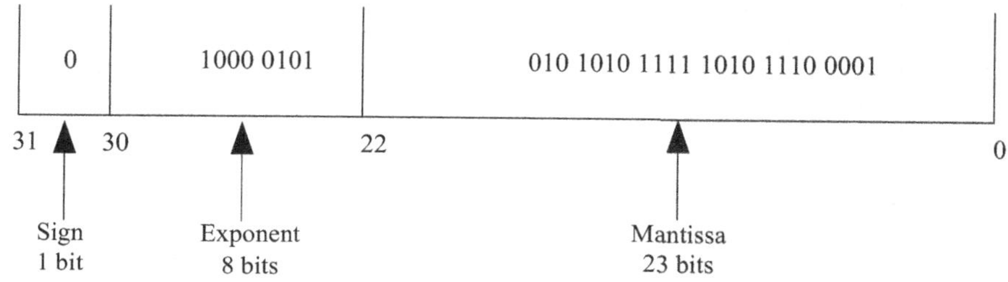


**Fig. 2.2** The binary single precision floating-point representation of a decimal number 85.49

approximated floating-point representations of real numbers. He proposed a new way to express an irrational number by using two nearest floating-point numbers where one is the upper bound and the other one is the lower bound. The irrational number is hence confined between these two floating-point numbers. Together Moore defined the corresponding interval calculation rules, which ensures that the interval containing the accurate computational result can be obtained. So far, interval number has become an active research area. The calculation rules of interval number and functions with respect to interval numbers were developed and formatted, leading to the corresponding theoretical system [14]. In a way, it was the demand for computing science that drove the early development of interval analysis theory.

## 2.2 The Basic Conceptions of Interval Mathematics

As shown in Fig. 2.3, a bounded real number closed set is called interval or interval number [10]:

$$A^I = \left[A^L, A^R\right] = \left\{x \middle| A^L \le x \le A^R, x \in R\right\} \tag{2.1}$$

where the superscripts $I$, $L$, $R$ represent the interval, the lower bound of interval and the upper bound of interval, respectively. An interval can be viewed as a pair of ordered real numbers consisting of two bounds, and it can also be called the interval number. The set containing all the interval numbers is denoted as $IR$. The sets that satisfy $A^L \ge 0$ or $A^R \le 0$ are denoted as $IR^+$ and $IR^-$, respectively.

Two interval numbers are equal if they have the same bounds. That is to say, when $A^L = B^L$ and $A^R = B^R$, we have $A^I = B^I$. An interval number $A^I = \left[A^L, A^R\right]$ is called point interval number when $A^L = A^R = A$. In such a situation, the interval number degenerates a real number. The inequality relation of a real number can also be extended to the interval number, i.e., $A^I < B^I$ if and only if $A^R < B^L$; $A^I > B^I$ if and only if $A^L > B^R$.

The width of an interval number $A^I$ is denoted as $\omega\left(A^I\right)$:

$$\omega\left(A^I\right) = A^R - A^L \tag{2.2}$$

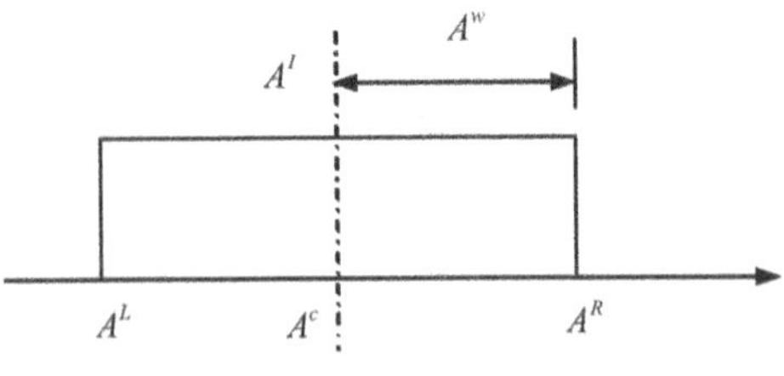


**Fig. 2.3** Geometric description of interval number

The midpoint of $A^I$, $m(A^I)$, represents the mean of $A^I$, and it can also be denoted as $A^c$:

$$A^c = \frac{A^L + A^R}{2} \tag{2.3}$$

The radius of $A^I$, which can also be called the deviation of $A^I$, is denoted as $A^w$:

$$A^w = \frac{A^R - A^L}{2} \tag{2.4}$$

An arbitrary interval $A^I = [A^L, A^R]$ can be represented by its midpoint $A^c$ and radius $A^w$ in the following form:

$$A^I = A^c \pm A^w \tag{2.5}$$

Equation (2.5) can also be written in the form of set:

$$A^I = \langle A^c, A^w \rangle = \{x | A^c - A^w \le x \le A^c + A^w\} \tag{2.6}$$

The absolute value of an interval number $|A^I|$ is defined as:

$$|A^I| = \max(|A^L|, |A^R|) \tag{2.7}$$

Therefore, $\forall x \in X^I$, we have:

$$|x| \le |A^I| \tag{2.8}$$

An ordered interval number array $\mathbf{A}^I = (A_1^I, A_2^I, \ldots, A_n^I)$ is called an interval vector. From the point of geometry, the interval vector is an $n$-dimensional cuboid in the variable space, and its width is defined as:

$$\omega(\mathbf{A}^I) = \max(\omega(A_1^I), \omega(A_2^I), \ldots, \omega(A_n^I)) \tag{2.9}$$

The norm of the interval vector is defined as:

$$\|\mathbf{A}^I\| = \max(|A_1^I|, |A_2^I|, \ldots, |A_n^I|) \tag{2.10}$$

The midpoint vector of the interval vector is defined as:

$$\mathbf{m}(\mathbf{A}^I) = (m(A_1^I), m(A_2^I), \ldots, m(A_n^I)) \tag{2.11}$$

Similarly, a matrix whose elements $A_{ij}$ are interval numbers, denoted as $\mathbf{A}^I = (A_{ij}^I)$, is called the interval matrix whose width, norm, and midpoint matrix are defined as:

$$\omega(\mathbf{A}^I) = \max_{i,j}(\omega(A_{ij}^I)) \tag{2.12}$$

$$\|\mathbf{A}^I\| = \max_i \sum_j |A_{ij}^I| \tag{2.13}$$

$$m(\mathbf{A}^I) = (m(A_{ij}^I)) \tag{2.14}$$

If any real matrix A $\in \mathbf{A}^I$ is a symmetric matrix, $\mathbf{A}^I$ is called a symmetric interval matrix.

## 2.3 The Basic Arithmetic Operations of Interval Number

There are four elementary arithmetic operations between two interval numbers $A^I = [A^L, A^R]$ and $B^I = [B^L, B^R]$ and they are defined as follows [10]:

$$\begin{cases} A^I + B^I = [A^L, A^R] + [B^L, B^R] = [A^L + B^L, A^R + B^R] \\ A^I - B^I = [A^L, A^R] - [B^L, B^R] = [A^L - B^R, A^R - B^L] \\ A^I \times B^I = [A^L, A^R] \times [B^L, B^R] = \begin{bmatrix} \min(A^L B^L, A^L B^R, A^R B^L, A^R B^R), \\ \max(A^L B^L, A^L B^R, A^R B^L, A^R B^R) \end{bmatrix} \\ A^I \div B^I = [A^L, A^R] \div [B^L, B^R] = [A^L, A^R] \times \left[\frac{1}{B^R}, \frac{1}{B^L}\right], \quad 0 \notin [B^L, B^R] \end{cases} \tag{2.15}$$

If $0 \in [B^L, B^R]$ and $B^L \neq B^R$, we have:

$$A^I \div B^I = \begin{cases} [A^R/B^L, \infty), & (A^R \le 0, B^R = 0) \\ (-\infty, A^R/B^R] \cup [A^R/B^R, \infty), & (A^R \le 0, B^L < 0 < B^R) \\ (-\infty, A^R/B^R], & (A^R \le 0, B^L = 0) \\ (-\infty, \infty), & (A^L \le 0 \le A^R) \\ (-\infty, A^L/B^L], & (A^L \ge 0, B^R = 0) \\ (-\infty, A^L/B^L] \cup [A^L/B^R, \infty), & (A^L \ge 0, B^L < 0 < B^R) \\ [A^L/B^R, \infty), & (A^L \ge 0, B^L = 0) \end{cases} \tag{2.16}$$

where $\left(A^R / B^L, \infty\right)$ and its similar forms are called the semi-infinite interval, and $(-\infty, \infty)$ is called the infinite interval.

The power operations of interval number are defined as follows:

$$\left(A^I\right)^n = \begin{cases} \left[0, \max\left(\left(A^L\right)^n, \left(A^R\right)^n\right)\right], & \left(n = 2k, 0 \in A^I\right) \\ \left[\min\left(\left(A^L\right)^n, \left(A^R\right)^n\right), \max\left(\left(A^L\right)^n, \left(A^R\right)^n\right)\right], & \left(n = 2k, 0 \notin A^I\right) \\ \left[\left(A^L\right)^n, \left(A^R\right)^n\right], & (n = 2k + 1) \end{cases} \tag{2.17}$$

where $k$ is the nonnegative integer.

Besides, only parts of the arithmetic operations for the real number are also suitable for interval number, and others are in the weak forms. For example, the following operations are also suitable for interval numbers $A^I$, $B^I$, and $C^I$.

The commutative law:

$$A^I + B^I = B^I + A^I \tag{2.18}$$

$$A^I \times B^I = B^I \times A^I \tag{2.19}$$

The associative law:

$$\left(A^I + B^I\right) \pm C^I = A^I + \left(B^I \pm C^I\right) \tag{2.20}$$

$$\left(A^I \times B^I\right) \times C^I = A^I \times \left(B^I \times C^I\right) \tag{2.21}$$

The identity law:

$$A^I + 0 = 0 + A^I; \qquad A^I \times 1 = 1 \times A^I \tag{2.22}$$

$$A^I - B^I = A^I + \left(-B^I\right) = -B^I + A^I \tag{2.23}$$

$$A^I / B^I = A^I \times \left(B^I\right)^{-1} = \left(B^I\right)^{-1} \times A^I, \qquad 0 \notin \left[B^L, B^R\right] \tag{2.24}$$

$$-\left(A^I - B^I\right) = B^I - A^I \tag{2.25}$$

$$A^I \times \left(-B^I\right) = \left(-A^I\right) \times B^I = -\left(A^I \times B^I\right) \tag{2.26}$$

$$A^I - \left(B^I \pm C^I\right) = \left(A^I - B^I\right) \pm C^I \tag{2.27}$$

$$\left(-A^I\right) \times \left(-B^I\right) = A^I \times B^I \tag{2.28}$$

Besides, there are weak-form arithmetic operations for interval numbers as follows:

The sub distributive law:

$$A^I \times \left(B^I \pm C^I\right) \subseteq A^I \times B^I \pm A^I \times C^I \tag{2.29}$$

$$\left(A^I \pm B^I\right) \times C^I \subseteq A^I \times C^I \pm B^I \times C^I \tag{2.30}$$

The cancellation law:

$$A^I - B^I \subseteq \left(A^I + C^I\right) - \left(B^I + C^I\right) \tag{2.31}$$

$$A^I / B^I \subseteq \left(A^I \times C^I\right) / \left(B^I \times C^I\right) \tag{2.32}$$

$$0 \in A^I - A^I, \qquad 1 \in A^I / A^I \tag{2.33}$$

## 2.4 The Overestimation Problem in Interval Arithmetic

The interval function is an extension of a real-valued function, namely, replacing the real variables in the real-valued function with corresponding interval variables to obtain an interval-valued function. When estimating the range of value for an interval function through the above interval arithmetic operations, this range value, in general, will be amplified or overestimated. For example, when calculating the range of value for an interval function $F\left(X^I\right) = X_1^I - X_1^I$, where $X_1^I = [0, 1]$, according to the subtraction rule of interval number as shown in Eq. (2.15), we can obtain $F\left(X^I\right) = [0, 1] - [0, 1] = [-1, 1]$. However, the actual result should be 0. This example demonstrates that the interval arithmetic generally results in an overestimation problem. The reason is the dependency between interval variables. The degree of overestimation is closely related to the number of times that the same interval variable appears. The more the number of an interval variable appearing in the interval function, the severer the interval overestimation phenomenon. For example, assuming $X^I = [0, 1]$, we can modify a function in three different forms, and calculate the range of value for these functions by the interval arithmetic operations:

$$\begin{cases} F_1(X^I) = \frac{1}{4} - \left(X^I - \frac{1}{2}\right)^2 = \left[0, \frac{1}{4}\right] \\ F_2(X^I) = X^I(1 - X^I) = [0, 1] \\ F_3(X^I) = X^I - (X^I)^2 = [-1, 1] \end{cases} \tag{2.34}$$

Obviously, the results are quite different. In $F_1(X^I)$, $X^I$ appears only once and hence the result is not expanded. However, $X^I$ appears twice in both $F_2(X^I)$ and $F_3(X^I)$, which hence causes the interval overestimation phenomenon.

At present, how to create more accurate computing methods to overcome the interval overestimation problem has become an important research direction in interval analysis field. Current methods for reducing the interval overestimation effects include the truncated method [15, 16], the perturbation method [17], the subinterval method [18, 19], the modified interval arithmetic operations considering the dependency [20], etc. However, such methods are only suitable for some relatively simple problems. How to develop a new interval analysis method that is universally applicable for complex nonlinear interval functions, and how to develop accurate interval analysis methods for differential equation problems will be the research emphasis in this field in future.

## 2.5 Summary

Firstly, this chapter briefly introduces the origin of interval number and demonstrates the necessity of interval analysis through an example of floating number calculation. Secondly, the basic conceptions of interval mathematics, the arithmetic operations of interval number, and the interval overestimation phenomenon are introduced, which provide some necessary and important theoretical foundations for the construction of interval optimization methods in subsequent chapters.

## References


1. Young RC (1931) The algebra of many-valued quantities. Math Ann 104(1):260–290
2. Grell H, Maruhn K, Rinow W (1966) Enzyklopadie der elementar mathematik, Band I Arithmetik, Dritte Auage. VEB Deutscher Verlag der Wissenschaften, Berlin
3. Dwyer P (1951) Linear computations. Wiley, New York
4. Warmus M (1956) Calculus of approximations. Bulletin De l'Academie Polonaise De Sciences 4(5):253–257
5. Sunaga T (1956) Geometry of numerals. Master thesis, University of Tokyo, Tokyo
6. Moore RE, Yang CT (1958) Theory of an interval algebra and its application to numeric analysis. RAAG Memories II, Gaukutsu Bunken Fukeyu-kai, Tokyo
7. Moore RE, Yang CT (1959) Interval analysis I. Technical document LMSD-285875, Lockheed Missiles and Space Division, Sunnyvale, CA, USA.

8. Moore RE (1962) Interval arithmetic and automatic error analysis in digital computing. PhD thesis, Stanford University, Stanford
9. Moore RE (1966) Interval analysis. Prentice-Hall, New Jersey
10. Moore RE (1979) Methods and applications of interval analysis. Prenice-Hall, London
11. Alefeld G, Mayer G (2000) Interval analysis: theory and applications. J Comput Appl Math 121(1–2):421–464
12. Stevenson D (1985) IEEE standard for binary floating point arithmetic. Technical report, IEEE/ANSI 754–1985, IEEE
13. Hu CY, Xu SY, Yang XG (2003) A brief introduction to the interval methods. Syst Eng Theory Pract 4(12):59–62
14. Sui JB (2006) Research on engineering structure uncertainty interval analysis method and its application. PhD thesis, Hohai University, Nanjing
15. Lv ZZ, Feng YW, Yue ZF (2002) A advanced interval-truncation approach and non-probabilistic reliability analysis based on interval analysis. Chinese J Computat Mech 3:260–264
16. Rao SS, Berke L (1997) Analysis of uncertain structural systems using interval analysis. AIAA J 35(4):727–735
17. Qiu ZP, Wang XJ (2005) Parameter perturbation method for dynamic responses of structures with uncertain-but-bounded parameters based on interval analysis. Int J Solids Struct 42(18–19):4958–4970
18. Qiu ZP (1994) Interval analysis method of static response of structures with uncertain parameters and the eigenvalue problem. PhD thesis, Jilin Industrial University, Changchun
19. Zhou YT, Jiang C, Han X (2006) Interval and subinterval analysis methods of the structural analysis and their error estimations. Int J Comput Methods 3(2):229–244
20. Jiang C, Fu CM, Ni BY, Han X (2016) Interval arithmetic operations for uncertainty analysis with correlated interval variables. Acta Mech Sin 32(4):743–752

# Chapter 3
# Mathematical Transformation Models of Nonlinear Interval Optimization

**Abstract** This chapter proposes two types of mathematical transformation models for general nonlinear interval optimization problems, i.e., the transformation model based on order relation of interval number and the transformation model based on possibility degree of interval number. Therefore, the uncertain optimization problem is transformed into a deterministic optimization. Furthermore, a two-layer optimization algorithm is established to solve the transformed deterministic optimization problem.

Generally, the uncertain optimization problem needs to be transformed into a deterministic optimization problem using a mathematical transformation model. In stochastic programming and fuzzy programming, such transformations are based on the probability theory and the fuzzy set theory, respectively. In interval optimization, the transformation is generally based on the order relation of interval number [1–3] or the minimax regret criterion [4–6]. The construction of a suitable mathematical transformation model is a primary and crucial task in interval optimization. Most current interval optimization methods aim at linear problems where the intervals of objective function and constraints can be explicitly obtained in the optimization procedure. Therefore, the transformation model of linear problems is much simpler than that of nonlinear problems. In previous research on nonlinear interval optimization [7], transformation models for general nonlinear interval optimization problems are deficient, which, to some degree, hinders the research and engineering application of interval optimization.

In this chapter, two types of mathematical transformation models are hence proposed for general nonlinear interval optimization problems, in which the objective function and constraints are both nonlinear and uncertain, and both inequality constraints and equality constraints appear. In these two models, the treatments to constraints are the same. According to the methods for handling the objective function, these two models are called *the mathematical transformation model based on order relation of interval number* and *the mathematical transformation model based on possibility degree of interval number*, respectively. For the convenience of expression, sometimes they are also called *the order relation transformation model* and *the possibility degree transformation model* in this book, respectively. The basic idea



C. Jiang et al., *Nonlinear Interval Optimization for Uncertain Problems*, Springer Tracts in Mechanical Engineering, https://doi.org/10.1007/978-981-15-8546-3_3

of the first transformation model is to extend the order relation of interval number that is often used in linear interval optimization into the nonlinear interval optimization problems, and then convert the original problem into a deterministic multi-objective optimization problem. The second transformation model, however, introduces a performance interval to maximize the possibility degree of objective function and then converts the uncertain objective function into a deterministic objective function. No matter which model we adopt, a deterministic nested optimization will be obtained after the transformation. Thus, we also present a two-layer optimization algorithm based on the intergeneration projection genetic algorithm (IP-GA) [8], for solving the nested optimization. In the end of this chapter, these two mathematical transformation models are applied to a numerical example.

## 3.1 The Description of a General Nonlinear Interval Optimization Problem

In traditional optimization problems, the parameters or coefficients in the optimization models are generally given by precise values. Thus, the values of the objective function and constraints at any given design point can be accurately calculated. However, many parameters are uncertain in practical engineering problems, which may lead to uncertainty in calculating the values of objective function and constraints. Therefore, the important uncertain optimization problem arises. According to the different mathematical tools used to model the uncertainties, the uncertain optimization methods can be classified into different categories such as the stochastic programming, the fuzzy programming and the interval optimization. A general nonlinear interval optimization problem, i.e., the research object of this book, can be expressed as follows:

$$\begin{cases} \min\limits_{\mathbf{X}} f(\mathbf{X}, \mathbf{U}) \\ \text{s.t. } g_i(\mathbf{X}, \mathbf{U}) \leq (=, \geq) b_i^I = \left[b_i^L, b_i^R\right],\ i = 1, 2, \ldots, l,\ \mathbf{X} \in \Omega^n \\ \quad \mathbf{U} \in \mathbf{U}^I = \left[\mathbf{U}^L, \mathbf{U}^R\right],\ U_i \in U_i^I = \left[U_i^L, U_i^R\right],\ i = 1, 2, \ldots, q \end{cases} \tag{3.1}$$

where $\mathbf{X} \in \Omega^n$ represents the $n$-dimensional design vector; $\mathbf{U}$ represent the $q$-dimensional uncertain vector that is described using a $q$-dimensional interval vector $\mathbf{U}^I$; $f$ and $g$ represent the objective function and the constraint, respectively; $l$ denotes the number of the constraints; $b_i^I$ represents an allowable interval of the $i$-th constraint, and in practical problems $b_i^I$ is allowed to be a real number. Here, $f$ and $g$ are all continuous and differentiable functions with respect to $\mathbf{X}$ and $\mathbf{U}$, and at least one function among the objective function and all the constraints are nonlinear with respect to $\mathbf{X}$. In addition, in this book, the objective function and constraint that contain interval parameters are also called interval objective function or interval constraint, respectively.

The values of the objective function $f(\mathbf{X}, \mathbf{U})$ or the constraint $g_i(\mathbf{X}, \mathbf{U})$ at a specific design vector are both intervals since $f(\mathbf{X}, \mathbf{U})$ and $g_i(\mathbf{X}, \mathbf{U})$ are continuous functions with respect to the interval vector $\mathbf{U}$. For this reason, Eq. (3.1), cannot be solved by traditional optimization methods that need precise values of the objective function and constraints at design vector. The rest of this chapter will first propose the treatment for the uncertain constraints. Afterwards, based on the different strategies to deal with the objective function, two different transformation models for nonlinear interval optimization will be constructed.

## 3.2 Possibility Degree of Interval Number and Transformation of Uncertain Constraints

Two real numbers can be compared regarding their values. However, two intervals cannot be compared directly since an interval is a set of real numbers. Therefore, a type of mathematical tool called the ranking model must be developed to compare the intervals. To clarify the conception and make the expression more convenient, the ranking models to compare the intervals are classified into two categories in this book. The first category is called the order relation of interval number, which is used to qualitatively judge if an interval is greater or superior to another interval. The other is called the possibility degree of interval number (or satisfaction degree, acceptability index, etc.), which is used to quantitatively describe the degree that an interval is greater or superior to another one.

### *3.2.1 An Improved Possibility Degree of Interval Number*

At present, several methods for constructing the possibility degree of interval number have been proposed. A possibility degree of interval number based on the fuzzy set was developed by Nakahara et al. [9]. Kundu [10] proposed a kind of possibility degree of interval number using the min transitivity of fuzzy leftness relationship. Sengupta and Pal [11] proposed the conception of "acceptability index" based on the mean and radius of the interval, and gave a utility function to compare interval numbers in optimistic or pessimistic attitudes. Liu and Da [12] presented a modified method to construct the possibility degree of interval number based on [13–15]. Xu and Da [16] proposed a kind of possibility degree of interval number to obtain a possibility degree matrix for interval ranking, and they further proved the equivalence between this model and those given in [17, 18].

Essentially, the above methods are all based on the fuzzy set to construct the possibility degree. However, it is known that the subjectivity will be inevitable in the selection of membership function when constructing a fuzzy set. In order to provide a more strict and objective mathematical explanation for possibility degree

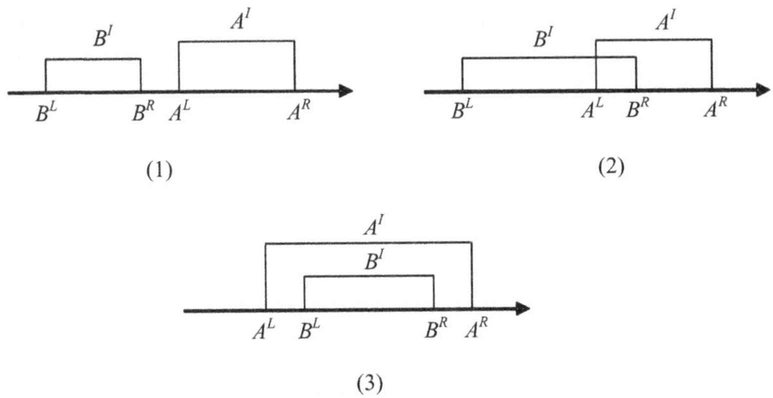


**Fig. 3.1** Three position relationships between intervals $A^I$ and $B^I$ [19]

of interval number, Zhang et al. [19], proposed a new approach to construct the possibility degree by introducing the probability theory. For the three cases as shown in Fig. 3.1, the possibility degree that the interval $A^I$ is greater or superior to another interval $B^I$, i.e., $P\left(A^I \geq B^I\right)$, is constructed as follows [19]:

$$P\left(A^I \geq B^I\right) = \begin{cases} 1, & A^L \geq B^R \\ \frac{A^R - B^R}{A^R - A^L} + \frac{B^R - A^L}{A^R - A^L} \cdot \frac{A^L - B^L}{B^R - B^L} + 0.5 \cdot \frac{B^R - A^L}{A^R - A^L} \cdot \frac{B^R - A^L}{B^R - B^L}, & B^L \leq A^L < B^R \leq A^R \\ \frac{A^R - B^R}{A^R - A^L} + 0.5 \cdot \frac{B^R - B^L}{A^R - A^L}, & A^L < B^L < B^R < A^R \end{cases} \tag{3.2}$$

Here the interval numbers $A^I$ and $B^I$ are regarded as random variables $\tilde{A}$ and $\tilde{B}$ with a uniform distribution in their intervals. The possibility degree $P\left(A^I \geq B^I\right)$ can then be obtained by calculating the probability that the random variable $\tilde{A}$ is greater than $\tilde{B}$. Taking the third case shown in Fig. 3.1, as an example, the probability that $\tilde{A}$ is between $A^R$ and $B^R$ is $\frac{A^R - B^R}{A^R - A^L}$. In this situation, no matter what value of $\tilde{B}$ is, the probability for $\tilde{A} \geq \tilde{B}$ is always 1. The probability that $\tilde{A}$ is between $B^L$ and $B^R$ is $\frac{B^R - B^L}{A^R - A^L}$, and in this situation the probability for $\tilde{A} \geq \tilde{B}$ is 0.5. The probability that $\tilde{A}$ is between $A^L$ and $B^L$ is $\frac{B^L - A^L}{A^R - A^L}$, and the probability for $\tilde{A} \geq \tilde{B}$ in this situation is 0. Finally, the probability for $\tilde{A} \geq \tilde{B}$ in this case is obtained as $\frac{A^R - B^R}{A^R - A^L} + 0.5 \cdot \frac{B^R - B^L}{A^R - A^L}$. Similarly, the possibility degree $P\left(B^I \geq A^I\right)$ can be also obtained through the above method [19]

$$P\left(B^I \geq A^I\right) = \begin{cases} 0, & A^L \geq B^R \\ 0.5 \cdot \frac{B^R - A^L}{A^R - A^L} \cdot \frac{B^R - A^L}{B^R - B^L}, & B^L \leq A^L < B^R \leq A^R \\ \frac{B^L - A^L}{A^R - A^L} + 0.5 \cdot \frac{B^R - B^L}{A^R - A^L}, & A^L < B^L < B^R < A^R \end{cases} \tag{3.3}$$

It can be seen that the mathematical meanings and objectivity of the possibility degree of interval number have been strengthened by introducing the probability method. Through such a treatment, the possibility degree of interval number becomes more intuitional, which helps the decision maker understand and use it. However, this method still has its limitations in the following two aspects:

(1) The construction of the possibility degree is based on three position cases as shown in Fig. 3.1, while these three cases are only part of all the possible cases. Therefore, we need to use two formulas, i.e., Eqs. (3.2) and (3.3), to compare the same two interval numbers, which whereby affects the convenience of the method.
(2) The situation that one of the intervals degenerates into a real number is not taken into consideration. However, such a situation is common in practical interval optimization problems, which also affects the practicality of the method.

Aiming at overcoming these shortcomings, an improved possibility degree of interval number is proposed in this chapter. Figure 3.2 lists all the possible six positions of $A^I$ and $B^I$, and by also introducing the above probability approach $P\left(A^I \leq B^I\right)$ can be expressed based on the positions [20]

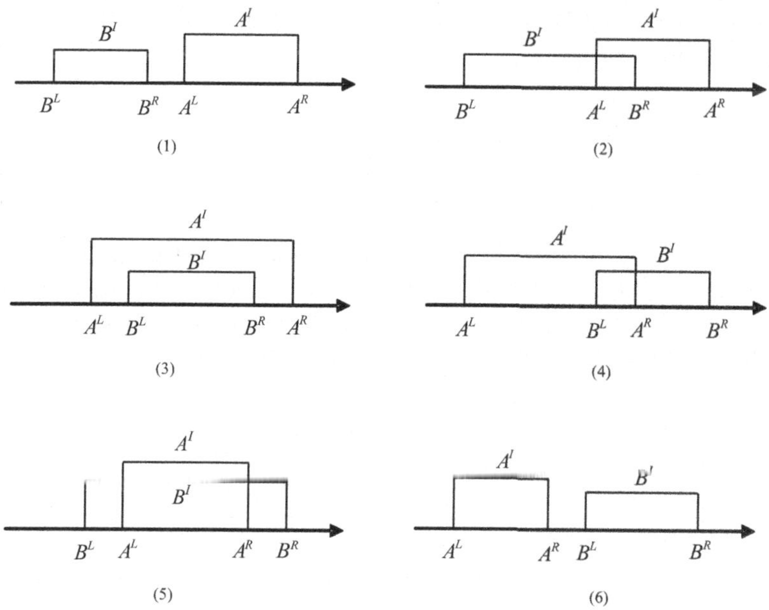


**Fig. 3.2** Six possible position relationships between intervals $A^I$ and $B^I$ [20]

$$P\left(A^I \le B^I\right) = \begin{cases} 0, & A^L \ge B^R \\ 0.5 \cdot \frac{B^R - A^L}{A^R - A^L} \cdot \frac{B^R - A^L}{B^R - B^L}, & B^L \le A^L < B^R \le A^R \\ \frac{B^L - A^L}{A^R - A^L} + 0.5 \cdot \frac{B^R - B^L}{A^R - A^L}, & A^L < B^L < B^R \le A^R \\ \frac{B^L - A^L}{A^R - A^L} + \frac{A^R - B^L}{A^R - A^L} \cdot \frac{B^R - A^R}{B^R - B^L} + 0.5 \cdot \frac{A^R - B^L}{A^R - A^L} \cdot \frac{A^R - B^L}{B^R - B^L}, & A^L < B^L \le A^R < B^R \\ \frac{B^R - A^R}{B^R - B^L} + 0.5 \cdot \frac{A^R - A^L}{B^R - B^L}, & B^L \le A^L < A^R < B^R \\ 1, & A^R < B^L \end{cases} \tag{3.4}$$

$P\left(A^I \le B^I\right)$ quantitatively gives the degree that the interval $A^I$ is less or inferior to another interval $B^I$, and it has the following properties:

(1) $0 \le P\left(A^I \le B^I\right) \le 1$;
(2) $A^I = B^I$ if $P\left(A^I \le B^I\right) = P\left(B^I \le A^I\right)$;
(3) $P\left(A^I \le B^I\right) = 0$ represents it is impossible for $A^I \le B^I$, i.e., $A^I$ must be totally greater than or equal to $B^I$;
(4) $P\left(A^I \le B^I\right) = 1$ represents $A^I$ must be totally less than or equal to $B^I$;
(5) $P\left(B^I \le A^I\right) = 1 - a$ if $P\left(A^I \le B^I\right) = a$.

If $B^I$ degenerates into a real number, denoted as $b$, the possible positions of $A^I$ and $b$ can be seen in Fig. 3.3. Based on these positions, the possibility degree $P\left(A^I \le b\right)$ has the following form:

$$P\left(A^I \le b\right) = \begin{cases} 0, & b \le A^L \\ \frac{b - A^L}{A^R - A^L}, & A^L < b \le A^R \\ 1, & b > A^R \end{cases} \tag{3.5}$$

In Eq. (3.5), only $A^I$ is regarded as a random variable $\tilde{A}$ with the uniform distribution in its interval. The probability for $\tilde{A} \le b$ is then regarded as the possibility degree $P\left(A^I \le b\right)$. Similarly, when $A^I$ degenerates into a real number $a$ (see Fig. 3.4), the possibility degree $P\left(a \le B^I\right)$ has the following form:

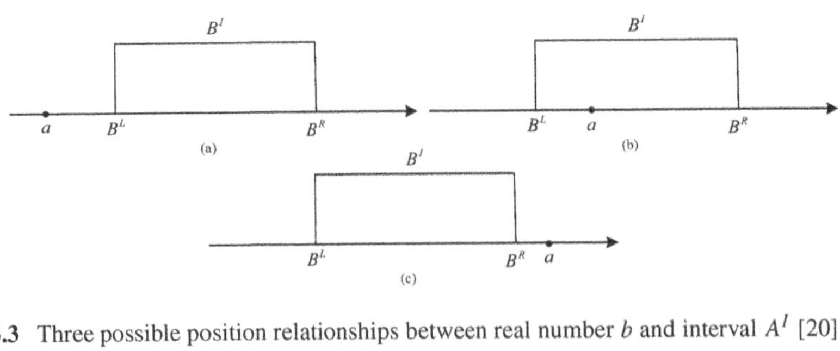


**Fig. 3.3** Three possible position relationships between real number $b$ and interval $A^I$ [20]

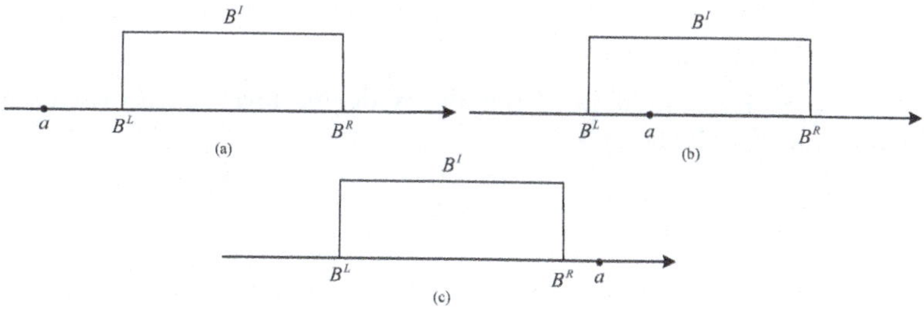


**Fig. 3.4** Three possible position relationships between real number $a$ and interval $B^I$ [20]

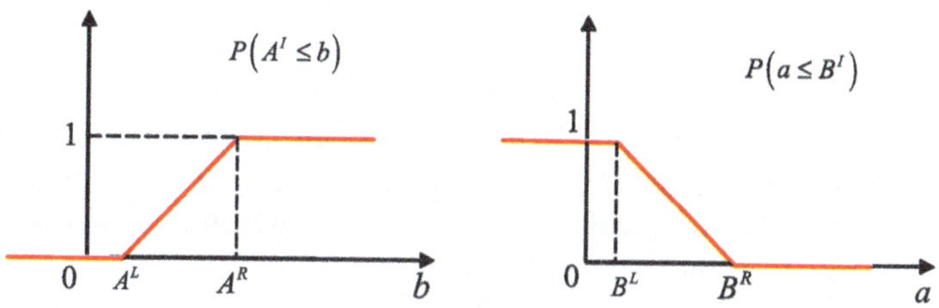


**Fig. 3.5** Geometric description for the interval possibility degree $P(A^I \le b)$ and $P(a \le B^I)$ [21]

$$P\left(a \le B^I\right) = \begin{cases} 1, & a \le B^L \\ \frac{B^R - a}{B^R - B^L}, & B^L < a \le B^R \\ 0, & a > B^R \end{cases} \tag{3.6}$$

Figure 3.5 gives the geometric description of $P\left(A^I \le b\right)$ and $P\left(a \le B^I\right)$. The possibility degree is a linear function with respect to $b$ or $a$ when $b \in \left[A^L, A^R\right]$ or $a \in \left[B^L, B^R\right]$.

### 3.2.2 Transformation of Uncertain Constraints Based on Possibility Degree of Interval Number

As mentioned in Chap. 1, in chance-constrained stochastic programming [22, 23], the random constraints can be transformed into deterministic constraints by satisfying the constraints under a given confidence level. In a similar way, we can make an interval constraint to be satisfied under a given possibility degree level and whereby transform it into a deterministic constraint. Actually, such treatment is often adopted for inequality constraints in linear interval optimization. Here, we extend it to the nonlinear interval optimization problems. For a "≤" constraint in Eq. (3.1), such as $g_i(\mathbf{X}, \mathbf{U}) \le b_i^I$, it thus can be converted into the following deterministic inequality constraint

$$P\left(g_i^I(\mathbf{X}) \le b_i^I\right) \ge \lambda_i \tag{3.7}$$

where $0 \le \lambda_i \le 1$ is a predefined possibility degree level. $g_i^I(\mathbf{X})$ represents the possible interval of $g_i(\mathbf{X}, \mathbf{U})$ at $\mathbf{X}$:

$$g_i^I(\mathbf{X}) = \left[g_i^L(\mathbf{X}), g_i^R(\mathbf{X})\right] \tag{3.8}$$

Unlike in linear interval optimization, $g_i^I(X)$ cannot be explicitly obtained. Here we use the approach in [23] to obtain $g_i^I(X)$ through the optimization method:

$$\begin{aligned} &g_i^L(\mathbf{X}) = \min_{\mathbf{U}} g_i(\mathbf{X}, \mathbf{U}),\ g_i^R(\mathbf{X}) = \max_{\mathbf{U}} g_i(\mathbf{X}, \mathbf{U}) \\ &\mathbf{U} \in \Gamma = \left\{\mathbf{U} \middle| U_i^L \le U_i \le U_i^R,\ i = 1, 2, \ldots, q\right\} \end{aligned} \tag{3.9}$$

Once $g_i^I(\mathbf{X})$ is obtained, the possibility degree of interval number of the constraint $P\left(g_i^I(\mathbf{X}) \le b_i^I\right)$ can be calculated through Eq. (3.4) or Eq. (3.5), according to whether $b_i^I$ is an interval number or a real number, and hence whether the given possibility degree level is satisfied can be determined.

For the "≥" constraints, such as $g_i(\mathbf{X}, \mathbf{U}) \ge b_i^I$, it can be converted into the following deterministic inequality constraint:

$$P\left(g_i^I(\mathbf{X}) \ge b_i^I\right) = P\left(b_i^I \le g_i^I(\mathbf{X})\right) \ge \lambda_i \tag{3.10}$$

where $P\left(b_i^I \le g_i^I(\mathbf{X})\right)$ can be calculated by Eq. (3.4) or Eq. (3.6).

The uncertain equality constraints can be transformed into inequality constraints. For example, the equality constraint $g_i(\mathbf{X}, \mathbf{U}) = b_i^I$ can be transformed into the following form:

$$b_i^L \le g_i(\mathbf{X}, \mathbf{U}) \le b_i^R \tag{3.11}$$

The above equation can be further converted into the following two inequality constraints:

$$\begin{cases} b_i^L \le g_i(\mathbf{X}, \mathbf{U}) \\ g_i(\mathbf{X}, \mathbf{U}) \le b_i^R \end{cases} \tag{3.12}$$

Since the inequality constraints are obtained, we can use the above transformation approach for inequality constraints to deal with Eq. (3.12), and obtain

$$\begin{cases} P\left(b_i^L \le g_i^I(\mathbf{X})\right) \ge \lambda_{i1} \\ P\left(g_i^I(\mathbf{X}) \le b_i^R\right) \ge \lambda_{i2} \end{cases} \tag{3.13}$$

where $P\left(b_i^L \le g_i^I(\mathbf{X})\right)$ and $P\left(g_i^I(\mathbf{X}) \le b_i^R\right)$ can be calculated by Eq. (3.6) and Eq. (3.5), respectively.

Therefore, the uncertain constraints in Eq. (3.1), are converted into deterministic constraints and can be expressed in a unified form:

$$P\left(M_i^I \leq N_i^I\right) \geq \lambda_i,\ i = 1, 2, \ldots, k \tag{3.14}$$

where $k > l$ if there exist the uncertain equality constraints. Whether $M_i^I$ and $N_i^I$ are intervals or real numbers are determined by not only by the form of $b_i^I$ but also by the conversion procedure of the uncertain constraints.

## 3.3 The Mathematic Transformation Model Based on Order Relation of Interval Number

### *3.3.1 Order Relation of Interval Number and Transformation of Uncertain Objective Function*

The order relation of interval number is used to qualitatively judge if an interval is greater or lesser than another, which is often used in linear interval optimization to deal with the uncertain objective function. As mentioned above, the value of the interval objective function at a specific design vector is also an interval, rather than a real number. Therefore, in interval optimization, we should compare the values of the objective function at different design vectors to judge which design vector is better to find the optimal one. The order relation generally has different expressions in maximization and minimization optimization problems since the evaluation indexes in these two problems are different. For example, a larger objective function value is better in the maximization optimization, while it is the opposite in the minimization optimization.

Some frequently used order relations of interval number are summarized in [7], and their expressions for the maximization and minimization optimization problems are as follows:

(1) The order relation of interval number $\leq_{LR}$

$$\begin{cases} A^I \leq_{LR} B^I \text{ if and only if } A^L \leq B^L \text{ and } A^R \leq B^R \\ A^I <_{LR} B^I \text{ if and only if } A^I \leq_{LR} B^I \text{ and } A^I \neq B^I \end{cases} \quad \text{(for maximization optimization)} \tag{3.15}$$

$$\begin{cases} A^I \leq_{LR} B^I \text{ if and only if } A^L \geq B^L \text{ and } A^R \geq B^R \\ A^I <_{LR} B^I \text{ if and only if } A^I \leq_{LR} B^I \text{ and } A^I \neq B^I \end{cases} \quad \text{(for minimization optimization)} \tag{3.16}$$

This type of order relation of interval number represents the decision maker's preference on the lower and upper bounds of the interval.

(2) The order relation of interval number $\leq_{cw}$

$$\begin{cases} A^I \le_{cw} B^I \text{ if and only if } A^c \le B^c \text{ and } A^w \ge B^w \\ A^I <_{cw} B^I \text{ if and only if } A^I \le_{cw} B^I \text{ and } A^I \ne B^I \end{cases} \quad \text{(for maximization optimization)} \quad (3.17)$$

$$\begin{cases} A^I \le_{cw} B^I \text{ if and only if } A^c \ge B^c \text{ and } A^w \ge B^w \\ A^I <_{cw} B^I \text{ if and only if } A^I \le_{cw} B^I \text{ and } A^I \ne B^I \end{cases} \quad \text{(for minimization optimization)} \quad (3.18)$$

This type of order relation of interval number represents the decision maker's preference on the midpoint and radius of the interval.

(3) The order relation of interval number $\le_{Lc}$

$$\begin{cases} A^I \le_{Lc} B^I \text{ if and only if } A^L \le B^L \text{ and } A^c \le B^c \\ A^I <_{Lc} B^I \text{ if and only if } A^I \le_{Lc} B^I \text{ and } A^I \ne B^I \end{cases} \quad \text{(for maximization optimization)} \quad (3.19)$$

$$\begin{cases} A^I \le_{Lc} B^I \text{ if and only if } A^L \ge B^L \text{ and } A^c \ge B^c \\ A^I <_{Lc} B^I \text{ if and only if } A^I \le_{Lc} B^I \text{ and } A^I \ne B^I \end{cases} \quad \text{(for minimization optimization)} \quad (3.20)$$

This type of order relation of interval number represents the decision maker's preference on the midpoint and lower bound of the interval.

(4) The order relation of interval number $\le_L$

$$\begin{cases} A^I \le_L B^I \text{ if and only if } A^L \le B^L \\ A^I <_L B^I \text{ if and only if } A^I \le_L B^I \text{ and } A^I \ne B^I \end{cases} \quad \text{(for maximization optimization)} \quad (3.21)$$

$$\begin{cases} A^I \le_L B^I \text{ if and only if } A^L \ge B^L \\ A^I <_L B^I \text{ if and only if } A^I \le_L B^I \text{ and } A^I \ne B^I \end{cases} \quad \text{(for minimization optimization)} \quad (3.22)$$

This type of order relation of interval number represents the decision maker's preference on the lower bound of the interval.

(5) The order relation of interval number $\le_R$

$$\begin{cases} A^I \le_R B^I \text{ if and only if } A^R \le B^R \\ A^I <_R B^I \text{ if and only if } A^I \le_R B^I \text{ and } A^I \ne B^I \end{cases} \quad \text{(for maximization optimization)} \quad (3.23)$$

$$\begin{cases} A^I \le_R B^I \text{ if and only if } A^R \ge B^R \\ A^I <_R B^I \text{ if and only if } A^I \le_R B^I \text{ and } A^I \ne B^I \end{cases} \quad \text{(for minimization optimization)} \quad (3.24)$$

This type of order relation of interval number represents the decision maker's preference on the upper bound of the interval.

The definitions of the order relation of interval number are not limited by the above types. In this book we also define the other two types of order relations as follows:

(6) The order relation of interval number $\le_c$

$$\begin{cases} A^I \le_c B^I \text{ if and only if } A^c \le B^c \\ A^I <_c B^I \text{ if and only if } A^I \le_c B^I \text{ and } A^I \ne B^I \end{cases} \quad \text{(for maximization optimization)} \quad (3.25)$$

$$\begin{cases} A^I \leq_c B^I \text{ if and only if } A^c \geq B^c \\ A^I <_c B^I \text{ if and only if } A^I \leq_c B^I \text{ and } A^I \neq B^I \end{cases} \quad \text{(for minimization optimization)} \tag{3.26}$$

This type of order relation of interval number represents the decision maker's preference for the midpoint of the interval.

(7) The order relation of interval number $\leq_w$

$$\begin{cases} A^I \leq_w B^I \text{ if and only if } A^w \geq B^w \\ A^I <_w B^I \text{ if and only if } A^I \leq_w B^I \text{ and } A^I \neq B^I \end{cases} \quad \text{(for maximization and minimization optimizations)} \tag{3.27}$$

This type of order relation of interval number represents the decision maker's preference for the radius of the interval.

In our formulation, the order relation "$\leq_{cw}$" is chosen to handle the uncertain objective function in Eq. (3.1). The possible value of the objective function $f(\mathbf{X}, \mathbf{U})$ at a specific design vector $\mathbf{X}$ is also an interval due to the existence of $\mathbf{U}$:

$$f^I(\mathbf{X}) = \left[f^L(\mathbf{X}), f^R(\mathbf{X})\right] = \left\langle f^c(\mathbf{X}), f^w(\mathbf{X})\right\rangle \tag{3.28}$$

where

$$f^c(\mathbf{X}) = \frac{f^L(\mathbf{X}) + f^R(\mathbf{X})}{2} \tag{3.29}$$

$$f^w(\mathbf{X}) = \frac{f^R(\mathbf{X}) - f^L(\mathbf{X})}{2} \tag{3.30}$$

Based on the order relation of interval number "$\leq_{cw}$" as shown in Eq. (3.18), the design vectors can be compared regarding their midpoint and radius of the objective function interval: if the interval of the objective function at $\mathbf{X}^1$ is better than that at $\mathbf{X}^2$, i.e., $f^c(\mathbf{X}^1) \leq f^c(\mathbf{X}^2)$ and $f^w(\mathbf{X}^1) \leq f^w(\mathbf{X}^2)$, the design vector $\mathbf{X}^1$ is regarded to be better than $\mathbf{X}^2$. Therefore, we hope to find the optimal solution that minimizes the midpoint value and radius of the objective function interval, and hence the uncertain objective function in Eq. (3.1), can be transformed into a deterministic multi-objective optimization:

$$\min_{\mathbf{X}} \left(f^c(\mathbf{X}), f^w(\mathbf{X})\right) \tag{3.31}$$

In Eq. (3.31), the midpoint and radius of the objective function interval should be calculated at $\mathbf{X}$. Here the objective function interval can still be calculated by using the optimization method:

$$\begin{aligned} & f^L(\mathbf{X}) = \min_{\mathbf{U}} f(\mathbf{X}, \mathbf{U}), \quad f^R(\mathbf{X}) = \max_{\mathbf{U}} f(\mathbf{X}, \mathbf{U}) \\ & \mathbf{U} \in \Gamma = \left\{\mathbf{U} \middle| U_i^L \leq U_i \leq U_i^R, \ i = 1, 2, \ldots, q\right\} \end{aligned} \tag{3.32}$$

The two objective functions in Eq. (3.31), are just like the expectation and standard deviation of the uncertain objective function in stochastic programming. In some sense, the "average design performance" of the objective function under uncertainty is enhanced by optimizing $f^c(\mathbf{X})$. Besides, the minimization of $f^w(\mathbf{X})$ will decrease the sensitivity of the objective function to the parametric uncertainties, thus ensuring the robustness of the optimal design. The order relation of interval number "$\leq_{cw}$" is selected here to deal with the uncertain objective function since in many cases it shows more intuitive engineering meanings and better engineering practicality compared with other order relations. However, we can also select other order relations according to the practical situation. This will not influence the framework of the above mathematical transformation model, and only the deterministic objective functions after transformation will be different.

### 3.3.2 The Transformed Deterministic Optimization

Through the above treatment, the nonlinear interval optimization as shown in Eq. (3.1), can be converted into a deterministic optimization problem:

$$\begin{cases} \min\limits_{\mathbf{X}} \left( f^c(\mathbf{X}), f^w(\mathbf{X}) \right) \\ \text{s.t. } P\left(M_i^I \leq N_i^I\right) \geq \lambda_i,\ i = 1, 2, \ldots, k \\ \quad \mathbf{X} \in \Omega^n \end{cases} \tag{3.33}$$

where

$$\begin{aligned} f^c(\mathbf{X}) &= \frac{\min\limits_{\mathbf{U}} f(\mathbf{X}, \mathbf{U}) + \max\limits_{\mathbf{U}} f(\mathbf{X}, \mathbf{U})}{2} \\ f^w(\mathbf{X}) &= \frac{\max\limits_{\mathbf{U}} f(\mathbf{X}, \mathbf{U}) - \min\limits_{\mathbf{U}} f(\mathbf{X}, \mathbf{U})}{2} \\ \mathbf{U} \in \Gamma &= \left\{\mathbf{U} \middle| U_i^L \leq U_i \leq U_i^R,\ i = 1, 2, \ldots, q \right\} \end{aligned} \tag{3.34}$$

So far, the transformation based on the order relation of interval number is finished, and a deterministic optimization problem is constructed. However, in order to make the obtained deterministic optimization more convenient to solve, we use the linear combination method [24], to deal with the multi-objective problem. In multi-objective optimization, applying the linear combination method [24] to integrate the objective functions is a relatively easy and convenient way. Therefore, Eq. (3.33), is further converted into a single-objective optimization problem:

$$\begin{cases} \min_{\mathbf{X}} f_d(\mathbf{X}) = (1-\beta)\big(f^c(\mathbf{X})+\xi\big)/\phi + \beta\big(f^w(\mathbf{X})+\xi\big)/\psi \\ \text{s.t. } P\big(M_i^I \le N_i^I\big) \ge \lambda_i,\ i = 1, 2, \ldots, k \\ \quad \mathbf{X} \in \Omega^n \end{cases} \tag{3.35}$$

where $f_d$ represents the multi-objective evaluation function, $0 \le \beta \le 1$ is a weighting factor, $\xi$ is a number which ensures that $f^c(\boldsymbol{X})+\xi$ and $f^w(\boldsymbol{X})+\xi$ are nonnegative. $\phi$ and $\psi$ are two normalization factors, which can be obtained through

$$\phi = \min_{\mathbf{X}}\big(f^c(\mathbf{X})+\xi\big),\ \psi = \min_{\mathbf{X}}\big(f^w(\mathbf{X})+\xi\big),\ \mathbf{X} \in \Omega^n \tag{3.36}$$

In practical application, $\phi$ and $\psi$ can be also approximately determined according to the order of magnitudes of $f^c(\boldsymbol{X})$ and $f^w(\boldsymbol{X})$.

By introducing the penalty function method [25], to deal with the constraints, Eq. (3.35), can be further converted into a non-constraint and single-objective optimization problem:

$$\begin{cases} \min_{\mathbf{X}} f_p(\mathbf{X}) = f_d(\mathbf{X}) + \sigma \sum_{i=1}^{k} \varphi\big(P\big(M_i^I \le N_i^I\big) - \lambda_i\big) \\ \qquad = (1-\beta)\big(f^c(\boldsymbol{X})+\xi\big)/\phi + \beta\big(f^w(\boldsymbol{X})+\xi\big)/\psi \\ \qquad\quad + \sigma \sum_{i=1}^{k} \varphi\big(P\big(M_i^I \le N_i^I\big) - \lambda_i\big) \\ \text{s.t. } \mathbf{X} \in \Omega^n \end{cases} \tag{3.37}$$

where $f_p$ denotes the penalty function, $\sigma$ is the penalty factor which is usually taken as a large value and $\varphi$ is a penalty function with the following form:

$$\varphi\big(P\big(M_i^I \le N_i^I\big) - \lambda_i\big) = \Big(\max\big(0, -\big(P\big(M_i^I \le N_i^I\big) - \lambda_i\big)\big)\Big)^2 \tag{3.38}$$

## 3.4 The Mathematic Transformation Model Based on Possibility Degree of Interval Number

As mentioned in Chap. 1, in the P-model method of stochastic programming, a deterministic objective function is obtained by maximizing the probability that the objective function is lower than a given value (for the minimization problem). Here we introduced this idea into nonlinear interval optimization to construct another mathematical transformation model, i.e., the mathematic transformation model based on possibility degree of interval number. In this model, we transform the uncertain objective function into a deterministic one by maximizing the possibility degree that the objective function is less than a given interval (for the minimization problem). On the other hand, we also use the possibility degree of interval number given in Sect. 3.2, to deal with the uncertain constraints. Therefore, Eq. (3.1), can be converted into the following deterministic optimization problem:

$$\begin{cases} \max\limits_{\mathbf{X}} P\left(f^I(\mathbf{X}) \leq V^I\right) \\ \text{s.t. } P\left(M_i^I \leq N_i^I\right) \geq \lambda_i,\ i = 1, 2, \ldots, k \\ \quad \mathbf{X} \in \Omega^n \end{cases} \tag{3.39}$$

where $V^I = \left[V^L, V^R\right]$ is called performance interval, which is determined according to the decision maker's requirements for the optimization and it can also be set to be a real number. For the maximization problem, the deterministic objective function should be changed to $\max\limits_{\mathbf{X}} P\left(f^I(\mathbf{X}) \geq V^I\right)$. Through such optimization, the given performance requirements can be satisfied at the maximum level under uncertainties.

By solving Eq. (3.39), an optimum design vector $\mathbf{X}^*$ that maximizes the possibility degree of the uncertain objective function to $P_{\max} = P\left(f^I(\mathbf{X}^*) \leq V^I\right)$ can be obtained. When $P_{\max} = 1$, it means that the objective function will absolutely satisfy the performance requirements at $\mathbf{X}^*$. That is to say, whatever the uncertain vector is, it is impossible for the objective function to be greater than the performance interval, which is the most desirable situation to the decision maker. However, in this case, $\mathbf{X}^*$ is generally not unique and there exists a set that makes $P_{\max} = 1$. Theoretically speaking, any $\mathbf{X}^*$ in this set will satisfy the design requirements. However, if the decision maker hopes to select the best design in this set, the robustness criterion can be introduced to construct the following optimization problem:

$$\begin{cases} \min\limits_{\mathbf{X}} f^w(\mathbf{X}) \\ \text{s.t. } P\left(f^I(\mathbf{X}) \leq V^I\right) = 1 \\ \quad P\left(M_i^I \leq N_i^I\right) \geq \lambda_i,\ i = 1, 2, \ldots, k \\ \quad \mathbf{X} \in \Omega^n \end{cases} \tag{3.40}$$

Through the above optimization, the design vector that minimizes the objective function interval, i.e., the design vector with the best robustness, can be found in the set that makes $P_{\max} = 1$. It should be mentioned that the optimization in Eq. (3.40), is not necessary for the proposed mathematic transformation model based on possibility degree of interval number. It is just an option for the user to improve their optimization result if the computational cost permits.

Similarly, Eqs. (3.39) and (3.40), can be further transformed into the non-constraint optimization problems by using the penalty function approach. For Eq. (3.39), the non-constraint optimization problem has the following form:

$$\begin{cases} \max_{\mathbf{X}} f_p(\mathbf{X}) = P\left(f^I(\mathbf{X}) \leq V^I\right) - \sigma \sum_{i=1}^{k} \varphi\left(P\left(M_i^I \leq N_i^I\right) - \lambda_i\right) \\ \text{s.t. } \mathbf{X} \in \Omega^n \end{cases} \tag{3.41}$$

And for Eq. (3.40), it has the following form:

$$\begin{cases} \min_{\mathbf{X}} f_p(\mathbf{X}) = f^w(\mathbf{X}) + \sigma \left[ \sum_{i=1}^{k} \varphi\left(P\left(M_i^I \leq N_i^I\right) - \lambda_i\right) + \left(P\left(f^I(X) \leq V^I\right) - 1\right)^2 \right] \\ \text{s.t. } \mathbf{X} \in \Omega^n \end{cases} \tag{3.42}$$

Compared with the mathematic transformation model based on order relation of interval number, the model based on possibility degree shows the following advantages and disadvantages:

(1) The treatments of both objective function and constraints are based on the same approach, i.e., the possibility degree of interval number, which unifies and simplifies the mathematical form of the optimization problem, and hence makes the model more convenient to use;
(2) By using the possibility degree of interval number to deal with the uncertain objective function, a deterministic single-objective optimization problem is obtained, which is easier to solve compared with the multi-objective optimization problem;
(3) The decision maker needs to be very familiar with the optimization problem to determine a reasonable performance interval, which thereby may limit its applicability.

## 3.5 A Two-Layer Optimization Algorithm Based on IP-GA

The deterministic optimization problems obtained through the above two transformation models are both double-loop nested optimization problems with the outer loop for updating the design vector and the inner loop for calculating the uncertain objective function and constraints. Although the original uncertain optimization problem is continuous and differentiable, the converted optimization problem is usually not because of the involvement of nested optimization. Therefore, traditional gradient-based optimization methods are incompetent in such a situation. As one of the intelligent optimization algorithms, the genetic algorithm (GA) is a kind of random search algorithm, which only needs the information of function value rather than the gradient during the optimization process. Therefore, a two-layer optimization algorithm based on IP-GA [8] is given in this chapter to solve the above-mentioned optimization problem. Both the inner loop and outer loop adopt IP-GA as solver that shows excellent performance on global optimization, which, on one hand, overcomes the difficulties caused by the discontinuity and disdifferentiability of the outer loop, and on the other hand, tries to avoid the local optimum in the inner loop. However, it must be pointed out that this algorithm generally could obtain ideal optimization results at the cost of the computational efficiency, since the IP-GA needs much more calculations compared with the traditional gradient-based algorithms when dealing with the inner loop optimization. Therefore, in the subsequent researches of this book, the calculation results of this algorithm will be used to measure the accuracy and computational efficiency of other developed algorithms. Besides, its transformation form is also embedded in other algorithms in subsequent chapters to construct more efficient and practical nonlinear interval optimization methods.

### 3.5.1 A Brief Introduction of IP-GA

IP-GA is an improved algorithm of the micro-genetic algorithm ($\mu$ GA) [26], which is extended from the traditional genetic algorithm. The $\mu$ GA can avoid the prematurity and find out the optimal area more quickly compared with traditional GA. The population size of $\mu$ GA is usually small, about 5–8 in most instances. Such a small population can quickly converge to a local optimum. Meanwhile, a restart strategy is adopted to replace the mutation operation in traditional GA. Once the current generation converges, a new generation with the same population size is randomly generated, and the new generation contains the optimal individual in the last generation. By adding an IP operator into $\mu$ GA, the IP-GA can quickly generate a better individual through constructing a moving direction based on the optimal individuals in two consecutive generations, which dramatically promotes the convergence speed. Assume $p_j^b$ and $p_{j-1}^b$ are the optimal individuals of current and the last generations, respectively, three new offspring individuals can be obtained by the following IP operator [8]:

$$\begin{cases} c_1 = p_j^b + r(p_j^b - p_{j-1}^b) \\ c_2 = p_{j-1}^b + s(p_j^b - p_{j-1}^b) \\ c_3 = p_j^b - t(p_j^b - p_{j-1}^b) \end{cases} \tag{3.43}$$

where $0 \le r \le 1, 0 \le s \le 1$ and $0 \le t \le 1$ are three nonnegative searching parameters that are used to adjust the distances between the three new individuals and $p_j^b$, $p_{j-1}^b$. The new individuals will replace the worst three individuals in the next generation. Six test functions (including two complex multimodal functions) are analyzed in [8], and the results demonstrate that IP-GA shows prominent global optimization performance. For any combinations of $r$, $s$, and $t$, the number of generations needed for searching the global optimum in IP-GA are only 6.4−74.4% of that in $\mu$ GA.

### 3.5.2 Procedure of the Algorithm

The two-layer optimization algorithm based on IP-GA is used to solve the deterministic optimization problems involving penalty functions, and the algorithm flowchart is shown in Fig. 3.6. In the outer layer of IP-GA, individuals of the design vector **X** are generated, and for each individual, the inner layer of IP-GA will be called several times to compute the intervals of the uncertain objective function and constraints. Based on these intervals, the values of the transformed deterministic objective function and constraints can be obtained. Thus, the value of the penalty function can also be calculated. For both the inner and outer layers of IP-GA, the maximum generation number is used as a stopping criterion. In the process of IP-GA, a design vector with large fitness value is regarded as a good design vector. Therefore, in the inner layer of IP-GA, the function value is used as fitness value when solving the upper bounds of the objective function and constraints, while the negative function value is used as fitness value when solving the lower bounds. Besides, in the outer layer IP-GA, the value or the negative value of the penalty function is used as fitness value according to whether the transformed deterministic optimization is a maximization problem or a minimization problem.

When using penalty function to deal with the constraints, a big penalty factor will often lead to a large condition number of the Hessian matrix of the penalty function [25], and it seems difficult to solve the problem by using traditional gradient-based optimization methods. In this case, interior or exterior point penalty function methods will be a good choice, in which the computational accuracy can be ensured by solving a sequence of optimization problems with different penalty factors for obtaining an ideal factor. However, the proposed algorithm adopts GA as the optimization solver that does not require gradient information in the whole optimization process. Therefore, an ill-condition Hessian matrix caused by a large penalty factor generally does not influence the solution of GA. Therefore, it is unnecessary to update the penalty factor $\sigma$ in the proposed algorithm. The decision maker only needs to set a relatively large penalty factor to guarantee the accuracy of the optimization solution.

## 3.6 Numerical Example and Discussions

Consider an interval optimization problem as follows [21]:

$$\begin{cases} \min\limits_{\mathbf{X}} f(\mathbf{X}, \mathbf{U}) = 130.0 - U_1^2(X_1 + 2) - U_2 X_2^2 - U_3^2 X_3^2 \\ \text{s.t. } g_1(\mathbf{X}, \mathbf{U}) = U_1 X_1^2 - U_2^2 X_2 + U_3 X_3 = [6.2, 7.3] \\ g_2(\mathbf{X}, \mathbf{U}) = U_1 X_1 + U_2 X_2 + U_3^2 X_3^2 + 1.0 \geq [15.0, 20.0] \\ -1.0 \leq X_1 \leq 5.0,\ -3.0 \leq X_2 \leq 6.0,\ -2.0 \leq X_3 \leq 7.0 \\ U_1 \in [1.0, 1.3],\ U_2 \in [0.9, 1.1],\ U_3 \in [1.2, 1.4] \end{cases} \quad (3.44)$$

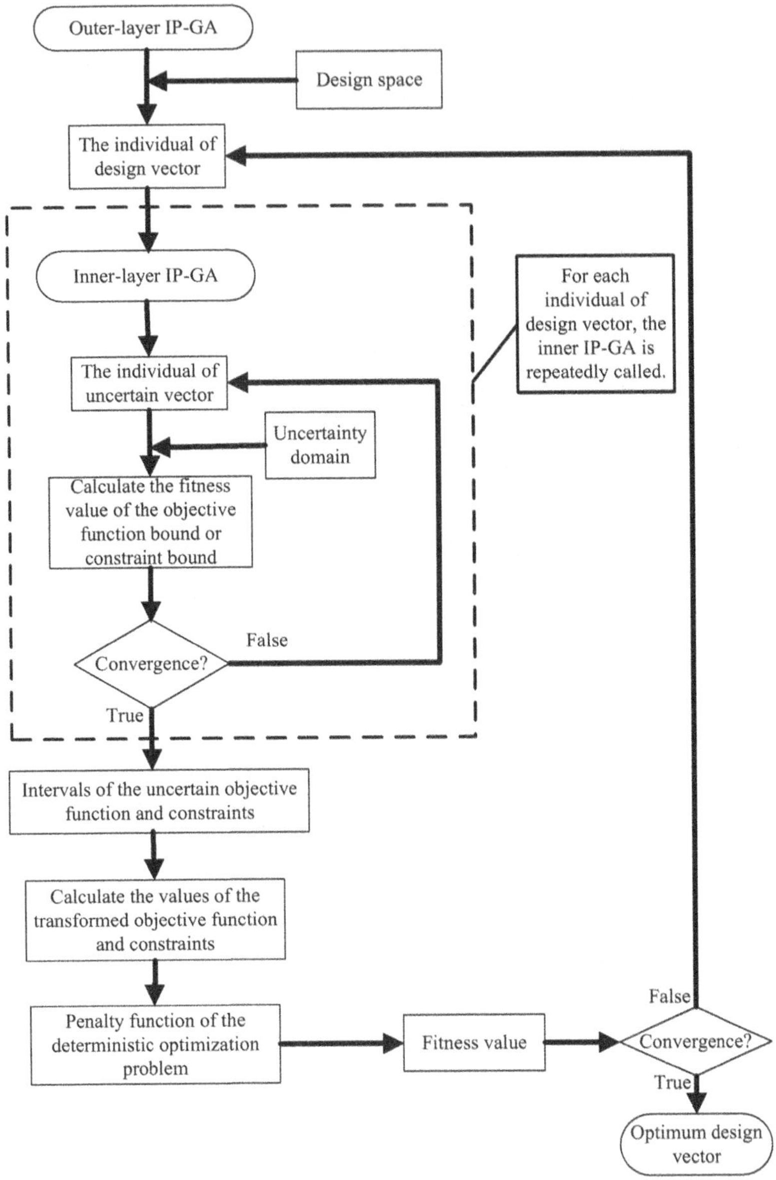


**Fig. 3.6** The two-layer optimization algorithm based on IP-GA [21]

where the uncertainty levels of $U_1$, $U_2$, and $U_3$ are 13.0%, 10.0%, and 7.7%, respectively. Here, the uncertainty level of interval is defined as the ratio of its radius to the absolute value of its midpoint. In the following text, we will use the two mathematical transformation models to deal with the problem and then adopt the two-layer optimization algorithm based on IP-GA to solve the obtained deterministic optimization problems.

### 3.6.1 By Using the Mathematic Transformation Model Based on Order Relation of Interval Number

The parameters in the optimization are set as follows: $\xi$ is set as 0.0, the penalty factor $\sigma$ is set as 100,000, and the normalization factors, $\phi$ and $\psi$ that are calculated through Eq. (3.36), are 1.33 and 0.34, respectively. For IP-GA, three searching factors, *r*, *s*, and *t*, are all set as 0.6 and the population size and crossover rate are set as 5 and 0.5, respectively. The parameters will be the same in the subsequent chapters of this book when using IP-GA. Also, the maximum generations in outer layer and inner layer of IP-GA are set as 300 and 100, respectively.

Firstly, the problem is optimized under different weighting factors for a given possibility degree for the constraints. When the constraint $g_1$ is converted into two inequality constraints, the possibility degree levels of these two constraints, $\lambda_1$ and $\lambda_2$, are both set as 0.8. Besides, the possibility degrees level of $g_2$, is also set as $\lambda_3 = 0.8$. The problem is optimized under six different weighting factors and the optimization results are shown in Table 3.1. It can be seen that the design requirements, i.e., the possibility degree of the constraints should be greater than the given possibility degree level 0.8, are satisfied at the optimum design vector for all the six cases. The optimum design vector and the intervals of the objective function and constraints are different under different $\beta$. As $\beta$ decreases, the midpoint of the objective function interval at the optimum design vector has a decreasing tendency, while the radius of the interval shows the opposite tendency (except for $\beta = 0$ and $\beta = 0.2$, the optimization results in these two situations are same). The reason is that a large $\beta$ means the preference on the radius of the objective function increases, while the preference on the midpoint decreases, which leads to an optimum design vector with a relatively small radius and relatively large midpoint of the objective function interval.

Secondly, for a given weighting factor $\beta = 0.5$, four problems are optimized under different possibility degree levels of the constraints. The optimization results can be seen in Table 3.2. Four optimum design vectors are obtained in these cases, and the possibility degrees of the constraints for each optimum design vector satisfy the design requirements. As the possibility degree level decreases, the value of the penalty function $f_p$, which here is equal to the value of multi-objective evaluation function $f_d$, has a decreasing tendency. The reason is that a low possibility degree level generally leads to a large feasible region in the deterministic optimization problem Eq. (3.33). Therefore, the algorithm can find the design vector with better

**Table 3.1** Optimization results under six different weighting factors [21]

| $\beta$ | Optimum design vector | Interval of objective function | Middle point of objective function | Radius of objective function | Interval of equality constraint | Interval of inequality constraint | Possibility degrees of constraints | Penalty function |
|---|---|---|---|---|---|---|---|---|
| 1.0 | (−0.60, −1.39, 3.77) | [97.62, 06.27] | 101.95 | 4.33 | [6.01, 7.42] | [19.18, 27.02] | 0.86, 0.92, 0.99 | 12.71 |
| 0.8 | (−0.96, −0.34, 4.19) | [93.65, 03.52] | 98.59 | 4.94 | [6.24, 7.50] | [24.69, 34.19] | 1.00,0.84, 1.00 | 26.44 |
| 0.6 | (−0.53, 0.37, 5.31) | [72.08, 87.79] | 79.94 | 7.86 | [6.21, 7.50] | [41.27, 56.16] | 1.00, 0.85, 1.00 | 37.90 |
| 0.4 | (0.56, 1.21, 5.87) | [56.47, 76.36] | 66.42 | 9.95 | [5.90, 7.65] | [52.34, 70.69] | 0.83, 0.80, 1.00 | 41.66 |
| 0.2 | (0.34, 1.20, 6.04) | [52.95, 73.71] | 63.33 | 10.38 | [5.92, 7.64] | [54.98, 74.29] | 0.84, 0.80, 1.00 | 44.37 |
| 0.0 | (0.34, 1.20, 6.04) | [52.95, 73.71] | 63.33 | 10.38 | [5.92, 7.64] | [54.98, 74.29] | 0.84, 0.80, 1.00 | 47.62 |

**Table 3.2** Optimization results under different possibility degree levels of the constraints [21]

| Possibility degree level of constraints | Optimum design vector | Interval of objective function | Interval of equality constraint | Interval of inequality constraint | Possibility degrees of constraints | Penalty function |
|---|---|---|---|---|---|---|
| 1.0 | (−0.13, 0.10, 5.24) | [73.01, 88.58] | [6.19, 7.28] | [40.47, 54.80] | 1.00,1.00, 1.00 | 50.12 |
| 0.8 | (−0.34, −0.37, 5.31) | [71.77, 87.60] | [6.04, 7.29] | [41.53, 56.35] | 0.87,1.00, 1.00 | 41.60 |
| 0.4 | (1.81, 6.00, 7.00) | [−12.01, 23.21] | [4.44, 9.20] | [78.77, 105.99] | 0.63,0.60, 1.00 | 28.00 |
| 0.0 | (0.31, 6.00, 7.00) | [−9.40, 24.60] | [1.24, 5.06] | [77.27, 104.04] | 0.00,1.00, 1.00 | 27.86 |

objective function performance, i.e., lower value of the multi-objective evaluation function. Figure 3.7 gives the convergence curves of the four cases, where the fitness value is the negative value of the penalty function. As seen in Fig. 3.7, the IP-GA shows a good convergence performance since the algorithm generally fast converges to a stable fitness value.

In practical application, the selection of the weighting factor $\beta$ and the possibility degree level mainly depends on two factors: the practical problem itself and the decision maker's preference. If the fluctuation of the objective function under

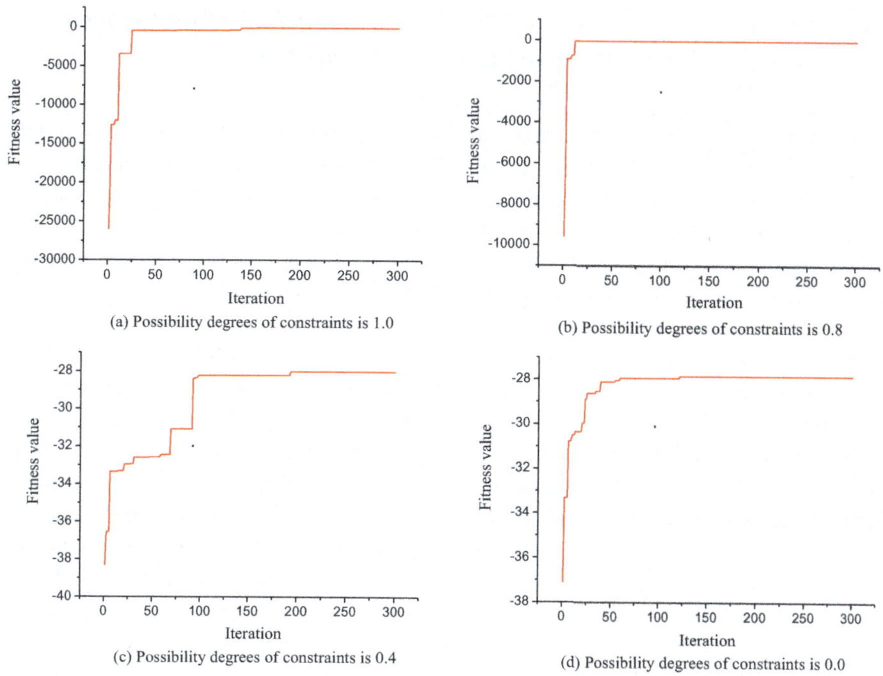


**Fig. 3.7** Convergence curves under different possibility degree levels of the constraints [21]

uncertainties, i.e., the robustness of the system, has little influence on the problem, a small weighting factor is recommended to increase the preference on the midpoint of the objective function to achieve better average design performance. If the stability or robustness of the problem is the priority, the weighting factor is recommended to be large. Besides, if we have a high priority on reliability, a large possibility degree level of the constraints should be selected. On the contrary, if the design objective is very important, and a certain degree of violation of the constraints will not result in a serious consequence, the possibility degree level can be given a relatively small value. In addition, the decision procedure will inevitably be influenced by the preference of the decision maker once the uncertainties exist, and the interval optimization method is no exception. An optimistic decision maker is prone to small weighting factor and possibility degree level to pursue the design objective. However, the risks that the fluctuation of the objective function is too large or the constraints are violated will be higher at the same time. Conversely, a pessimistic decision maker is more likely to select large weighting factor and possibility degree level to reduce the risk even though the design objective is sacrificed. Therefore, to some degree, the decision maker's design experience and the degree how familiar the decision maker is with the problem are very important since a decision maker with rich experience is likely to have a reasonable decision-making attitude.

### 3.6.2 By Using the Mathematic Transformation Model Based on Possibility Degree of Interval Number

In the optimization procedure, the possibility degree level $\lambda$ of the constraints and the penalty factor $\sigma$ are set to be 0.8 and 100,000, respectively. The maximum generation number is set to be 300 and 100 for the outer layer of IP-GA and inner layer of IP-GA, respectively. Firstly, let us consider the situation that the performance interval of the objective function is $V^I = [69, 79]$ and establish the optimization model as Eq. (3.39), to maximize the possibility degree of the objective function. The results, as shown in Table 3.3, indicate that the possibility degree of the objective function at the optimum design vector is 0.98. Meanwhile, the possibility degrees of the constraints are 0.80, 0.83, and 1.00, respectively, which all satisfy the predetermined requirements.

Secondly, consider the situation that the performance interval of the objective function is $V^I = [79, 89]$ and construct the corresponding optimization problem. Table 3.4 gives the optimization results. It can be seen in that the possibility degree of the objective function at the optimum design vector reaches 1.0 and the possibility degree requirements are satisfied for all the constraints. Since the possibility degree of the objective function is 1.0, we construct an optimization problem as Eq. (3.40), considering the robustness for further optimization. The optimization results are listed in Table 3.5. As shown in the table, the optimum design vector still satisfies the design requirements, while the radius of the uncertain objective function is reduced to 9.2 from 9.5, which means the fluctuation of the objective function caused by uncertainties is alleviated and thus the robustness is enhanced.

**Table 3.3** Optimization results for $V^I = [69, 79]$ [21]

| Optimum design vector | Interval of objective function | Possibility degree of objective function | Interval of equality constraint | Interval of inequality constraint | Possibility degrees of constraints | Penalty function |
|---|---|---|---|---|---|---|
| (−0.17, 1.29, 6.16) | [50.80, 72.10] | 0.98 | [5.85, 7.61] | [56.52, 76.52] | 0.80, 0.83, 1.00 | 0.98 |

**Table 3.4** Optimization results maximizing the possibility degree of the objective function for $V^I = [79, 89]$ [21]

| Optimum design vector | Interval of objective function | Radius of objective function | Possibility degree of objective function | Interval of equality constraint | Interval of inequality constraint | Possibility degrees of constraints | Penalty function |
|---|---|---|---|---|---|---|---|
| (0.60, 1.13, 5.72) | [60.00, 79.02] | 9.5 | 1.00 | [5.86, 7.56] | [49.81, 67.25] | 0.80, 0.84, 1.00 | 1.00 |

**Table 3.5** Optimization results considering the robustness of the objective function for $V^I = [79, 89]$ [21]

| Optimum design vector | Interval of objective function | Radius of objective function | Possibility degree of objective function | Interval of equality constraint | Interval of inequality constraint | Possibility degrees of constraints | Penalty function |
|---|---|---|---|---|---|---|---|
| (−0.59,1.16, 5.76) | [61.21,79.61] | 9.2 | 1.00 | [5.86, 7.57] | [49.08, 66.69] | 0.80, 0.84, 1.00 | 9.2 |

In practical application, a reasonable expected performance interval of the objective function is important, which requires that the decision maker be familiar with practical problems and have engineering experience. A too low performance interval generally makes the possibility degree of the objective function reach 1.0 very easily. Although we can construct an optimization problem as shown in Eq. (3.4), to further improve the robustness, the new constraint (the possibility degree that the value of the objective function is less than the performance interval should be 1.0) has little effect on the optimization. On the other hand, a too high performance interval will lead to the possibility degree of the objective function being 0, which deprives the meaning of the optimization. Therefore, the decision maker should avoid selecting a too low or too high performance interval. Besides, in practical problems, the performance interval can be replaced by a specific performance value.

## 3.7 Summary

This chapter proposes two mathematical transformation models for nonlinear interval optimization to convert the uncertain optimization problem into a deterministic problem. The uncertain objective function is converted into a deterministic two objective optimization problem by the mathematic transformation model based on order relation of interval number. In the transformation model based on possibility degree of interval number, the uncertain objective function is converted to a deterministic single-objective optimization problem. Compared to the transformation model based on order relation, the transformation model based on possibility degree adopts the same strategy to deal with the objective function and constraints, which leads to a more unified mathematical form. Besides, the single-objective optimization is easier to solve compared with the multi-objective optimization. However, the transformation model based on possibility degree requires that the decision maker should be more familiar with the practical problem to determine a suitable performance interval. In addition, a two-layer optimization algorithm based on IP-GA is also proposed in this chapter for solving the deterministic optimization problems obtained through the two transformation models. In the proposed algorithm, the IP-GA that has an excellent global optimization performance is used to solve both the outer and inner layer optimizations, which, on one hand, overcomes the difficulties caused by the discontinuity and disdifferentiability of the outer loop optimization, and on the other hand, avoids the constraint violation caused by the local optimum of the inner loop optimization to ensure the safety.

## References


1. Tanaka H, Okuda T, Asai K (1973) On Fuzzy-mathematical programming. J Cybern 3(4):37–46
2. Ishibuchi H, Tanaka H (1990) Multiobjective programming in optimization of the interval objective function. Eur J Oper Res 48(2):219–225
3. Lai KK, Wang SY, Xu JP, Zhu SS, Fang Y (2002) A class of linear interval programming problems and its application to portfolio selection. IEEE Trans Fuzzy Syst 10(6):698–704
4. Inuiguchi M, Sakawa M (1995) Minimax regret solution to linear programming problems with an interval objective function. Eur J Oper Res 86(3):526–536
5. Inuiguchi M, Sakawa M (1996) Maximum regret analysis in linear programs with an interval objective function. In: Proceedings of IWSCI'96
6. Averbakh I, Lebedev V (2005) On the complexity of minmax regret linear programming. Eur J Oper Res 160(1):227–231
7. Ma LH (2002) Research on method and application of robust optimization for uncertain system. PhD thesis, Zhejiang University, Hangzhou
8. Liu GR, Han X (2003) Computational inverse techniques in nondestructive evaluation. CRC Press, Florida
9. Nakahara Y, Sasaki M, Gen M (1992) On the linear programming problems with interval coefficients. Comput Ind Eng 23(1):301–304
10. Kundu S (1997) Min-transitivity of fuzzy leftness relationship and its application to decision making. Fuzzy Sets Syst 86(3):357–367
11. Sengupta A, Pal TK (2000) On comparing interval numbers. Eur J Oper Res 127(1):28–43
12. Liu XW, Da QL (1999) A satisfactory solution for interval linear programming. J Syst Eng 14(2):123–128
13. Ramik J, Rommelfanger H (1993) A single- and a multi-valued order on fuzzy numbers and its use in linear programming with fuzzy coefficients. Fuzzy Sets Syst 57(2):203–208
14. Dubois D, Prade H (1983) Ranking fuzzy numbers in the setting of possibility theory. Inf Sci 30(3):183–224
15. Ohta H, Yamaguchi T (1996) Linear fractional goal programming in consideration of fuzzy solution. Eur J Oper Res 92(1):157–172
16. Xu ZS, Da QL (2003) Possibility degree method for ranking interval numbers and its application. J Syst Eng 18(1):67–70
17. Da QL, Liu XW (1999) Interval number linear programming and its satisfactory solution. Syst Eng Theory Pract 19(4):3–7
18. Facchinetti G, Ricci RG, Muzzioli S (1998) Note on ranking fuzzy triangular numbers. Int J Intell Syst 13(7):613–622
19. Zhang Q, Fan ZP, Pan DH (1999) A ranking approach for interval numbers in uncertain multiple attribute decision making problems. Syst Eng Theory Pract 19(5):129–133
20. Jiang C, Han X, Liu GR, Liu GP (2008) A nonlinear interval number programming method for uncertain optimization problems. Eur J Oper Res 188(1):1–13
21. Jiang C (2008) Theories and algorithms of uncertain optimizationbased on interval. PhD thesis, Hunan University, Changsha
22. Liu BD, Zhao RQ, Wang G (2003) Uncertain programming with applications. Tsinghua University Press, Beijing
23. Liu BD, Zhao RQ (1998) Stochastic programming and fuzzy programming. Tsinghua University Press, Beijing

24. Hu YD (1990) Practical multi-objective optimization. Shanghai Technological Press, Shanghai
25. Chen BL (2003) Optimization theory and algorithm. Tsinghua University Press, Beijing
26. Krishnakumar K (1990) Micro-genetic algorithms for stationary and non-stationary function optimization. In: 1989 symposium on visual communications, image processing, and intelligent robotics systems, Philadelphia, PA, United States, February 1990

# Chapter 4
# Interval Optimization Based on Hybrid Optimization Algorithm

**Abstract** By combining the genetic algorithm (GA) with the artificial neural network (ANN), this chapter establish two hybrid optimization algorithms to solve the nested optimization problem after transformation, based on which two efficient nonlinear interval optimization methods are developed.

The previous chapter proposes two mathematical transformation models to convert the interval optimization problem into a deterministic double-loop nested optimization problem. Although a two-layer optimization algorithm based on IP-GA is proposed to solve the problem, it is only suitable for some simple problems. For most practical engineering problems, the optimization model is generally based on some numerical analysis models that are very computationally expensive, such as the finite element model, the multi-rigid-body dynamic model, etc. The nested optimization based on the numerical analysis models will generally lead to an extremely low computational efficiency, which limits its application to many engineering problems.

Based on the transformation model based on order relation of interval number, two types of hybrid optimization algorithms based on the genetic algorithm (GA) and the artificial neural network (ANN) are proposed in this chapter to solve the double-loop nested optimization problem, and thus the computational efficiency of nonlinear interval optimization is greatly improved. These two algorithms are then applied to two practical engineering problems: the optimization of the variable binder force in U-shaped forming and the locator optimization of automobile welding fixture.

## 4.1 The Nonlinear Interval Optimization with Uniformly Expressed Constraints

It can be seen in the previous chapter that the uncertain equality constraints and inequality constraints with "$\geq$" in nonlinear interval optimization problems can both be converted to inequality constraints with "$\leq$". Therefore, the subsequent chapters in this book will adopt a simpler and more unified nonlinear interval optimization model for convenience:



C. Jiang et al., *Nonlinear Interval Optimization for Uncertain Problems*, Springer Tracts in Mechanical Engineering, https://doi.org/10.1007/978-981-15-8546-3_4

$$\begin{cases} \min_{\mathbf{X}} f(\mathbf{X}, \mathbf{U}) \\ \text{s.t. } g_i(\mathbf{X}, \mathbf{U}) \leq b_i^I = \left[b_i^L, b_i^R\right],\ i = 1, 2, \ldots, l,\ \mathbf{X} \in \Omega^n \\ \qquad \mathbf{U} \in \mathbf{U}^I = \left[\mathbf{U}^L, \mathbf{U}^R\right],\ U_i \in U_i^I = \left[U_i^L, U_i^R\right],\ i = 1, 2, \ldots, q \end{cases} \tag{4.1}$$

The optimization problem can be converted into a deterministic optimization problem as follows using the mathematic transformation model based on order relation of interval number:

$$\begin{cases} \min_{\mathbf{X}} \left(f^c(\mathbf{X}), f^w(\mathbf{X})\right) \\ \text{s.t. } P\left(g_i^I(\mathbf{X}) \leq b_i^I\right) \geq \lambda_i,\ i = 1, 2, \ldots, l \\ \qquad \mathbf{X} \in \Omega^n \end{cases} \tag{4.2}$$

Equation (4.2) can be further converted into:

$$\begin{cases} \min_{\mathbf{X}} f_p(\mathbf{X}) = f_d(\mathbf{X}) + \sigma \sum_{i=1}^{l} \varphi\left(P\left(g_i^I(\mathbf{X}) \leq b_i^I\right) - \lambda_i\right) \\ \qquad = (1-\beta)\left(f^c(\mathbf{X}) + \xi\right)/\phi + \beta\left(f^w(\mathbf{X}) + \xi\right)/\psi \\ \qquad + \sigma \sum_{i=1}^{l} \varphi\left(P\left(g_i^I(X) \leq b_i^I\right) - \lambda_i\right) \\ \text{s.t. } \mathbf{X} \in \Omega^n \end{cases} \tag{4.3}$$

Hereinafter, the ANN model is to be introduced, and then two types of hybrid optimization algorithms are constructed to solve the above optimization problem.

## 4.2 The ANN Model

ANN model is a nonlinear system constructed by many simple computing units (neurons). To some extent, it simulates some functions of the human brain, such as information processing, information storage and information retrieval. Therefore, ANN has the intelligent functions of learning, memory, and calculation. At present, there are dozens of neural network models and these models are widely applied in many fields. In this chapter, multilayer feed-forward network [1], as shown in Fig. 4.1, is adopted. The network typically consists of a set of neurons that constitute the input layer, one or more hidden layers, and an output layer, meanwhile, there are weights (and bias) corresponding to the connection between the layers. Although the number of the hidden layers is adjustable, the research in [2], indicates that the ANN model with two hidden layers is adequate to deal with most structural problems. The training of most ANN models is based on the back propagation (BP) learning algorithm that is generally inefficient. This chapter adopts an improved BP

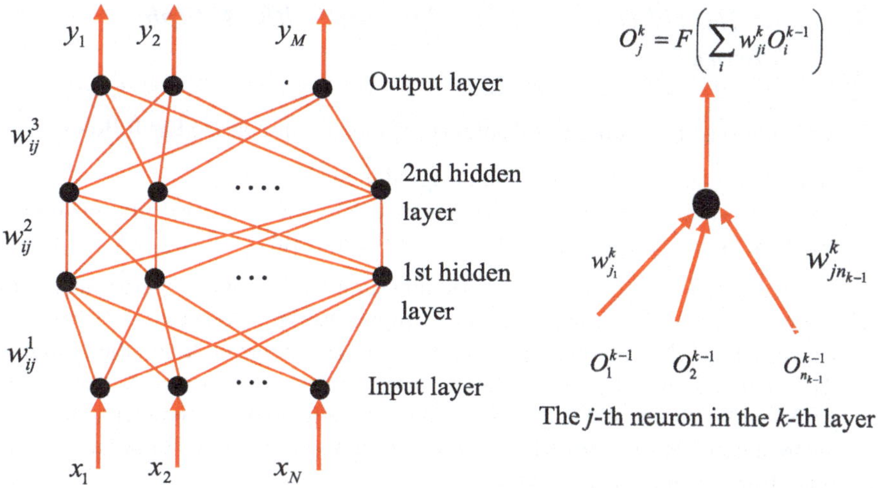


**Fig. 4.1** Multilayer feed-forward network connected by neighboring layers with $N$ inputs and $M$ outputs [1]

learning algorithm with adjustable learning rate and jump factor that is proposed in [3]. This improved BP learning algorithm has two important characteristics: firstly, the learning rate is adjusted every few generations, rather than each generation; secondly, the jump factor is used to avoid stagnation of the weight coefficient matrix.

## 4.3 Construction of the Hybrid Optimization Algorithms

In this section, two methods, i.e., the hybrid optimization algorithm with multiple networks and the hybrid optimization algorithm with a single network, are constructed based on the IP-GA and the ANN model to efficiently solve the deterministic optimization problem in Eq. (4.3). In the algorithm with multiple networks, several ANN models are created to establish the mapping relationship between the design variables and the intervals of the objective function and constraints. On the contrary, in the algorithm with a single network, only one ANN model is created to create the mapping relationship between the design variables, interval variables, and the values of the objective function and constraints. In the whole optimization procedure, the ANN model is used to replace the time-consuming original computational model, and hence the computational efficiency can be dramatically promoted.

### 4.3.1 The Hybrid Optimization Algorithm with Multiple Networks

The computational procedure of this algorithm can be summarized as follows:

(1) Select a set sample of $\mathbf{X}$ from the design domain $\Omega^n$.
(2) For each sample $\mathbf{X}'$, call the original model to carry out two optimizations, $\min_{\mathbf{U}} f(\mathbf{X}', \mathbf{U})$, and $\max_{\mathbf{U}} f(\mathbf{X}', \mathbf{U})$ to obtain the interval $[f^L(\mathbf{X}'), f^R(\mathbf{X}')]$. $\mathbf{X}'$ and $[f^L(\mathbf{X}'), f^R(\mathbf{X}')]$ will constitute a training sample. Repeat the operations for all samples, and thus a group of training samples can be obtained.
(3) Construct an ANN model whose input is the design vector $\mathbf{X}$ and the output is the upper and lower bounds of the uncertain objective function at $\mathbf{X}$, denoted as $f^R(\mathbf{X})$ and $f^L(\mathbf{X})$. Use the training samples to train the ANN model, and thereby create a nonlinear mapping between the design vector and the upper and lower bounds of the objective function.
(4) Repeat the above operations for all the constraints, and $l$ ANN models can be obtained. Each ANN model has a nonlinear mapping between the design vector and the upper and lower bounds of the corresponding constraint.
(5) Combine all the trained ANN models with IP-GA, and solve the optimization problem in Eq. (4.3). The optimization procedure can be seen in Fig. 4.2. In the optimization process, the intervals of the objective function and the constraints can be quickly obtained through the ANN models for each design vector generated by IP-GA and then the multi-objective evaluation function, the possibility degree of the constraints, and the penalty function value can be calculated. The inner optimization is thus avoided since the interval can be directly obtained from ANN models.

In practical problems, generally not all the objective function and constraints need to be calculated through the time-consuming numerical analysis models. Actually, some of them will be explicit functions. Therefore, when using the above hybrid optimization algorithm, we only need to construct the ANN models for those functions that should be calculated by the numerical analysis models. For the explicit functions, we can adopt the optimization approach to obtain their bounds. Most computational costs of the above hybrid optimization algorithm come from the acquisition of the training samples of ANN models. In practical application, the acquisition of these training samples can be carried out offline and concurrently. Once all training samples are obtained and the training for all the ANN models is finished, we can put the original model aside and realize the fast optimization.

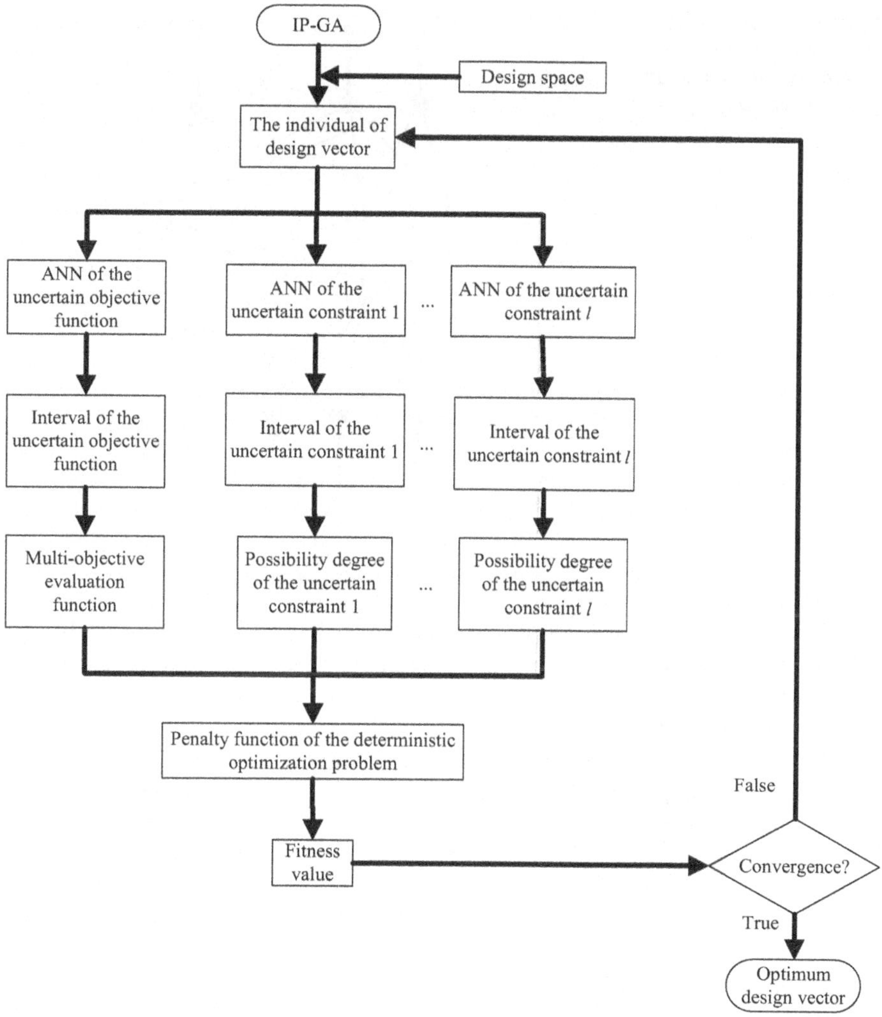


**Fig. 4.2** Interval optimization by using the hybrid optimization algorithm with multiple networks [4]

### 4.3.2 *The Hybrid Optimization Algorithm with a Single Network*

The computational procedure of this algorithm can be summarized as follows:

(1) Select a set of samples in the hybrid space composed of the design domain $\Omega^n$ and the uncertainty domain $\Gamma$. Each sample consists of the design vector and the uncertain vector.

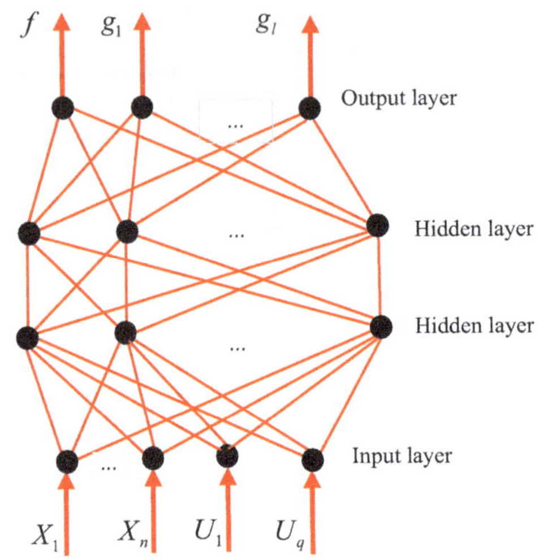


**Fig. 4.3** The ANN model created in the hybrid optimization algorithm with single network [4]

(2) For each sample $\left(\mathbf{X}', \mathbf{U}'\right)$, call the original model to calculate the objective function $f\left(\mathbf{X}', \mathbf{U}'\right)$, and the constraints $g_i\left(\mathbf{X}', \mathbf{U}'\right)$, $i = 1, 2, \ldots, l$, thus a training sample is obtained. Repeat the above operations for all the samples to obtain a group of training samples.

(3) Construct an ANN model as shown in Fig. 4.3, whose input is the design vector and uncertain vector, and the output is the corresponding values of the objective function and constraints.

(4) Use the training samples to train the ANN model and establish the nonlinear mapping between variables (design vector and uncertain vector) and functions (uncertain objective function and constraints).

(5) Combine the trained ANN model with the IP-GA, and solve the optimization problem in Eq. (4.3). The optimization procedure can be seen in Fig. 4.4. Like the nested optimization based on IP-GA in Chap. 3, the whole optimization process is composed of the outer loop for updating the design vector and the inner loop for calculating the intervals of the uncertain objective function and constraints. What is different is that the nested optimization is based on the ANN model, rather than the original model. In practical engineering application, the original model is generally a numerical analysis model that is time-consuming. By introducing the ANN model, the algorithm is able to quickly obtain the values of the objective function and constraints for any design vector and uncertain vector to avoid the original numerical analysis model. Therefore, the optimization efficiency is dramatically promoted.

In the above algorithm, the objective function and all the constraints are included in the output of the ANN model for better expression. However, as mentioned above, generally only part of the objective function and constraints are calculated by the

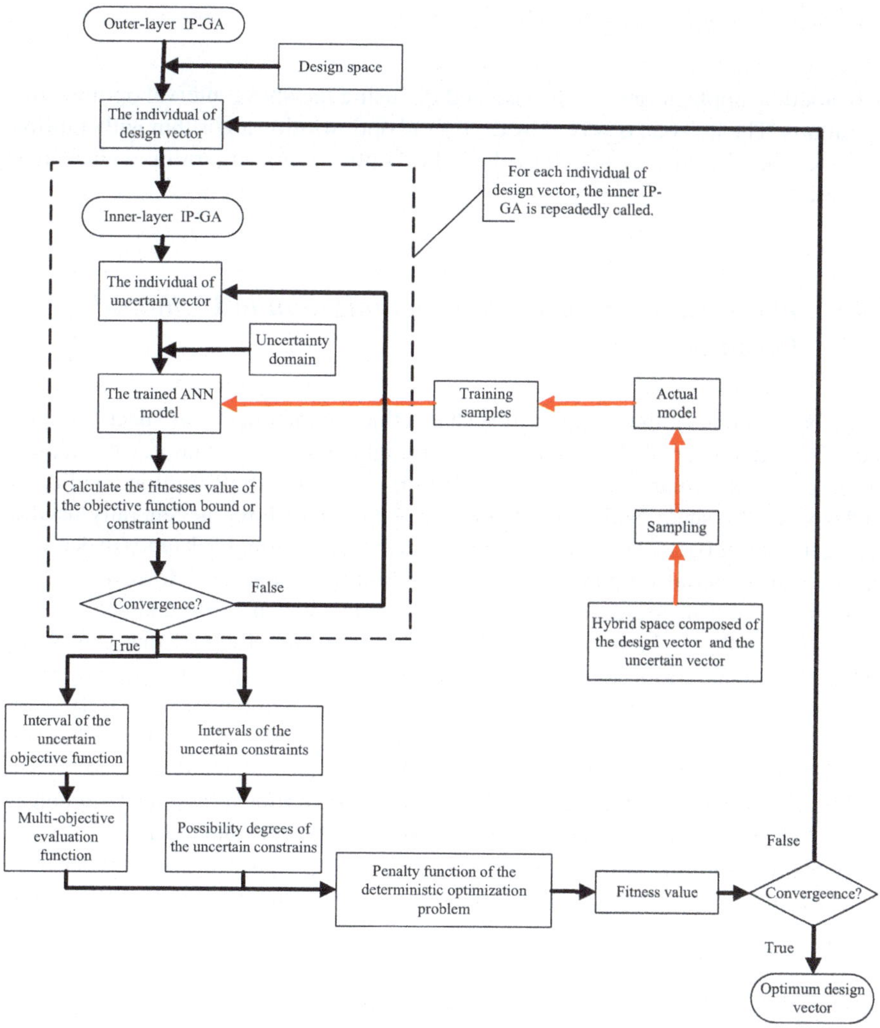


**Fig. 4.4** Interval optimization by using the hybrid optimization algorithm with single network [4]

numerical analysis models in practical optimization problems, and others are calculated by explicit functions. Therefore, when constructing the ANN model, only those based on the time-consuming numerical analysis models need to be included in the output of the ANN model. In the inner optimization, only these functions are calculated by the ANN model, while other functions can be obtained by directly calling the original explicit functions. Therefore, through such a strategy, the complexity of the ANN model is reduced, and a higher mapping accuracy can be ensured.

## 4.4 Engineering Applications

Two practical applications are then carried out using the above interval optimization algorithms. The first one is solved by the hybrid optimization algorithm with multiple networks, and the other is solved by the hybrid optimization algorithm with a single network.

### *4.4.1 The Variable Binder Force Optimization in U-Shaped Forming*

Springback is one of the major factors that influence the quality of sheet stamping and it is hard to control. The formation of springback involves complex mechanics, and it is related to many factors such as the material properties, friction coefficient, thickness of the sheet, and temperature, etc. The binder force is generally used to reduce the springback. As the binder force increases, the springback tends to decrease under given material property and other machining parameters. However, a large binder force will increase the strain in the material, which could damage the part [5]. To solve this problem, a stepped variable binder force model that can obtain a relatively small springback and strain at the same time was proposed by [6–8], as shown in Fig. 4.5. Different to the conventional constant binder force (CBF), the stepped variable binder force strategy firstly uses a low binder force (LBF) to help the material flow, and then a high binder force (HBF) replaces the LBF to cause plastic strains. This stepped variable binder force curve is determined by three parameters: $LBF$, $HBF$, and the percentage of the total punch displacements ($PPD$).

At present, most of the studies on the design of the stepped variable binder force use the finite element method (FEM) to simulate the stamping procedure [9, 10], in

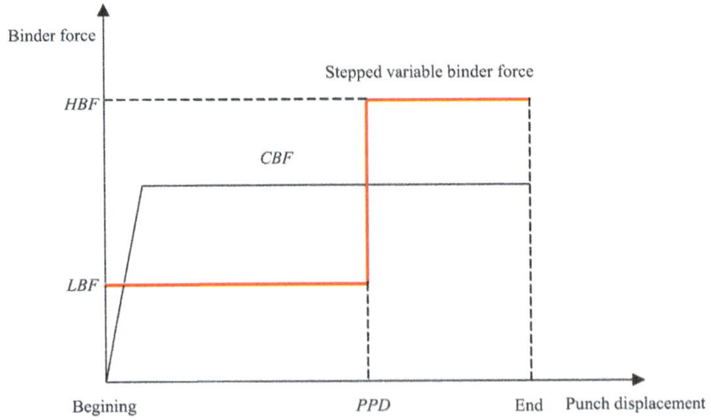


**Fig. 4.5** The pattern of the stepped variable binder force [9]

which the material properties and the machining parameters are all assumed to be deterministic. However, sheet stamping is a complex mechanical process and some parameters are generally uncertain. For example, the friction coefficient fluctuates according to the lubrication condition during the stamping processes. Besides, even in one stamping process, the friction coefficient is not constant. Considering that the friction coefficient is an important factor for the springback [11], in this application it is treated as an uncertain parameter and an interval optimization is thus formulated to find the optimal variable binder force curve.

1. Formulation of the interval optimization problem

As shown in Fig. 4.6, the stamping process of a U-shaped part of NUMISHEET'93 is selected as the research object. The geometric parameters and material properties are listed in Table 4.1. In the practical stamping, *LBF* should not be given a too small value to avoid wrinkling. However, there is no wrinkling problem for the U-shaped part [9], and hence a small and safe value can be selected. Therefore, the optimization of the variable binder force here is actually equivalent to the optimization of the two parameters, *HBF* and *PPD*.

Figure 4.7 gives the comparison between the ideal deformation (no springback) and actual deformation (with springback) after stamping, where $H$ denotes the

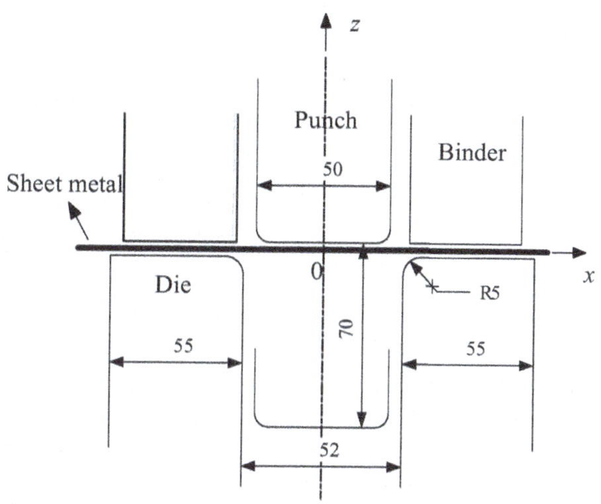


**Fig. 4.6** The U-shaped forming problem of NUMISHEET'93 (unit: mm) [9]

**Table 4.1** The geometric parameters and material properties of the part [12]

| Sheet length (mm) | Sheet width (mm) | Sheet thickness (mm) | Young's modulus (GPa) | Poisson's ratio | Relation of stress and strain (MPa) |
|---|---|---|---|---|---|
| 350 | 35 | 1.0 | 206 | 0.3 | $\sigma = 565.2(0.007117 + \varepsilon^p)^{0.2589}$ |

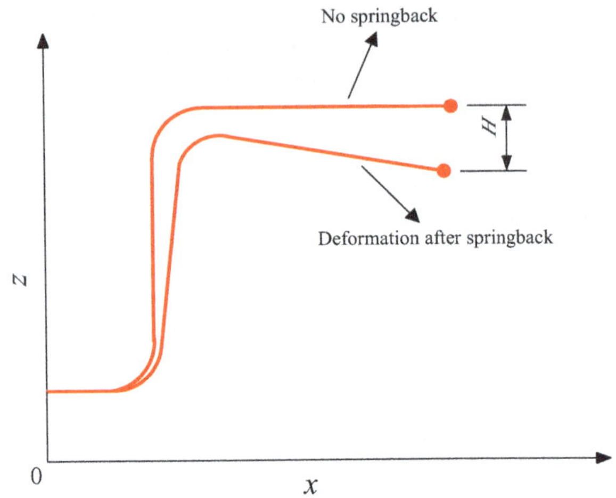


**Fig. 4.7** Comparison between the ideal deformation and actual deformation after stamping [12]

distance between the two curves in $z$ direction. It is obvious that $H$ can be used to represent the springback: the smaller $H$ indicates the small springback. Besides, the variation range of the friction coefficient $\mu$ can be estimated as an interval of $[0.1, 0.2]$ according to the practical lubrication condition and engineering experience. Therefore, a nonlinear interval optimization problem can be formulated as follows [12]:

$$\begin{cases} \min\limits_{HBF,PPD} H(HBF, PPD, \mu) \\ \text{s.t. } 2.45\text{kN} \le HBF \le 40\text{kN} \\ 55\% \le PPD \le 85\% \\ \mu \in \left[\mu^L, \mu^R\right] = [0.1, 0.2] \end{cases} \tag{4.4}$$

2. Construction of the FEM model and ANN model

Only 1/2 of the FEM model is built because of the symmetry. The two-dimensional degenerated shell element [13], is selected for meshing as shown in Fig. 4.8. The part is divided into 84 elements. The elements contacting the binder are sparse, while the elements contacting the tools are dense. The whole FEM simulation process is divided into two parts: stamping and springback. In the analysis procedure, the *LBF* is specified as a small value 2.45 kN and the total punch displacement is 70 mm.

In Eq. (4.4), only the objective function is calculated through FEM. Therefore, only one ANN model with one hidden layer is constructed. The inputs of this ANN model are *HBF* and *PPD*, the outputs are the upper and lower bounds of $H$ value, and the numbers of neuron in the input layer, hidden layer and output layer are 2, 8, 2, respectively. Let us uniformly select $5 \times 5 = 25$ sampling points in the design space and call the FEM model to perform optimization to obtain the upper and lower bounds for each sampling point. Thus, 25 training samples are obtained to train the

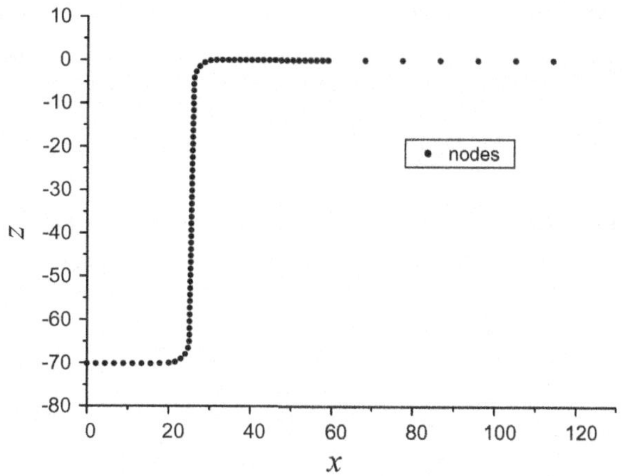


**Fig. 4.8** The FEM meshing of the part [12]

ANN model. In order to test the accuracy of the trained ANN model, we select 4 extra sampling points in the design domain and independently use the ANN model and the optimization method that is based on the actual FEM model to calculate the upper and lower bounds of the objective function. The calculation results are listed in Table 4.2. It can be found that the maximum variation at the 4 sampling points is only 4.9%, which to some extent indicates the good prediction accuracy of the ANN model.

3. Optimization results

Here IP-GA is combined with the trained ANN model to perform the nonlinear interval optimization. The parameters in the optimization are set as follows : $\beta = 0.5$, $\xi = 0.0$, $\phi = \psi = 1.0$, and the maximum generation in IP-GA is 100. The optimization results are shown in Table 4.3. As seen in the Table, the optimum is $HBF = 25.8$ kN and $PPD = 67.7\%$, and the corresponding interval of $H$ is [1.3 mm, 6.0 mm]. To verify the effectiveness of the results, the situation that the binder force is a constant 25.8 kN is also considered. Using the optimization method

**Table 4.2** Comparison of the calculation results on 4 sampling points by optimization method and the ANN model [4]

| Sample (*HBF*, *PPD*) | Optimization method | | ANN model (deviation from the optimization method) | |
|---|---|---|---|---|
| | Upper bound of *H* (mm) | Lower bound of *H* (mm) | Upper bound of *H* (mm) | Lower bound of *H* (mm) |
| (20 kN, 60%) | 3.53 | 9.12 | 3.39 (4.1%) | 8.76 (3.9%) |
| (30 kN, 70%) | 1.97 | 7.15 | 1.93 (2.1%) | 7.47 (4.5%) |
| (30 kN, 60%) | 2.51 | 7.58 | 2.59 (3.1%) | 7.80 (2.9%) |
| (20 kN, 70%) | 3.12 | 8.66 | 3.06 (1.8%) | 9.08 (4.9%) |

**Table 4.3** Optimization results of the variable binder force [4]

| *HBF* (kN) | *PPD* | Lower bound of *H* (mm) | Upper bound of *H* (mm) | Multi-objective evaluation function $f_d$ |
|---|---|---|---|---|
| 25.8 | 67.7% | 1.3 | 6.0 | 3.0 |

**Table 4.4** Comparison of the constant binder force and the optimal variable binder force [4]

| Parameters | Binder force (kN) | Maximum major strain (%) | | Maximum thickness reduction rate (%) | | *H* (mm) | | Multi-objective evaluation function $f_d$ |
|---|---|---|---|---|---|---|---|---|
| | | Lower bound | Upper bound | Lower bound | Upper bound | Lower bound | Upper bound | |
| Constant binder force | 25.8 | 18.2 | 22.9 | 11.0 | 13.3 | 4.3 | 7.4 | 3.7 |
| Variable binder force | *HBF* 25.8, *PPD* 67.7% | 13.8 | 16.4 | 8.1 | 9.5 | 1.3 | 6.0 | 3.0 |

to calculate the bounds of the mechanical parameters about the deformation under uncertain friction coefficient for these two cases (stepped variable binder force and constant binder force), we can obtain the computational results as shown in Table 4.4. It can be seen that the maximum major strain and the maximum thickness reduction rate with constant hinder force are obviously larger than those with stepped variable binder force, which means the part is more difficult to be damaged with the optimal variable binder force. Besides, a better interval of *H* value (smaller evaluation function value) is achieved using the optimal variable binder force, which further indicates that the optimal variable binder force can help control the springback and promote the stamping quality.

### 4.4.2 The Locator Optimization in Welding Fixture

The cold stamping process widely exists in manufacturing industries of vehicles, air planes, and domestic appliances for its high productivity and material utilization. The welding assembly of stampings is the critical process, and the performance of welding fixture influences not only the productivity, but also the product quality [14]. The key of the welding fixture design for the thin sheet is to determine the optimal locator number and their positions and to realize the localization and the minimum deformation under the influence of gravity and processing force to improve the accuracy of welding fixture and product quality.

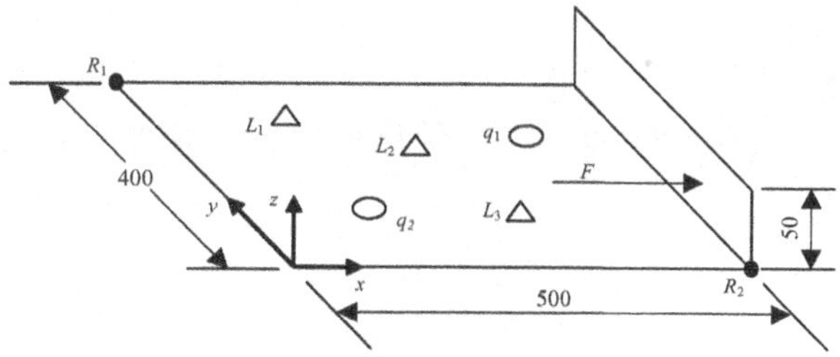


**Fig. 4.9** Locating model of an automobile welding fixture (mm) [15]

1. Formulation of the interval optimization problem

As shown in Fig. 4.9, $q_1$ and $q_2$ are two location pinholes with a diameter of 15 mm and the central points are (80 mm, 80 mm, 0 mm) and (420 mm, 320 mm, 0 mm), respectively. $q_1$ restrains the translational degrees of freedom in $x$ and $y$ directions and $q_2$ restrains the rotational degree of freedom around $z$-axis. $L_1$, $L_2$, and $L_3$ are three locators of the fixture on the datum plane ($z = 0$) to restrain the other three degrees of freedom. The force that the welding gun imposes on the part in $x$-direction is $F = 50\text{N}$ at (420 mm, 320 mm, 0 mm). The thickness of the sheet metal is 1 mm. Due to manufacturing error, the Young's modulus $E$ and Poisson's ratio $\nu$ are uncertain parameters, and the intervals are $E^I = \left[1.8 \times 10^5\text{MPa}, 2.2 \times 10^5\,\text{MPa}\right]$ and $\nu^I = [0.3, 0.4]$, respectively. The objective of this application is to optimize the locations of the three locators to minimize the whole deformation of the part under gravity and machining force. Select five key points on the datum plane: (250 mm, 0 mm), (500 mm, 200 mm), (250 mm, 400 mm), (0 mm, 200 mm) and (250 mm, 200 mm). The whole deformation of the part is described by the total displacements along the $z$-direction of these five key points. Besides, the displacements along the $z$-direction of $R_1$ and $R_2$ should satisfy the constraints to satisfy the following assembly accuracy requirements. Therefore, the uncertain optimization problem can be constructed as follows [4]:

$$\begin{cases} \min\limits_{\mathbf{X}} \sum\limits_{i=1}^{5} d_i(\mathbf{X}, E, \nu) \\ \text{s.t. } d_{R_1}(\mathbf{X}, E, \nu) \le [0.3\,\text{mm}, 0.4\,\text{mm}] \\ \quad d_{R_2}(\mathbf{X}, E, \nu) \le [0.76\,\text{mm}, 0.85\,\text{mm}] \\ \quad \mathbf{X} \in \Omega^6 \\ \quad E \in \left[1.8 \times 10^5\,\text{MPa}, 2.2 \times 10^5\,\text{MPa}\right],\ \nu \in [0.3, 0.4] \end{cases} \tag{4.5}$$

where $\mathbf{X}$ is a six-dimensional design vector composed of $x$ and $y$ coordinates of three locators; $\Omega^6$ represents a 500 mm × 400 mm rectangular field of the sheet metal (excluding the two location pinholes); $d_i$ represents the absolute value of the displacement along the $z$-direction of the $i$-th key point; $d_{R_1}$ and $d_{R_2}$ represent the absolute value of the displacement along the $z$-direction of $R_1$ and $R_2$.

2. Construction of the FEM model and ANN model [16]

Here the objective function and constraints in the optimization problem are both calculated through FEM. The four-node shell element [16], which combines the two-dimensional solid element with the plane element is used to mesh the part. The total number of the elements is 1912. The contact between the locators and part is simplified as point contact, and all the degrees of freedom of the nodes at the contact points are confined. The translational freedom degrees of $x$ and $y$ directions of all nodes on the circumference of the location pinholes are confined, and the translational degrees of freedom of the $z$-direction and all the rotational degrees of freedom are retained.

An ANN model with two hidden layers is constructed, and the inputs are six design variables and two uncertain variables. The objective function and two constraints are the outputs of the ANN model since both of them are calculated through FEM. The numbers of neuron in the input layer, the first hidden layer, the second hidden layer, and the output layer are 8, 18, 9, 3, respectively. Afterwards, the orthogonal design method [17], is used to select 97 sampling points, and each of them requires an FEM calculation. For verifying the prediction accuracy of the trained ANN model, five extra sampling points are selected, and the objective function and constraints are calculated at these five points using FEM and ANN model, respectively. The calculation results are listed in Table 4.5. The maximum error between ANN model and FEM model is only 7.1%, which indicates that the ANN model is accurate enough to satisfy the engineering requirements.

3. Optimization results

Here the IP-GA is combined with the trained ANN model to perform nonlinear interval optimization. The parameters in the optimization are set as follows: $\beta = 0.5$, $\xi = 0.0$, $\phi = 1.2$, $\psi = 0.3$, the penalty factor is 1000, the possibility degree levels of two constraints are both 0.9, and the maximum generation in inner and outer IP-GA are 500 and 100, respectively. The optimization results are shown in Table 4.6, and the optimal locator arrangement is shown in Fig. 4.10. Under the optimal arrangement, the possible interval of the whole deformation is [2.68 mm, 3.42 mm], the penalty function value is $f_p = 1.89$, the intervals of the deformations at $R_1$ and $R_2$ are [0.19 mm, 0.24 mm] and [0.65 mm, 0.81 mm] respectively, and the possibility degree of the constraints are 1.0 and 0.91, respectively, both of which are larger than the possibility degree level 0.9. Besides, the number of FEM model calls in the optimization procedure is only 97.

**Table 4.5** Comparison of calculation results with FEM and ANN models [4]

| Input (design variables and uncertain variable) | | | FEM model | | | ANN model (deviation from FEM model) | | |
|---|---|---|---|---|---|---|---|---|
| Positions of the three locators ($x$, $y$) (mm) | $E$ (MPa) | $\nu$ | Objective function (mm) | Constraint 1 (mm) | Constraint 2 (mm) | Objective function (mm) | Constraint 1 (mm) | Constraint 2 (mm) |
| (160.1,315.2), (468.5, 346.7), (370.3, 90.4) | $2.0 \times 10^5$ | 0.32 | 1.32 | 0.80 | 0.45 | 1.41(6.5%) | 0.82(2.1%) | 0.42(5.8%) |
| (71.8, 270.5), (180.8, 45.2), (321.3, 352.8) | $1.9 \times 10^5$ | 0.34 | 4.00 | 0.14 | 5.69 | 3.85(3.8%) | 0.15(5.9%) | 5.34(6.2%) |
| (480.6, 231.9), (18.0, 190.7), (220.0, 21.4) | $1.8 \times 10^5$ | 0.30 | 5.46 | 2.12 | 2.81 | 5.35(2.1%) | 2.03(4.2%) | 2.89(3.0%) |
| (19.7, 175.9), (259.8, 25.2), (494.1, 320.5) | $2.1 \times 10^5$ | 0.36 | 5.78 | 2.50 | 3.91 | 6.14(6.2%) | 2.58(3.1%) | 3.86(1.2%) |
| (493.1, 125.6), (328.6, 6.1), (214.8, 128.1) | $2.2 \times 10^5$ | 0.38 | 7.72 | 5.28 | 1.19 | 7.17(7.1%) | 5.06(4.2%) | 1.25(5.4%) |

**Table 4.6** Optimization results of the locators [4]

| Positions of the three locators (*x*, *y*) (mm) | Interval of objective function (mm) | Interval of constraint 1 (mm) | Interval of constraint 2 (mm) | Possibility degrees of constraints | Penalty function $f_p$ |
|---|---|---|---|---|---|
| (169.5, 78.6), (485.3, 50.2), (109.5, 340.2) | [2.68, 3.42] | [0.19, 0.24] | [0.65, 0.81] | 1.0, 0.91 | 1.89 |

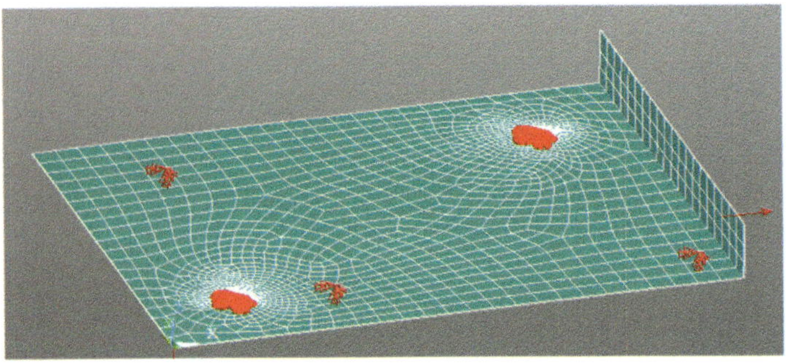

**Fig. 4.10** Optimal arrangement of the locators [4]

## 4.5 Summary

Based on the IP-GA and ANN model, this chapter proposed the two hybrid optimization algorithms to solve the transformed nested optimization problem, and hence two corresponding efficient nonlinear interval optimization methods were developed. The characteristics and application scopes of these two algorithms can be summarized as follows:

(1) In the hybrid optimization algorithm with multiple networks, the inner-layer optimization is avoided by directly obtaining the intervals of the objective function and constraints through the ANN model. On the contrary, nested optimization still exists in the hybrid optimization algorithm with a single network. However, the computational cost caused by the nested optimization in the single-network algorithm is still cheap due to the use of efficient ANN model, though its solution procedure seems more complex than the multi-network algorithm.

(2) For the same problem, the ANN model in the multi-network algorithm is simpler (less inputs and outputs) than that in the single-network algorithm, and hence it is generally easier to ensure the prediction accuracy. However, the acquisition of the training samples needs multiple optimizations, which generally requires a relatively large number of calls of the actual computational models. Therefore, the computational efficiency of the multi-network algorithm is generally lower than the single-network algorithm. Besides, it should be noted that although the construction of the training samples is more efficient in the single-network

algorithm, the structure is more complex (more inputs and outputs). It is known that the training accuracy is harder to ensure when there are more input and output parameters for an ANN model. Therefore, the computational efficiency of the single-network algorithm is generally higher while the adaptability of the multi-network algorithm seems better.

Lastly, the two algorithms can also be combined with the mathematic transformation model based on possibility degree of interval number to construct corresponding efficient interval optimization methods.

## References


1. Liu GR, Han X (2003) Computational inverse techniques in nondestructive evaluation. CRC Press, Florida
2. Masi S (1993) Identification of nonlinear dynamic system using neural networks. ASME J Appl Mech 60:123–133
3. Han X (2001) Elastic waves in functionally graded materials and its applications to material characterization. PhD thesis, National University of Singapore, Singapore
4. Jiang C (2008) Optimization theory and algorithm based on interval uncertainty. PhD thesis, Hunan University, Changsha
5. Cao J, Kinsey B, Solla SA (2000) Consistent and minimal springback using a stepped binder force trajectory and neural network control. J Eng Mater Technol 122:113
6. Ayres RA (1984) SHAPESET: a process to reduce sidewall curl springback in high-strength steel rails. J Appl Metalwork 3(2):127–134
7. Hishida Y, Wagoner RH (1993) Experimental analysis of blank holding force control in sheet forming. Technical report, SAE Technical Paper, p 930285
8. Sunseri M, Cao J, Karafillis AP, Boyce MC (1996) Accommodation of springback error in channel forming using active binder force control: numerical simulations and experiments. J Eng Mater Technol 118(3):426–435
9. Liu G, Lin ZQ, Xu W, Bao Y (2002) Variable blankholder force in U-shaped part forming for eliminating springback error. J Mater Process Technol 120(1–3):259–264
10. Han X, Jiang C, Li GY, Zhong ZH, Hu DB (2006) An inversion procedure for determination of variable binder force in U-shaped forming. Inverse Probl Sci Eng 14(3):301–312
11. Papeleux L, Ponthot J-P (2002) Finite element simulation of springback in sheet metal forming. J Mater Process Technol 125–126:785–791
12. Jiang C, Han X, Liu GR, Li GY (2007) The optimization of the variable binder force in U-shaped forming with uncertain friction coefficient. J Mater Process Technol 182(1–3):262–267
13. Hughes TJR, Liu WK (1981) Nonlinear finite element analysis of shells-Part II. Two-dimensional shells. Comput Methods Appl Mech Eng 27(2):167–181
14. Lai XM, Chen GL, Lin ZQ, Luo CY (2000) On fixture design of sheet stamping assembly. Mech Sci Technol Aerosp Eng 19(5):785–787
15. Jiang C, Han X, Liu GR, Liu GP (2008) A nonlinear interval number programming method for uncertain optimization problems. Eur J Oper Res 188(1):1–13
16. Liu GR, Quek SS (2003) The finite element method: a practical course. Elsevier Science Ltd., England
17. Besterfield DH, Besterfiels-Michna C, Besterfield GH, Besterfield-Sacre M (1999) Total quality management. Prentice-Hall Inc, Englewood Cliffs, New Jersey

# Chapter 5
# Interval Optimization Based on Interval Structural Analysis

**Abstract** This chapter first introduces the conventional interval structural analysis method with small uncertainties. Based on the interval set theory and the subinterval technology, it is then extended to problems with large uncertainties. Consequently, an efficient interval optimization method based on the interval structural analysis is established.

The two hybrid optimization algorithms proposed in Chap. 4, are able to improve the computation efficiency of the nonlinear interval optimizations and make it possible for obtaining the solutions of many practical engineering problems. However, the accuracy of the hybrid optimization algorithms depends on the mapping accuracy of the ANN model. In some complicated design problems where the number of design variables is large or the nonlinearity of optimization models is high, constructing accurate ANN models can be difficult, thus restricting the application of those algorithms. If some advanced analysis methods can be employed to efficiently obtain the response intervals under uncertainties, the inner layer of the interval optimization could be avoided and the two-layer nested optimization problem will be transformed into a conventional single-layer optimization problem, which will be more efficient and practical for engineering problems. The interval structural analysis can be such a numerical analysis method. It combines the interval mathematics and numerical simulation models, such as FEM, and it is able to obtain the response intervals of a structure with only a few simulations. So far, a series of approaches [1–6], about the interval structural analysis have been proposed, and some of them have been applied to solve problems in statics, dynamics, and eigenvalue problem, etc.

This chapter will first give an introduction to the interval set theory, the interval extension theory [7], and the conventional interval structural analysis method under small uncertainties [1, 2]. Based on the interval set theory and the subinterval technology, the existing interval structural analysis is then extended to problems with large uncertainties. Consequently, an efficient interval optimization method based on the interval structural analysis is established.



C. Jiang et al., *Nonlinear Interval Optimization for Uncertain Problems*, Springer Tracts in Mechanical Engineering, https://doi.org/10.1007/978-981-15-8546-3_5

## 5.1 Interval Set Theory and Interval Extension

If $A^L > B^R$ or $B^L > A^R$, the intersection of $A^I$ and $B^I$ is a null set, or $A^I \cap B^I = \emptyset$, otherwise their union set will also be an interval [7]:

$$A^I \cup B^I = \left[\min\left(A^L, B^L\right), \max\left(A^R, B^R\right)\right] \tag{5.1}$$

The union operation of intervals can be extended to cases with an arbitrary number of intervals. For $n$ interval numbers $A_r^I = \left[A_r^L, A_r^R\right]$, $r \in n$, if there exists $t \in \{1, 2, ..., n\}$, $t \neq r$ that makes $A_r^I \cap A_t^I \neq \emptyset$ for any $r \in \{1, 2, ..., n\}$, namely, these interval numbers are singly connected, then $\bigcup\limits_{r\in\{1,2,...,n\}} A_r^I$ is also an interval number

$$\bigcup_{r\in\{1,2,...,n\}} A_r^I = \left[\min_{r\in\{1,2,...,n\}}\left\{A_r^L\right\}, \max_{r\in\{1,2,...,n\}}\left\{A_r^R\right\}\right] \tag{5.2}$$

Assuming that $F\left(\mathbf{A}^I\right) = F\left(A_1^I, A_2^I, \ldots, A_n^I\right)$ is an interval function with respect to the interval vector $\mathbf{A}^I = \left(A_1^I, A_2^I, \ldots, A_n^I\right)^{\mathrm{T}}$, if $A_i^I \subseteq B_i^I$, $i = 1, 2, \ldots, n$, we have

$$F\left(A_1^I, A_2^I, \ldots, A_n^I\right) \subseteq F\left(B_1^I, B_2^I, \ldots, B_n^I\right) \tag{5.3}$$

then $F\left(\mathbf{A}^I\right) = F\left(A_1^I, A_2^I, \ldots, A_n^I\right)$ is called to have inclusion monotonicity with respect to $\mathbf{A}^I = \left(A_1^I, A_2^I, \ldots, A_n^I\right)^{\mathrm{T}}$.

The interval arithmetic operations and the rounded interval arithmetic operations have the inclusion monotonicity, hence the rational interval function is inclusion monotonic. If the interval of a function can be determined by finite times of interval arithmetic operations, the function is called a rational interval function.

Assuming $f(a_1, a_2, \ldots, a_n)$ is a real rational function with respect to $a_1, a_2, \ldots, a_n$, if the real variables in $f(a_1, a_2, \ldots, a_n)$ are replaced by corresponding interval variables, and the arithmetic operation of real number is replaced by corresponding interval arithmetic operation, then the obtained rational interval function $F\left(A_1^I, A_2^I, \ldots, A_n^I\right)$ is obviously inclusion monotonic and becomes a natural interval extension of $f$.

If $F$ is an inclusion monotonic interval extension of $f$, then:

$$f\left(A_1^I, A_2^I, \ldots, A_n^I\right) \subseteq F\left(A_1^I, A_2^I, \ldots, A_n^I\right) \tag{5.4}$$

where $f\left(A_1^I, A_2^I, \ldots, A_n^I\right)$ represents the value domain of $f(a_1, a_2, \ldots, a_n)$:

$$f\left(A_1^I, A_2^I, \ldots, A_n^I\right) = \left\{f(a_1, a_2, \ldots, a_n) | a_i \in A_i^I, i = 1, 2, \ldots, n\right\} \tag{5.5}$$

Therefore, the inclusion monotonic interval extension $F(A_1^I, A_2^I, \ldots, A_n^I)$ with respect to $(A_1^I, A_2^I, \ldots, A_n^I)^{\mathrm{T}}$ contains the value domain of the real function $f(a_1, a_2, \ldots, a_n)$ with respect to $(A_1^I, A_2^I, \ldots, A_n^I)^{\mathrm{T}}$, providing a method to obtain the bounds of a real rational function defined on an $n$-dimensional cuboid [8].

## 5.2 The Interval Structural Analysis Method

The interval structural analysis method can be used to obtain the response bounds of a structure involving uncertainties. In this section, we first introduce the interval structural analysis method for problems with small uncertainties [1, 2], and then extend it to problems with large uncertainties.

### *5.2.1 Interval Structural Analysis for Small Uncertainties*

Based on FEM, the static displacement response of a structure can be obtained via solving the following governing equation [9]:

$$\mathbf{Kd} = \mathbf{F} \tag{5.6}$$

where $\mathbf{K}$ is the $n \times n$ symmetric global stiffness matrix, $\mathbf{d}$ and $\mathbf{F}$ are $n$-dimensional displacement response vector and load vector on the nodes, respectively. Assuming there is an $n$-dimensional dimensional uncertain vector $\mathbf{U}$ existing in the structure and loads, the above equation can be rewritten as:

$$\mathbf{K}(\mathbf{U})\mathbf{d}(\mathbf{U}) = \mathbf{F}(\mathbf{U}) \tag{5.7}$$

Here we use an interval vector $\mathbf{U}^I$ to describe the uncertainty of $\mathbf{U}$

$$\mathbf{U} \subset \mathbf{U}^I = \left[\mathbf{U}^L, \mathbf{U}^R\right],\ U_j \in U_j^I = \left[U_j^L, U_j^R\right],\ j = 1, 2, \ldots, q \tag{5.8}$$

$\mathbf{U}^I$ can be rewritten as

$$\mathbf{U}^I = \mathbf{U}^c + [-1, 1]\mathbf{U}^w = U_j^c + [-1, 1]U_j^w,\ j = 1, 2, \ldots, q \tag{5.9}$$

where

$$\mathbf{U}^c = \frac{\mathbf{U}^L + \mathbf{U}^R}{2},\ U_j^c = \frac{U_j^L + U_j^R}{2},\ j = 1, 2, \ldots, q \tag{5.10}$$

$$\mathbf{U}^w = \frac{\mathbf{U}^R - \mathbf{U}^L}{2},\ U_j^w = \frac{U_j^R - U_j^L}{2},\ j = 1, 2, \ldots, q \tag{5.11}$$

Based on Eqs. (5.8) and (5.9), **U** can be rewritten as

$$\mathbf{U} = \mathbf{U}^c + \delta\mathbf{U} \tag{5.12}$$

where

$$\delta\mathbf{U} \in [-1, 1]\mathbf{U}^w,\ \delta U_j \in [-1, 1]U_j^w,\ j = 1, 2, \ldots, q \tag{5.13}$$

Assuming the uncertainty levels of the variables are all relatively small, $\boldsymbol{d}(\mathbf{U})$ can be expanded into the following expression with respect to $\mathbf{U}^c$, using the first-order Taylor expansion [1]:

$$\mathbf{d}(\mathbf{U}) = \mathbf{d}(\mathbf{U}^c + \delta\mathbf{U}) \approx \boldsymbol{d}(\mathbf{U}^c) + \sum_{j=1}^{q} \frac{\partial \mathbf{d}(\mathbf{U}^c)}{\partial U_j}\delta U_j \tag{5.14}$$

Because $\delta\mathbf{U}$ is an interval vector defined in Eq. (5.13), the intervals of the displacement responses can be obtained via the natural interval extension of Eq. (5.14):

$$\mathbf{d}^I(\mathbf{U}^I) = \mathbf{d}(\mathbf{U}^c) + \sum_{j=1}^{q} \left| \frac{\partial \mathbf{d}(\mathbf{U}^c)}{\partial U_j} \right| [-1, 1]U_j^w \tag{5.15}$$

Therefore, the lower bound vector $\mathbf{d}^L(\mathbf{U}^I)$ and the upper bound vector $\mathbf{d}^R(\mathbf{U}^I)$ of the displacement responses can be explicitly obtained:

$$\mathbf{d}^L(\mathbf{U}^I) = \mathbf{d}(\mathbf{U}^c) - \sum_{j=1}^{q} \left| \frac{\partial \mathbf{d}(\mathbf{U}^c)}{\partial U_j} \right| U_j^w \tag{5.16}$$

$$\mathbf{d}^R(\mathbf{U}^I) = \mathbf{d}(\mathbf{U}^c) + \sum_{j=1}^{q} \left| \frac{\partial \mathbf{d}(\mathbf{U}^c)}{\partial U_j} \right| U_j^w \tag{5.17}$$

The gradients of the displacement response in the above two equations can be obtained by the following sensitivity analysis technique. Differentiating both sides of Eq. (5.7), yields:

$$\mathbf{K}(\mathbf{U}^c)\frac{\partial \mathbf{d}(\mathbf{U}^c)}{\partial U_j} = \frac{\partial \mathbf{F}(\mathbf{U}^c)}{\partial U_j} - \frac{\partial \mathbf{K}(\mathbf{U}^c)}{\partial U_j}\mathbf{d}(\mathbf{U}^c),\ j = 1, 2, \ldots, q \tag{5.18}$$

In FEM analysis, the elemental stiffness matrixes and the elemental load vectors are computed first, and then they are assembled into the global stiffness matrix and the global load vector:

$$\mathbf{K}(\mathbf{U}) = \sum_{i=1}^{s} \mathbf{K}_i(\mathbf{U}) \tag{5.19}$$

$$\mathbf{F}(\mathbf{U}) = \sum_{i=1}^{s} \mathbf{F}_i(\mathbf{U}) \tag{5.20}$$

where $\mathbf{K}_i(\mathbf{U})$ and $\mathbf{F}_i(\mathbf{U})$ represent the $i$-th elemental stiffness matrix and the $i$-th elemental load vector, respectively. $s$ is the number of the elements and $\sum$ represents the assembly of the matrixes according to FEM instead of a simple sum. The gradients of the stiffness matrix and the load vector with respect to the uncertain parameters can be obtained via the following equations:

$$\frac{\partial \mathbf{K}(\mathbf{U}^c)}{\partial U_j} = \frac{\partial}{\partial U_j}\left(\sum_{i=1}^{s} \mathbf{K}_i(\mathbf{U}^c)\right) = \sum_{i=1}^{s} \frac{\partial \mathbf{K}_i(\mathbf{U}^c)}{\partial U_j}, \; j = 1, 2, \ldots, q \tag{5.21}$$

$$\frac{\partial \mathbf{F}(\mathbf{U}^c)}{\partial U_j} = \frac{\partial}{\partial U_j}\left(\sum_{i=1}^{s} \mathbf{F}_i(\mathbf{U}^c)\right) = \sum_{i=1}^{s} \frac{\partial \mathbf{F}_i(\mathbf{U}^c)}{\partial U_j}, \; j = 1, 2, \ldots, q \tag{5.22}$$

In practical FEM applications, the element stiffness matrix is usually obtained by numerical integration. Therefore, it is unable to be explicitly represented with corresponding structural parameters. In such a case, the gradient of an implicit element stiffness matrix with respect to the structural parameters can be obtained with numerical integration after differential operations. In addition, the finite difference method can also be employed to compute the gradient by two evaluations of the element stiffness matrix.

After getting the gradients of the global stiffness matrix and the nodal load vector with respect to all uncertain variables, the gradient vectors of the displacement response with respect to all uncertain variables can be obtained with $q$ FEM evaluations based on Eq. (5.18). Then the bounds of structural displacement responses can be obtained via Eqs. (5.16) and (5.17). From the above analysis, it is clear that $q + 1$ FEM evaluations are needed to obtain the interval of the structural displacement response by the interval structural analysis method, among which $q$ evaluations are consumed by the structural sensitivity analysis and the other one is applied to compute the structural response on the middle point of the uncertain variables. In conclusion, compared to the optimization method to solve the response bounds, the interval structural analysis method can greatly improve the computation efficiency.

### 5.2.2 Interval Structural Analysis for Large Uncertainties

The conventional interval structural analysis method introduced above is based on the linear approximation (first-order Taylor expansion) of the displacement response. Hence, theoretically, it is only suitable for problems involving variables with relatively small uncertainties. In this chapter, based on the combination of the subinterval technology and the interval set theory, the interval structural analysis method is extended to problems with large uncertainties [10]. It is worth mentioning that the concept of subinterval has been proposed in reference [8], which is combined with interval perturbation method to develop a subinterval perturbation method. Here the subinterval technology is employed to extend the interval structural analysis method in order to provide solution to the problems involving large uncertainties.

Considering that the large uncertainties will in general cause large error in the interval structural analysis, the intervals with large uncertainties are divided into some small subintervals as follows:

$$\left(U_j^I\right)_i = \left[U_j^L + \frac{(i-1)2U_j^w}{t_j}, U_j^L + \frac{i2U_j^w}{t_j}\right],\ i = 1, 2, \ldots, t_j,\ j = 1, 2, \ldots, q \tag{5.23}$$

where $\left(U_j^I\right)_i$ represents the $i$-th subinterval of the $j$-th interval variable, and $t_j$ is the number of the subintervals of $U_j^I$. The number of subintervals can be determined according to the uncertainty of each variable. Since we can take out one subinterval from each uncertain variable, $r$ combinations of the subintervals can be obtained:

$$r = t_1 t_2 \ldots t_q \tag{5.24}$$

The FEM governing functions for all the subinterval combinations can be expressed as follows:

$$\mathbf{K}\left(\mathbf{U}^I_{G_1G_2\ldots G_q}\right)\mathbf{d}\left(\mathbf{U}^I_{G_1G_2\ldots G_q}\right) = \mathbf{F}\left(\mathbf{U}^I_{G_1G_2\ldots G_q}\right),\ G_j = 1, 2, \ldots, t_j,\ j = 1, 2, \ldots, q \tag{5.25}$$

where $\mathbf{U}^I_{G_1G_2\ldots G_q}$ denotes the subinterval vector composed of the $G_j$-th subinterval of the $j$-th uncertain variable where $j = 1, 2, \ldots, q$. Equation (5.25) thus represents $r = t_1 t_2 \ldots t_q$ FEM equations, each of which is under small uncertainties and can be solved using the conventional interval structural analysis method introduced in Sect. 5.2.1. For any subinterval vector $\mathbf{U}^I_{G_1G_2\ldots G_q}$, the displacement response interval vector $\mathbf{d}^I\left(\mathbf{U}^I_{G_1G_2\ldots G_q}\right)$ (denoted as $\mathbf{d}^I_{G_1G_2\ldots G_q}$ for convenience) can be obtained via the proposed method. Besides, for any two subinterval vectors $\mathbf{U}^I_{G_1G_2\ldots G_q}$ and $\mathbf{U}^I_{(G_1+1)G_2\ldots G_q}$, the following equations always hold:

$$\mathbf{K}\left(\mathbf{U}^I_{G_1G_2\ldots G_q}\right) \cap \mathbf{K}\left(\mathbf{U}^I_{(G_1+1)G_2\ldots G_q}\right) = \mathbf{K}\left(\mathbf{U}^R_{G_1} = U^L_{(G_1+1)}, U^I_{G_2}, U^I_{G_3}, \ldots, U^I_{G_q}\right) \tag{5.26}$$

$$\mathbf{F}\left(\mathbf{U}^I_{G_1G_2\ldots G_q}\right) \cap \mathbf{F}\left(\mathbf{U}^I_{(G_1+1)G_2\ldots G_q}\right) = \mathbf{F}\left(U^R_{G_1} = U^L_{(G_1+1)}, U^I_{G_2}, U^I_{G_3}, \ldots, U^I_{G_q}\right) \tag{5.27}$$

Therefore, there must be an intersection between $\mathbf{K}\left(\mathbf{U}^I_{G_1G_2\ldots G_q}\right)$, $\mathbf{K}\left(\mathbf{U}^I_{(G_1+1)G_2\ldots G_q}\right)$, and between $\mathbf{F}\left(\mathbf{U}^I_{G_1G_2\ldots G_q}\right)$, $\mathbf{F}\left(\mathbf{U}^I_{(G_1+1)G_2\ldots G_q}\right)$, hence

$$\mathbf{d}^I_{G_1G_2\ldots G_q} \cap \mathbf{d}^I_{(G_1+1)G_2\ldots G_q} \neq \emptyset \tag{5.28}$$

Similarly, the following equations also hold:

$$\mathbf{d}^I_{G_1G_2\ldots G_q} \cap \mathbf{d}^I_{G_1(G_2+1)\ldots G_q} \neq \emptyset,\ \mathbf{d}^I_{G_1G_2G_3\ldots G_q} \cap \mathbf{d}^I_{G_1G_2(G_3+1)\ldots G_q} \neq \emptyset, \ldots,$$
$$\mathbf{d}^I_{G_1G_2\ldots G_q} \cap \mathbf{d}^I_{G_1G_2\ldots(G_q+1)} \neq \emptyset \tag{5.29}$$

Equation (5.28) and Eq. (5.29), show that $\mathbf{d}^I\left(\mathbf{U}^I_{G_1G_2\ldots G_q}\right)$, $G_j = 1, 2, , t_j$, $j = 1, 2, \ldots, q$ are singly connected. Then the union set $\mathbf{d}^I_s$ of all these intervals can be obtained by Eq. (5.2)

$$\mathbf{d}^I_s = \left[\mathbf{d}^L_s, \mathbf{d}^R_s\right],\ \left(d^I_s\right)_i = \left[\left(d^L_s\right)_i, \left(d^R_s\right)_i\right],\ i = 1, 2, \ldots, n \tag{5.30}$$

where

$$\left(d^L_s\right)_i = \min_{G_j=1,2,\ldots,t_j\ \ j=1,2,\ldots,q} \left(d^L_{G_1G_2\ldots G_q}\right)_i,\ i = 1, 2, \ldots, n \tag{5.31}$$

$$\left(d^R_s\right)_i = \max_{G_j=1,2,\ldots,t_j\ \ j=1,2,\ldots,q} \left(d^R_{G_1G_2\ldots G_q}\right)_i,\ i = 1, 2, \ldots, n \tag{5.32}$$

$\mathbf{d}^I_s$ is just the displacement response interval vector obtained via the interval structural analysis method for problems with large uncertainties.

As shown above, the interval structural analysis method is extended to deal with problems with arbitrarily large uncertainties by dividing the large intervals into enough subintervals. However, it should be mentioned that the computation efficiency is sacrificed because the conventional interval structural analysis is performed many times according to the number of subintervals. For practical engineering problems, the term large uncertainties is a relative concept. Its absolute value is usually relatively small as the uncertainty is usually represented by a small perturbation around the nominal values of parameters, hence, in reality, the number of subintervals is generally not large. In addition, in practical problems only one or a few of uncertain variables need to be divided into subintervals, hence in problems with large uncertainties, we generally do not have to worry too much about the problem of too many subinterval combinations.

Although the above interval structural analysis method is formulated on the base of static FEM, it can be applied to dynamic problems. Based on dynamic FEM, Reference [2], developed a corresponding interval structural analysis method to compute the dynamic respond bounds of structures with small uncertainties. Similarly, this approach can also be conveniently extended to solve dynamic problems with large uncertainties. Also, interval structural analysis method is not confined to obtaining the response intervals based on FEM. It can be combined with other numerical algorithms, such as the meshless method and rigid multi-body dynamics, to develop corresponding structural analysis methods. On the other hand, besides the displacement response, it can be also extended to compute the other structural response intervals, such as stress and strain. The difference is that corresponding sensitivity analysis methods should be developed to compute the gradients of the structural responses. If the sensitivity information is difficult to get in practical computations, the finite difference method can be employed at the price of more computational costs.

The interval structural analysis method is actually a general numerical analysis method. It computes the bounds of a nonlinear function under disturbance by combining the Taylor expansion and the interval set theory. Therefore, in the viewpoint of mathematics, it should be called an interval analysis method. In this book, it is named as interval structural analysis from the viewpoint of structural mechanics.

### 5.2.3 Numerical Example and Discussions

In this example, a six-bar plane truss structure [11], is analyzed to demonstrate the effectiveness of interval structural analysis method. As shown in Fig. 5.1, the truss element is used to create the FEM mesh, in which each bar is an element, and hence there are a total of six elements and four nodes involved.

The elemental stiffness matrixes $\mathbf{K}_i$, $i = 1, 2, \ldots, 6$ and the nodal load vector $\mathbf{F}$ can be expressed as follows:

$$\mathbf{K}_1 = \frac{E_1 A_1}{L_1}\begin{bmatrix} 1 & 0 & -1 & 0 \\ 0 & 0 & 0 & 0 \\ -1 & 0 & 1 & 0 \\ 0 & 0 & 0 & 0 \end{bmatrix}, \mathbf{K}_2 = \frac{E_2 A_2}{L_2}\begin{bmatrix} 1 & 0 & -1 & 0 \\ 0 & 0 & 0 & 0 \\ -1 & 0 & 1 & 0 \\ 0 & 0 & 0 & 0 \end{bmatrix}$$

$$\mathbf{K}_3 = \frac{E_3 A_3}{L_3}\begin{bmatrix} 0 & 0 & 0 & 0 \\ 0 & 1 & 0 & -1 \\ 0 & 0 & 0 & 0 \\ 0 & -1 & 0 & 1 \end{bmatrix}, \mathbf{K}_4 = \frac{E_4 A_4}{L_4}\begin{bmatrix} 0 & 0 & 0 & 0 \\ 0 & 1 & 0 & -1 \\ 0 & 0 & 0 & 0 \\ 0 & -1 & 0 & 1 \end{bmatrix}$$

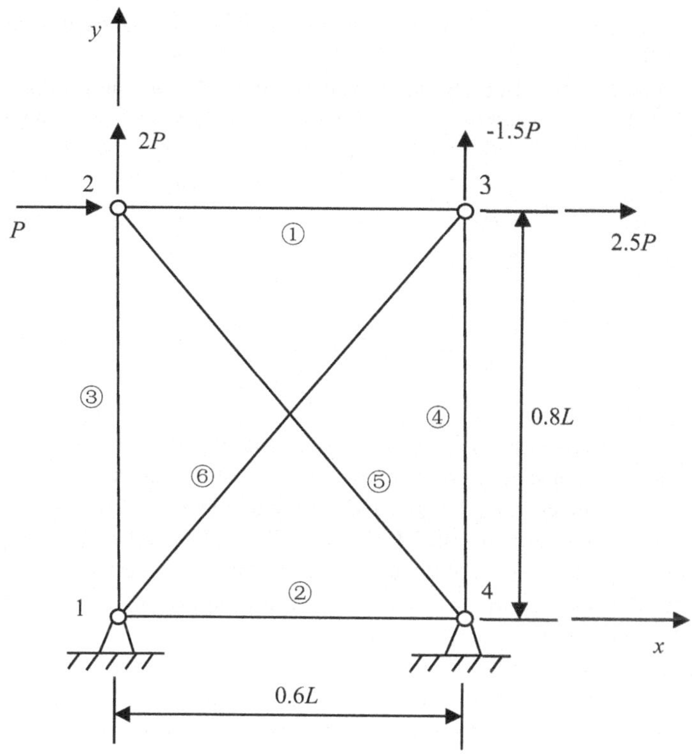


**Fig. 5.1** A six-bar plane truss [11]

$$\mathbf{K}_5 = \frac{E_5 A_5}{L_5}\begin{bmatrix} 0.36 & -0.48 & -0.36 & 0.48 \\ -0.48 & 0.64 & 0.48 & -0.64 \\ -0.36 & 0.48 & 0.36 & -0.48 \\ 0.48 & -0.64 & -0.48 & 0.64 \end{bmatrix}$$

$$\mathbf{K}_6 = \frac{E_6 A_6}{L_6}\begin{bmatrix} 0.36 & 0.48 & -0.36 & -0.48 \\ 0.48 & 0.64 & -0.48 & -0.64 \\ -0.36 & -0.48 & 0.36 & 0.48 \\ -0.48 & -0.64 & 0.48 & 0.64 \end{bmatrix} \tag{5.33}$$

$$\mathbf{F} = (P, 2P, 2.5P, -1.5P)^{\mathrm{T}} \tag{5.34}$$

where $E_i$, $A_i$, and $L_i$ respectively, denote the Young's modulus, the cross-section area and the length of the $i$-th bar.

The cross-section area of all bars is $A = 1.0 \times 10^3\,\mathrm{mm}^2$, the Young's modulus of all bars is $E = 2.0 \times 10^5$ MPa, the load is $P = 20$ kN, and the length of all bars is $L = 1000$ mm. Due to the manufacturing and measurement errors, $A$ and $E$ are uncertain parameters whose uncertainty levels are both 10%:

**Table 5.1** Displacement response intervals with interval structural analysis method for small uncertainties [12]

| Parameter | Low bound (mm) | Upper bound (mm) | Middle point (mm) | Radius (mm) | Uncertainty level |
|---|---|---|---|---|---|
| $d_1$ | 0.72 | 1.08 | 0.90 | 0.18 | 20.0% |
| $d_2$ | 0.27 | 0.40 | 0.34 | 0.07 | 20.6% |
| $d_3$ | 0.75 | 1.13 | 0.94 | 0.19 | 20.2% |
| $d_4$ | 0.25 | 0.38 | 0.32 | 0.07 | 21.9% |

$$A^I = \left[0.9 \times 10^3\,\text{mm}^2, 1.1 \times 10^3\,\text{mm}^2\right],\; E^I = \left[1.8 \times 10^5\,\text{MPa}, 2.2 \times 10^5\,\text{MPa}\right] \tag{5.35}$$

With the interval structural analysis method for small uncertainties, the intervals of absolute values of displacement responses at nodes 2 and 3 can be obtained as shown in Table 5.1. The uncertainty levels of the four displacement responses intervals caused by uncertain parameters are 20.0%, 20.6%, 20.2%, and 21.9%, respectively, basically equivalent to the sum of the uncertainty levels of two uncertain parameters, which means the uncertainties of $A$ and $E$ are not amplified in the displacement responses for this problem.

Via mechanical analysis, it is not difficult to know that the four displacement responses reach their upper bounds at $A^L$ and $E^L$, and their lower bounds at $A^R$ and $E^R$. Therefore, the accurate intervals, in the sense of FEM, of the four displacement responses can be obtained through once FEM computation for each case. The comparisons between the accurate results and the interval structural analysis method are listed in Table 5.2. We can see from the table that the results of the interval structural analysis method are quite close to the accurate results, with the maximum deviation of only 4% at the lower bound of $d_1$. Therefore, the computation accuracy of the conventional interval structural analysis method is very high when the uncertainties of $A$ and $E$ are only 10%. In addition, only 3 FEM evaluations are costed when performing the interval structural analysis method to obtain the results in this problem.

**Table 5.2** Comparison between the results calculated with interval structural analysis method for small uncertainties and the accurate values (under 10% uncertainty level) [12]

| Parameter | Accurate values | | Results with interval structural analysis method (Deviations) | |
|---|---|---|---|---|
| | Low bound(mm) | Upper bound(mm) | Low bound (mm) | Upper bound (mm) |
| $d_1$ | 0.75 | 1.11 | 0.72 (4.0%) | 1.08 (2.7%) |
| $d_2$ | 0.28 | 0.41 | 0.27 (3.6%) | 0.40 (2.4%) |
| $d_3$ | 0.78 | 1.16 | 0.75 (3.8%) | 1.13 (2.6%) |
| $d_4$ | 0.26 | 0.39 | 0.25 (3.8%) | 0.38 (2.6%) |

Then the uncertainties of $A$ and $E$ are set to be 20% and 30%, respectively, and their corresponding intervals are:

$$A^I = \left[0.8 \times 10^3\,\text{mm}^2, 1.2 \times 10^3\,\text{mm}^2\right],\ E^I = \left[1.4 \times 10^5\,\text{MPa}, 2.6 \times 10^5\,\text{MPa}\right] \tag{5.36}$$

Similarly, the conventional interval structural analysis is carried out again and the results are listed in Table 5.3 compared with the accurate results. It can be found that the deviations of the results obtained by the interval structural analysis noticeably increase, with the maximum deviation of 22.4%, also at the lower bound of $d_1$. It indicates that higher uncertainty levels of the parameters may lead to relatively larger errors when using the conventional interval structural analysis method. To improve the computation accuracy, the interval structural analysis method for large uncertainties is then employed to recalculate the response intervals. Here the intervals $A^I$ and $E^I$ in Eq. (5.36), are divided into two and three subintervals, respectively. Thus there are totally $r = 2 \times 3 = 6$ subinterval combinations involved, denoted by $s_i$, $i = 1, 2, \ldots, 6$ as listed in detail in Table 5.4. Firstly, applying the interval structural analysis method for small uncertainties to each subinterval combination, we can obtain the response intervals as listed in Tables 5.5, 5.6, 5.7, 5.8, 5.9 and 5.10. According to Tables 5.5, 5.6, 5.7, 5.8, 5.9 and 5.10, it can be found that the radiuses of response intervals, or the uncertainty levels of responses, are relatively small in each subinterval combination since use of the subinterval technique helps control the

**Table 5.3** Comparison between the results calculated with interval structural analysis method for small uncertainties and the accurate values under 20% and 30% uncertainty levels [12]

| Parameter | Accurate values | | Results with interval structural analysis method (Deviations) | |
|---|---|---|---|---|
| | Low bound (mm) | Upper bound (mm) | Low Bound (mm) | Upper bound (mm) |
| $d_1$ | 0.58 | 1.61 | 0.45 (22.4%) | 1.35 (16.1%) |
| $d_2$ | 0.21 | 0.60 | 0.17 (19.0%) | 0.50 (16.7%) |
| $d_3$ | 0.60 | 1.68 | 0.47 (21.7%) | 1.41 (16.1%) |
| $d_4$ | 0.20 | 0.57 | 0.16 (20.0%) | 0.48 (15.8%) |

**Table 5.4** Six subinterval combinations of the uncertain parameters $A$ and $E$ [12]

| Parameter | Subinterval of $A$ ($\text{mm}^2$) | Subinterval of $E$ (MPa) |
|---|---|---|
| $s_1$ | $[0.8 \times 10^3, 1.0 \times 10^3]$ | $[1.4 \times 10^5, 1.8 \times 10^5]$ |
| $s_2$ | $[0.8 \times 10^3, 1.0 \times 10^3]$ | $[1.8 \times 10^5, 2.2 \times 10^5]$ |
| $s_3$ | $[0.8 \times 10^3, 1.0 \times 10^3]$ | $[2.2 \times 10^5, 2.6 \times 10^5]$ |
| $s_4$ | $[1.0 \times 10^3, 1.2 \times 10^3]$ | $[1.4 \times 10^5, 1.8 \times 10^5]$ |
| $s_5$ | $[1.0 \times 10^3, 1.2 \times 10^3]$ | $[1.8 \times 10^5, 2.2 \times 10^5]$ |
| $s_6$ | $[1.0 \times 10^3, 1.2 \times 10^3]$ | $[2.2 \times 10^5, 2.6 \times 10^5]$ |

**Table 5.5** Displacement response intervals derived from subinterval combination $s_1$ (mm) [12]

| Parameter | Low bound (mm) | Upper bound (mm) | Middle point (mm) | Radius (mm) |
|---|---|---|---|---|
| $d_1$ | 0.96 | 1.55 | 1.26 | 0.30 |
| $d_2$ | 0.36 | 0.58 | 0.47 | 0.11 |
| $d_3$ | 1.00 | 1.62 | 1.31 | 0.31 |
| $d_4$ | 0.34 | 0.55 | 0.45 | 0.11 |

**Table 5.6** Displacement response intervals derived from subinterval combination $s_2$ (mm) [12]

| Parameter | Low bound (mm) | Upper bound (mm) | Middle point (mm) | Radius (mm) |
|---|---|---|---|---|
| $d_1$ | 0.79 | 1.21 | 1.00 | 0.21 |
| $d_2$ | 0.29 | 0.45 | 0.37 | 0.08 |
| $d_3$ | 0.82 | 1.27 | 1.05 | 0.23 |
| $d_4$ | 0.28 | 0.43 | 0.36 | 0.08 |

**Table 5.7** Displacement response intervals derived from subinterval combination $s_3$ (mm) [12]

| Parameter | Low bound (mm) | Upper bound (mm) | Middle point (mm) | Radius (mm) |
|---|---|---|---|---|
| $d_1$ | 0.67 | 1.00 | 0.84 | 0.17 |
| $d_2$ | 0.25 | 0.37 | 0.31 | 0.06 |
| $d_3$ | 0.70 | 1.04 | 0.87 | 0.17 |
| $d_4$ | 0.24 | 0.35 | 0.30 | 0.06 |

**Table 5.8** Displacement response intervals derived from subinterval combination $s_4$ (mm) [12]

| Parameter | Low bound (mm) | Upper bound (mm) | Middle point (mm) | Radius (mm) |
|---|---|---|---|---|
| $d_1$ | 0.80 | 1.25 | 1.03 | 0.23 |
| $d_2$ | 0.30 | 0.46 | 0.38 | 0.08 |
| $d_3$ | 0.84 | 1.30 | 1.07 | 0.23 |
| $d_4$ | 0.28 | 0.44 | 0.36 | 0.08 |

**Table 5.9** Displacement response intervals derived from subinterval combination $s_5$ (mm) [12]

| Parameter | Low bound (mm) | Upper bound (mm) | Middle point (mm) | Radius (mm) |
|---|---|---|---|---|
| $d_1$ | 0.66 | 0.98 | 0.82 | 0.16 |
| $d_2$ | 0.25 | 0.36 | 0.31 | 0.06 |
| $d_3$ | 0.69 | 1.02 | 0.86 | 0.17 |
| $d_4$ | 0.23 | 0.34 | 0.29 | 0.06 |

**Table 5.10** Displacement response intervals derived from subinterval combination $s_6$ (mm) [12]

| Parameter | Low bound (mm) | Upper bound (mm) | Middle point (mm) | Radius (mm) |
|---|---|---|---|---|
| $d_1$ | 0.56 | 0.80 | 0.68 | 0.12 |
| $d_2$ | 0.21 | 0.30 | 0.26 | 0.05 |
| $d_3$ | 0.59 | 0.84 | 0.72 | 0.13 |
| $d_4$ | 0.20 | 0.28 | 0.24 | 0.04 |

uncertainties of parameters at a small level. In addition, for each displacement component, all response intervals under six subinterval combinations are singly connected. It means that with the interval set operations defined by Eq. (5.2), the computation results of the proposed interval structural analysis method for large uncertainties can be obtained. The comparison between the results and the accurate values are listed in Table 5.11, which indicates that the results of the proposed interval structural analysis method are very close to the accurate values, with the maximum deviation of only 3.7% at the upper bound of $d_1$. Therefore, the interval structural analysis method for large uncertainties has a high accuracy even when relatively large uncertainties are involved in parameters of the structure. As for the computation cost, 3 FEM evaluations are performed for every subinterval combination, thus the number of all FEM evaluations is 18 in total, meaning that the computation efficiency of the interval structural analysis method for large uncertainties is lower compared with its counterpart.

This numerical example is relatively simple so that we only need to use the mechanical analysis to quickly obtain the accurate response bounds of the structure under parametric uncertainties. However, for many practical structures, the relationship between responses and uncertain parameters is so complicated that two optimization processes are in general necessary to solve the response intervals, leading to low computation efficiency. Therefore, compared to the optimization method, the interval structural analysis method for large uncertainties still has a higher computation efficiency.

**Table 5.11** Comparison between the results calculated with interval structural analysis method for large uncertainties and the accurate values [12]

| Parameter | Accurate values | | Results with interval structural analysis method (Deviations) | |
|---|---|---|---|---|
| | Low bound (mm) | Upper bound (mm) | Low bound (mm) | Upper bound (mm) |
| $d_1$ | 0.58 | 1.61 | 0.56 (3.4%) | 1.55 (3.7%) |
| $d_2$ | 0.21 | 0.60 | 0.21 (0.0%) | 0.58 (3.3%) |
| $d_3$ | 0.60 | 1.68 | 0.59 (1.7%) | 1.62 (3.6%) |
| $d_4$ | 0.20 | 0.57 | 0.20 (0.0%) | 0.55 (3.5%) |

## 5.3 An Efficient Interval Optimization Method

In this section, an efficient interval optimization method is proposed for structural design with combination of the interval structural analysis method and the mathematical transformation model based on order relation of interval number.

### 5.3.1 Algorithm Description

For a practical structure, a nonlinear interval optimization problem denoted by Eq. (4.1), can be constructed and then converted to a deterministic two-layer nested optimization problem denoted by Eq. (4.3). In structural optimization, the objective function or constraints are usually structural responses obtained with the numerical analysis method, such as displacement, stress, strain, etc. Therefore, when solving the converted deterministic optimization problem in nonlinear interval optimization, the interval structural analysis method can be employed to calculate the response bounds of objective function and constraints in each iteration to avoid the inner layer optimization. An efficient interval optimization method thus can be formulated and its computational process can be summarized as follows:

(1) IP-GA is still selected as the solver for outer layer optimization in consideration of the discontinuity and the disdifferentiability of the converted deterministic optimization problem.
(2) Calculate the response intervals of the uncertain objective function and constraints when a design vector $\boldsymbol{X}$ is given by IP-GA. If the uncertainty levels of all parameters are relatively small, use the interval structural analysis method for small uncertainties

$$f^L(\mathbf{X}, \mathbf{U}^I) = f(\mathbf{X}, \mathbf{U}^c) - \sum_{j=1}^{q} \left| \frac{\partial f(\mathbf{X}, \mathbf{U}^c)}{\partial U_j} \right| U_j^w \tag{5.37}$$

$$f^R(\mathbf{X}, \mathbf{U}^I) = f(\mathbf{X}, \mathbf{U}^c) + \sum_{j=1}^{q} \left| \frac{\partial f(\mathbf{X}, \mathbf{U}^c)}{\partial U_j} \right| U_j^w \tag{5.38}$$

$$g_i^L(\mathbf{X}, \mathbf{U}^I) = g_i(\mathbf{X}, \mathbf{U}^c) - \sum_{j=1}^{q} \left| \frac{\partial g_i(\mathbf{X}, \mathbf{U}^c)}{\partial U_j} \right| U_j^w, \ i = 1, 2, \ldots, l \tag{5.39}$$

$$g_i^R(\mathbf{X}, \mathbf{U}^I) = g_i(\mathbf{X}, \mathbf{U}^c) + \sum_{j=1}^{q} \left| \frac{\partial g_i(\mathbf{X}, \mathbf{U}^c)}{\partial U_j} \right| U_j^w, \ i = 1, 2, \ldots, l \tag{5.40}$$

If the uncertainty levels of parameters are relatively large, divide the intervals into subintervals and then obtain the intervals of the objective function and constraints by using the interval structural analysis method for large uncertainties

$$f_s^L(\mathbf{X}, \mathbf{U}^I) = \min f^L_{G_1G_2...G_q}(\mathbf{X}, \mathbf{U}^I),\ G_j = 1, 2, \ldots, t_j,\ j = 1, 2, \ldots, q \quad (5.41)$$

$$f_s^R(\mathbf{X}, \mathbf{U}^I) = \max f^R_{G_1G_2...G_q}(\mathbf{X}, \mathbf{U}^I),\ G_j = 1, 2, \ldots, t_j,\ j = 1, 2, \ldots, q \quad (5.42)$$

$$\left(g_s^L(\mathbf{X}, \mathbf{U}^I)\right)_i = \min \left(g^L_{G_1G_2...G_q}(\mathbf{X}, \mathbf{U}^I)\right)_i,\ G_j = 1, 2, \ldots, t_j,$$
$$j = 1, 2, \ldots, q,\ i = 1, 2, \ldots, l \quad (5.43)$$

$$\left(g_s^R(\mathbf{X}, \mathbf{U}^I)\right)_i = \max \left(g^R_{G_1G_2...G_q}(\mathbf{X}, \mathbf{U}^I)\right)_i,\ G_j = 1, 2, \ldots, t_j,$$
$$j = 1, 2, \ldots, q,\ i = 1, 2, \ldots, l \quad (5.44)$$

(3) Based on the interval of the objective function, obtain its midpoint and radius, and compute the multi-objective evaluation function $f_d(\mathbf{X})$; based on the constraint intervals, compute the possibility degrees $P\left(g_i^I(\mathbf{X}) \leq b_i^I\right),\ i = 1, 2, \ldots, l$.
(4) Compute the penalty function $f_p(\mathbf{X})$ expressed in Eq. (4.3), and whereby the fitness value of IP-GA.
(5) For each design vector generated by IP-GA, complete the above computations and obtain corresponding fitness value. Stop the optimization process when it comes to the maximum generation number. The design vector with the maximum fitness value is the optimal solution of the interval optimization problem.

The flowchart of the proposed algorithm is shown in Fig. 5.2. It can be seen that by means of the interval structural analysis, the objective function or the constraint bounds can be obtained for each design vector with only a few numerical analyses, thus avoiding the inner optimization. Therefore, the two-layer nested optimization problem is transformed into a single-layer optimization problem, and the computation efficiency of the nonlinear interval optimization is dramatically improved.

In the above analysis, the interval structural analysis was applied to the objective function and all the constraints, assuming that they all need to be obtained via numerical analysis models. In practical optimizations, however, there are only one or a few of objective function and constraints needing to be obtained via numerical analysis models. Therefore, in the practical use of the algorithm, the interval structural analysis is only applied to those objective function or constraints with numerical analysis models. For others based on explicit functions, optimization method will be better.

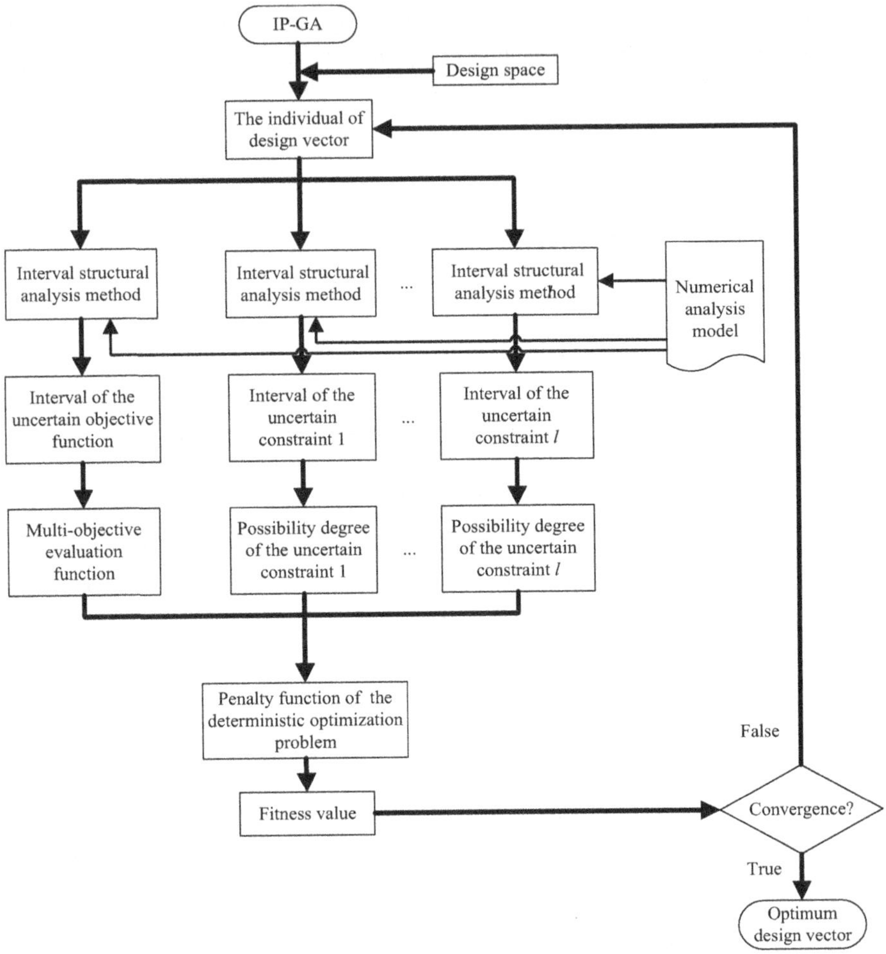


**Fig. 5.2** Interval optimization based on interval structural analysis method [12]

## *5.3.2 Engineering Applications*

The abovementioned interval optimization method will be illustrated by two numerical examples in this section, and only the interval structural analysis method for small uncertainties will be employed since the uncertainty levels in these two examples are relatively low.

1. Application to a 25-bar truss

This application analyzes a 25-bar plane truss as shown in Fig. 5.3 [13]. The cross-section areas of the bars need to be optimized to achieve the minimum volume

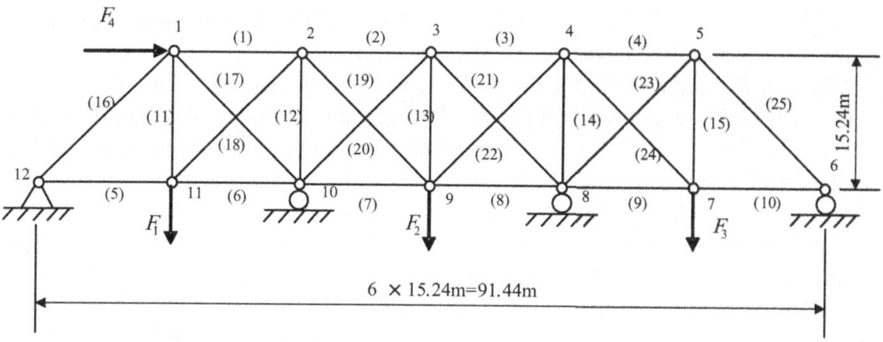


**Fig. 5.3** A 25-bar plane truss [13]

subject to several displacement constraints. The bars (1)–(4) have the same cross-section area denoted by $A_1$; the bars (16)–(25) have the same cross-section area denoted by $A_2$; the bars (11)–(15) have the same cross-section area denoted by $A_3$, and the bars (5)–(10) have the same cross-section area denoted by $A_4$. The length $L$ of each horizontal or vertical bar is 15.24 m. The Young's Modulus of the truss is 199949.2 MPa. Joint 12 is hinge-supported, and joints 6, 8, and 10 are roller-supported. Joints 7, 9, and 10 are subjected to the vertical loads $F_3 = 1779.2\,\text{kN}$, $F_2 = 2224\,\text{kN}$ and $F_1 = 1779.2\,\text{kN}$, respectively. Joint 1 is subjected to a horizontal load $F_4$. The horizontal displacement of joint 6 is denoted by $d_1$ and its allowable maximum is 23 mm. The vertical displacements of joints 7, 9, and 11 are denoted by $d_2$, $d_3$, and $d_4$, and their allowable maximums are 47 mm, 40 mm, and 48 mm, respectively. The four loads denoted by $\mathbf{F} = (F_1, F_2, F_3, F_4)^{\text{T}}$ are uncertain, all with the same uncertainty level of 10%. Therefore, an interval optimization problem can be modeled as follows [14]:

$$\begin{cases} \min\limits_{\mathbf{A}} Vol(\mathbf{A}) = \sum\limits_{i=1}^{25} (L_i A_i) = L\left(4A_1 + 10\sqrt{2}A_2 + 5A_3 + 6A_4\right) \\ \text{s.t.}\ d_1(\mathbf{A}, \mathbf{F}) \le 23\,\text{mm} \\ \quad d_2(\mathbf{A}, \mathbf{F}) \le 47\,\text{mm} \\ \quad d_3(\mathbf{A}, \mathbf{F}) \le 40\,\text{mm} \\ \quad d_4(\mathbf{A}, \mathbf{F}) \le 48\,\text{mm} \\ \quad 100\,\text{mm}^2 \le A_i \le 10,000\,\text{mm}^2,\ i = 1, 2, 3, 4 \\ \quad F_1 \in \left[F_1^L, F_1^R\right] = [1601.2\,\text{kN}, 1957.1\,\text{kN}] \\ \quad F_2 \in \left[F_2^L, F_2^R\right] = [2001.6\,\text{kN}, 2446.4\,\text{kN}] \\ \quad F_3 \in \left[F_3^L, F_3^R\right] = [1601.2\,\text{kN}, 1957.1\,\text{kN}] \\ \quad F_4 \in \left[F_4^L, F_4^R\right] = [1201.0\,\text{kN}, 1467.8\,\text{kN}] \end{cases} \tag{5.45}$$

where $Vol$ represents the volume of the truss. It is worth mentioning that in the above optimization, only the constraints involve uncertainties.

FEM is employed to compute the displacement responses, using the truss element to create the FEM model with totally 25 elements and 12 nodes. In the optimization process, the FEM-based interval structural analysis is applied to the constraints, while the objective function is dealt with by the conventional deterministic optimization method. The maximum generation number of IP-GA and the penalty factor $\sigma$ are set to be 300 and 10,000, respectively. Four different values are considered for the possibility degree level, and for each case, all the uncertain constraints are given the same value of possibility degree level. The optimization results can be obtained, as listed in Tables 5.12, 5.13, 5.14, 5.15, and 5.16, using the abovementioned interval optimization method. From the tables, it can be found that when the possibility level of constraints is 1.0, the truss volume corresponding to the optimum design vector is 2.58 $m^3$. Also, the volume decreases with the reduction of the possibility level to 2.13 $m^3$, when the level is 0.2. It demonstrates again that the possibility level of constraints and the objective design performance (lighter weight in this example) are usually contradictory in practical structural design. Although a lower possibility level can lead to a better design performance of the objective function, it will unavoidably cause a higher possibility of constraints violation and lead to lower reliability of the structure. For example, as shown in Table 5.13, when the possibility level is 0.8, the values of four displacement constraints at the optimum design vector are $d_1^I = [18.83\,\text{mm}, 23.91\,\text{mm}]$, $d_2^I = [36.49\,\text{mm}, 48.36\,\text{mm}]$, $d_3^I = [29.49\,\text{mm}, 42.04\,\text{mm}]$, and $d_4^I = [38.18\,\text{mm}, 50.41\,\text{mm}]$, respectively, which seem highly possible to violate their corresponding allowable maximums, i.e., 23 mm, 47 mm, 40 mm, and 48 mm, respectively.

**Table 5.12** Optimization results under the possibility degree level 1.0 (25-bar) [14]

| Optimal areas ($mm^2$) | Displacement intervals (mm) | Possibility degrees of constraints |
|---|---|---|
| $A_1$: 719.4 | $d_1$: [17.82, 22.39] | 1.00 |
| $A_2$: 6293.5 | $d_2$: [34.40, 44.76] | 1.00 |
| $A_3$: 5374.2 | $d_3$: [28.78, 39.46] | 1.00 |
| $A_4$: 8848.4 | $d_4$: [35.66, 46.26] | 1.00 |
| Volume of the truss is 2.58 $m^3$ | | |

**Table 5.13** Optimization results under the possibility degree level 0.8 (25-bar) [14]

| Optimal areas ($mm^2$) | Displacement intervals (mm) | Possibility degrees of constraints |
|---|---|---|
| $A_1$: 409.7 | $d_1$: [18.83, 23.91] | 0.82 |
| $A_2$: 5064.5 | $d_2$: [36.49, 48.36] | 0.89 |
| $A_3$: 6293.5 | $d_3$: [29.49, 42.04] | 0.84 |
| $A_4$: 8316.1 | $d_4$: [38.18, 50.41] | 0.80 |
| Volume of the truss is 2.32 $m^3$ | | |

**Table 5.14** Optimization results under the possibility degree level 0.6 (25-bar) [14]

| Optimal areas ($mm^2$) | Displacement intervals (mm) | Possibility degrees of constraints |
|---|---|---|
| $A_1$: 487.1 | $d_1$: [19.70, 24.89] | 0.64 |
| $A_2$: 5122.6 | $d_2$: [38.90, 51.12] | 0.66 |
| $A_3$: 5248.4 | $d_3$: [32.06, 44.85] | 0.62 |
| $A_4$: 7996.8 | $d_4$: [40.46, 52.99] | 0.60 |
| Volume of the truss is 2.23 $m^3$ | | |

**Table 5.15** Optimization results under the possibility degree level 0.4 (25-bar) [14]

| Optimal areas ($mm^2$) | Displacement intervals (mm) | Possibility degrees of constraints |
|---|---|---|
| $A_1$: 196.8 | $d_1$: [20.48, 26.27] | 0.44 |
| $A_2$: 5054.8 | $d_2$: [39.69, 53.53] | 0.53 |
| $A_3$: 5054.8 | $d_3$: [30.37, 45.44] | 0.64 |
| $A_4$: 7532.3 | $d_4$: [42.23, 56.63] | 0.40 |
| Volume of the truss is 2.14 $m^3$ | | |

**Table 5.16** Optimization results under the possibility degree level 0.2 (25-bar) [14]

| Optimal areas ($mm^2$) | Displacement intervals (mm) | Possibility degrees of constraints |
|---|---|---|
| $A_1$: 100.0 | $d_1$: [20.36, 26.29] | 0.45 |
| $A_2$: 5054.8 | $d_2$: [39.59, 54.12] | 0.51 |
| $A_3$: 5054.8 | $d_3$: [29.10, 45.26] | 0.68 |
| $A_4$: 7532.3 | $d_4$: [42.55, 57.72] | 0.36 |
| Volume of the truss is 2.13 $m^3$ | | |

The analyses for the four uncertain constraints are based on FEM, and the intervals of all the constraints can be obtained by an interval structural analysis with 5 FEM evaluations for each design vector **X**. The maximum generation number of IP-GA is 300 and the population size is 5. Thus $300 \times 5 \times 5 = 7500$ FEM evaluations are required for the interval optimization of this example. However, if the IP-GA-based two-layer nested optimization method mentioned in Chap. 3 is employed to cope with the optimization problem, and the maximum generation number of the inner and outer layers of IP-GA are both set to be 300, the optimization process will need $1.8 \times 10^7$ FEM evaluations, which is unacceptable in practical engineering. Therefore, through replacing the inner optimization with the interval structural analysis, the proposed method efficiently improves the optimization efficiency.

2. Application to the design of an automobile frame

Shown in Fig. 5.4, is an automobile frame model composed of two side beams and seven cross beams. The layout of the cross beams should be optimized to

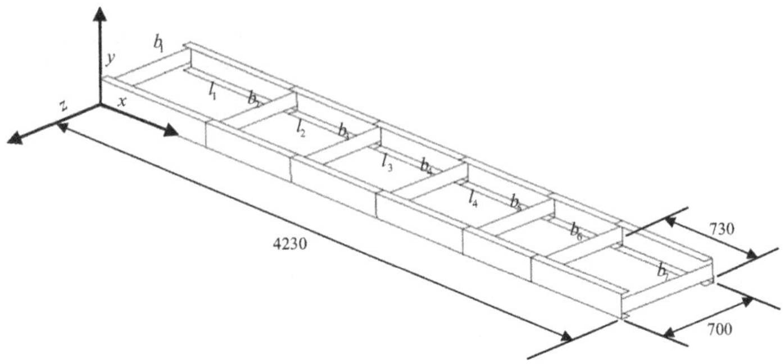


**Fig. 5.4** An automobile frame model (mm) [15]

achieve a maximum stiffness in $y$-direction. The side beams are manufactured by channel steel, and its cross section is shown in Fig. 5.5. The cross beams denoted by $b_i$, $i = 1, 2, \ldots, 7$ are made of steel plate with the size of 690 mm × 90 mm × 5 mm. Since most parts and unit assemblies of an automobile are fixed on the frame through connecting pieces, the frame is a base of the whole automobile. Through simplifying the loads applied on the frame, a static mechanical model of the frame can be constructed as shown in Fig. 5.6. The single arrow, double arrow, and small triangle represent concentrated force, bending moment, and fixed constraint, respectively. The concentrated forces $F_i$, $i = 1, 2, \ldots, 10$ are 201 N, 222.5 N, 500 N, 222.5 N, 300 N, 300 N, 300 N, 300 N, 200 N, and 200 N, respectively. The moments $M_i$, $i = 1, 2, \ldots, 10$ are 190.2 N·m, 190.2 N·m, 120 N·m, 120 N·m, 110.5 N·m, 120 N·m, 110.5 N·m, 120 N·m, 180 N·m, and 180 N·m, respectively. Besides, the density

**Fig. 5.5** Cross-section of the side beam (mm) [15]

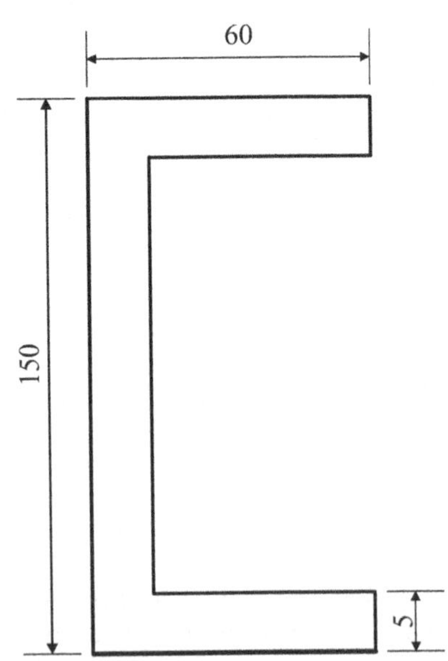

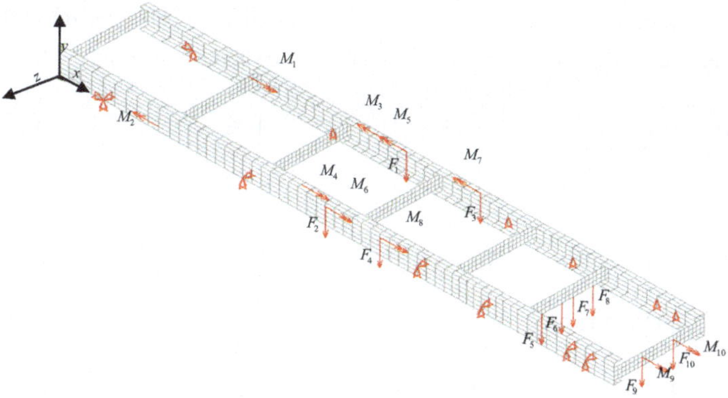


**Fig. 5.6** Loads and constraints of the frame model [15]

of the material is $\rho = 7.8 \times 10^{-6}\ \text{kg/mm}^3$, the Young's modulus is $E = 2.0 \times 10^5$ MPa and the Poisson's ratio is $\nu = 0.3$.

The cross beams $b_1$, $b_6$, and $b_7$ are fixed and the spans of other beams need to be optimized, thus with $\boldsymbol{l} = (l_1, l_2, l_3, l_4)^{\mathrm{T}}$ being the design vector. The stiffness of the frame can be equivalently represented by the maximum nodal displacement in $y$-direction. Due to the manufacturing and measurement errors, the Young's modulus $E$ and Poisson's ratio $\nu$ of the frame material are uncertain, both with an uncertainty level of 10%. Then an interval optimization model can be formulated as follows:

$$\begin{cases} \min\limits_{\mathbf{l}} d_{\max}(\mathbf{l}, E, \nu) \\ \text{s.t. } l_1 + l_2 + l_3 + l_4 \leq 3200\,\text{mm} \\ \quad 300\,\text{mm} \leq l_i \leq 1500\,\text{mm},\ i = 1, 2, 3, 4 \\ \quad E \in \left[E^L, E^R\right] = \left[1.8 \times 10^5\,\text{MPa}, 2.2 \times 10^5\,\text{MPa}\right] \\ \quad \nu \in \left[\nu^L, \nu^R\right] = [0.27, 0.33] \end{cases} \tag{5.46}$$

where $d_{\max}$ is the maximum nodal displacement in $Y$ direction and the uncertainties only exist in the objective function.

In the FEM model of the automobile frame, the four-node shell element combining 2-D solid element and plate element [9], is adopted. In the optimization process, the FEM-based interval structural analysis is only applied to the objective function, while the constraints are dealt with by conventional deterministic optimization method. The factors $\xi$, $\phi$ and $\psi$, and the maximum generation number of IP-GA are set to be 0.0, 1.16, 0.13, and 200, respectively. Listed in Table 5.17, are the optimization results under 6 different multi-objective weighting factors, and Fig. 5.7, shows the distribution of solutions in the two-dimensional objective function space. From the table and figure, it can be found that with the variation of $\beta$, the middle point and the radius of the objective function interval evaluated at the optimum design vector shows different trends. When $\beta = 1.0$, the objective function has the highest middle

**Table 5.17** Optimization results under different multi-objective weighting factors [12]

| $\beta$ | Optimum design vector **l** (mm) | Interval of objective function (mm) | Middle point (mm) | Radius (mm) |
|---|---|---|---|---|
| 1.0 | (1041.3, 569.8, 307.0, 300.0) | [1.45, 1.71] | 1.58 | 0.13 |
| 0.8 | (1069.5, 532.3, 300.0, 301.2) | [1.24, 1.62] | 1.43 | 0.18 |
| 0.6 | (1109.4, 494.7, 300.0, 300.0) | [1.11, 1.59] | 1.35 | 0.24 |
| 0.4 | (1071.8, 497.1, 302.3, 312.9) | [0.94, 1.56] | 1.25 | 0.31 |
| 0.2 | (1071.8, 497.1, 302.3, 312.9) | [0.94, 1.56] | 1.25 | 0.31 |
| 0.0 | (1081.5, 492.3, 304.6, 302.7) | [0.81, 1.51] | 1.16 | 0.35 |

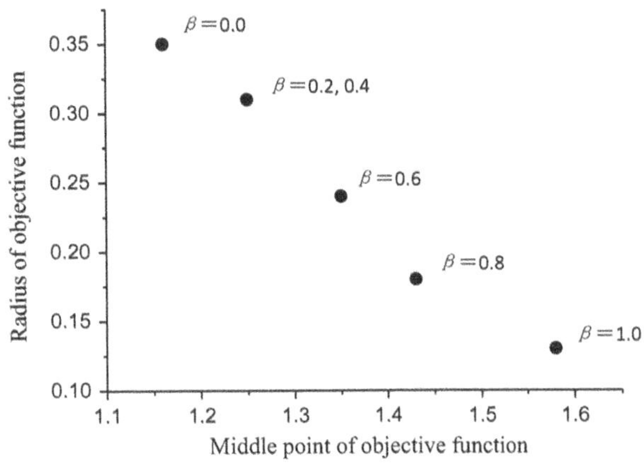


**Fig. 5.7** The optimum solution set under different multi-objective weighting factors [12]

point value of 1.58 mm and the lowest radius of 0.13 mm. In this case, the stiffness is least sensitive to the uncertain variables, and the robustness is well guaranteed, but the average stiffness of the frame under the influence of uncertainties is quite small. On the contrary, when $\beta = 0.0$, the average stiffness is the largest but the robustness is the worst. As is mentioned in Chap. 3, a reasonable multi-objective weighting factor should be set according to practical engineering problems and the experience of the designer in order to reach a good balance between the design robustness and the objective performance. Finally, in this example, each interval structural analysis needs 3 FEM evaluations and the total FEM evaluations needed in each time of interval optimization is 5000.

## 5.4 Summary

In this chapter, firstly the interval structural analysis method is improved and extended. The previous interval structural analysis method which is based on the first-order Taylor expansion is only suitable for problems with a relatively low uncertainty level of parameters. Therefore, based on the interval set theory and the subinterval technology, an interval structural analysis method for large uncertainties is proposed to calculate the response interval of structures with high uncertainty level. Additionally, in this chapter, an efficient nonlinear interval optimization method is developed by introducing the interval structural analysis. In this method, the inner layer of optimization is replaced by the interval structural analysis. Therefore, the original two-layer nested optimization problem is converted into a single-layer optimization problem and the computation cost of nonlinear interval optimizations is greatly reduced. Finally, the effectiveness of the proposed method is illustrated by the designs of a 25-bar plane truss and an automobile frame. The results show that the efficiency of the method is far higher than the two-layer nested optimization problem. Besides, in the formulation of the above method, we use the transformation model based on the order relation of interval number. Similarly, the interval structural analysis method can also be combined with the transformation model based on possibility degree of interval number to develop the corresponding efficient interval optimization algorithm.

## References


1. Qiu ZP (2003) Comparison of static response of structures using convex models and interval analysis method. Int J Numer Meth Eng 56(12):1735–1753
2. Qiu ZP, Wang XJ (2003) Comparison of dynamic response of structures with uncertain-but-bounded parameters using non-probabilistic interval analysis method and probabilistic approach. Int J Solids Struct 40(20):5423–5439
3. Qiu ZP, Elishakoff I (2001) Anti-optimization technique—a generalization of interval analysis for nonprobabilistic treatment of uncertainty. Chaos, Solitons Fractals 12(9):1747–1759
4. Han X, Jiang C, Gong S, Huang YH (2008) Transient waves in composite-laminated plates with uncertain load and material property. Int J Numer Meth Eng 75(3):253–274
5. Liu N, Gao W, Song CM, Zhang N, Pi Y-L (2013) Interval dynamic response analysis of vehicle-bridge interaction system with uncertainty. J Sound Vib 332(13):3218–3231
6. Wang C, Qiu ZP (2015) Interval analysis of steady-state heat convection–diffusion problem with uncertain-but-bounded parameters. Int J Heat Mass Transf 91:355–362
7. Moore RE (1979) Methods and applications of interval analysis. Prenice-Hall, London
8. Qiu ZP (1994) Interval analysis method of static response of structures with uncertain parameters and the eigenvalue problem. PhD thesis, Jilin Industrial University, Changchun
9. Liu GR, Quek SS (2003) The finite element method: a practical course. Elsevier Science Ltd., England
10. Zhou YT, Jiang C, Han X (2006) Interval and subinterval analysis methods of the structural analysis and their error estimations. Int J Comput Methods 3(2):229–244
11. Qiu ZP (2005) Convex method based on non-probabilistic set-theory and its application. National Defence Industry Press, Beijing

12. Jiang C (2008) Theories and algorithms of uncertain optimization based on interval. PhD thesis, Hunan University, Changsha
13. Au FTK, Cheng YS, Tham LG, Zeng GW (2003) Robust design of structures using convex models. Comput Struct 81(28–29):2611–2619
14. Jiang C, Han X, Liu GR (2007) Optimization of structures with uncertain constraints based on convex model and satisfaction degree of interval. Comput Methods Appl Mech Eng 196(49–52):4791–4800
15. Jiang C, Han X, Guan FJ, Li YH (2007) An uncertain structural optimization method based on nonlinear interval number programming and interval analysis method. Eng Struct 29(11):3168–3177

# Chapter 6
# Interval Optimization Based on Sequential Linear Programming

**Abstract** By introducing the sequential linear programming technique, this chapter develops an efficient nonlinear interval optimization algorithm. An iterative mechanism is also provided to ensure the convergence of the proposed algorithm.

In Chap. 5, a nonlinear interval optimization algorithm was proposed on the base of interval structural analysis, effectively eliminating the nested optimization and improving the computation efficiency. The algorithm linearizes the optimization model with respect to the uncertain variables using the interval structural analysis method, thereby obtaining the intervals of the uncertain objective function and the constraints in each iteration of the outer optimization with only a few evaluations of the actual model. However, in this algorithm, the linearization is only implemented to the uncertain variables, excluding the design variables. Therefore, the outer layer optimization still exists. Especially when GA is employed as the solver of the outer optimization, a large number of evaluations of the actual model are still necessary.

In this chapter, the sequential linear programming (SLP) [1–3], a conventional deterministic optimization method, is introduced to the nonlinear interval optimization to linearize the model with respect to both the uncertain variables and the design variables in each iteration. Also, an effective iteration strategy is established to guarantee the convergence. The inner and outer optimizations based on the actual model could be both eliminated, and hence a more efficient interval optimization method can be constructed. This chapter will first give the formulation of the proposed method, and then provides two test functions and also an engineering application to illustrate its effectiveness.

## 6.1 Formulation of the Algorithm

The problem this chapter studies is denoted by Eq. (4.1), where the uncertainty level of the variables **U** is relatively small, so the method proposed in this chapter can only apply to problems with small uncertainties. As mentioned above, such problem



C. Jiang et al., *Nonlinear Interval Optimization for Uncertain Problems*, Springer Tracts in Mechanical Engineering, https://doi.org/10.1007/978-981-15-8546-3_6

is usual in practical engineering because parametric uncertainties are usually small perturbations around the nominal values of parameters.

The whole optimization process is composed of iterations of a series of approximated interval optimizations. In each iteration, the objective function and constraints are linearized with respect to design variables and uncertain variables using the first-order Taylor expansion, and hence an approximated interval optimization problem of Eq. (4.1), can be constructed. In the $s$-th iteration, the approximated interval optimization problem can be expressed as [4]:

$$\begin{cases} \min_{\mathbf{X}} \tilde{f}(\mathbf{X},\mathbf{U}) \approx f(\mathbf{X}^{(s)},\mathbf{U}^c) + \sum_{j=1}^{n} \frac{\partial f(\mathbf{X}^{(s)},\mathbf{U}^c)}{\partial X_j}\left(X_j - X_j^{(s)}\right) + \sum_{j=1}^{q} \frac{\partial f(\mathbf{X}^{(s)},\mathbf{U}^c)}{\partial U_j}\left(U_j - U_j^c\right) \\ \text{s.t. } \tilde{g}_i(\mathbf{X},\mathbf{U}) \approx g_i(\mathbf{X}^{(s)},\mathbf{U}^c) + \sum_{j=1}^{n} \frac{\partial g_i(\mathbf{X}^{(s)},\mathbf{U}^c)}{\partial X_j}\left(X_j - X_j^{(s)}\right) + \sum_{j=1}^{q} \frac{\partial g_i(\mathbf{X}^{(s)},\mathbf{U}^c)}{\partial U_j}\left(U_j - U_j^{(c)}\right) \\ \quad \le b_i^I = \left[b_i^L, b_i^R\right],\ i = 1, 2, \ldots, l \\ \quad \max\left[\mathbf{X}_l, \mathbf{X}^{(s)} - \boldsymbol{\delta}^{(s)}\right] \le \mathbf{X} \le \min\left[\mathbf{X}_r, \mathbf{X}^{(s)} + \boldsymbol{\delta}^{(s)}\right] \\ \quad \mathbf{U} \in \mathbf{U}^I = \left[\mathbf{U}^L, \mathbf{U}^R\right],\ U_i \in U_i^I = \left[U_i^L, U_i^R\right],\ i = 1, 2, \ldots, q \end{cases} \tag{6.1}$$

where $\mathbf{X}_l$ and $\mathbf{X}_r$ represent the lower and upper bounds of design vector $\mathbf{X}$; $\tilde{f}$ and $\tilde{g}_i$ represent the approximate objective function and the $i$-th approximate constraint after linearization, respectively; $\boldsymbol{\delta}^{(s)}$ represents the step size vector of design vector that constitutes the current design space with $\mathbf{X}^{(s)}$. The step size vector varies during the optimization process. Considering the first-order Taylor expansion can only well approximate the function near the expansion point, especially when the function is highly nonlinear. The value ranges of the design variables thus should be confined by $\boldsymbol{\delta}^{(s)}$. In addition, both design variables and uncertain variable are used to linearize the objective and constraints, making $\tilde{f}$ and $\tilde{g}_i$, $i = 1, 2, \ldots, l$ in Eq. (6.1), become linear functions with respect to both design variables and uncertain variables.

### 6.1.1 Solution of the Linear Interval Optimization Problems

After the linearization, the approximated interval optimization expressed in Eq. (6.1), is, in fact, a linear interval optimization problem that is far easier to solve than the nonlinear one because its objective function interval and constraints intervals can be explicitly obtained, and hence the inner optimization could be avoided.

From Eqs. (5.12) and (5.13), it can be known that

$$\left(\mathbf{U} - \mathbf{U}^c\right) \in [-1, 1]\mathbf{U}^w,\ \left(U_i - U_i^c\right) \in [-1, 1]U_i^w,\ i = 1, 2, \ldots, q \tag{6.2}$$

Applying the natural interval extension to the objective function in Eq. (6.1), the explicit expression of the interval of the approximated objective function at any $\mathbf{X}$

can be obtained:

$$\begin{aligned}
\tilde{f}^L(\mathbf{X}) &= f(\mathbf{X}^{(s)}, \mathbf{U}^c) + \sum_{j=1}^{n} \frac{\partial f(\mathbf{X}^{(s)}, \mathbf{U}^c)}{\partial X_j}\left(X_j - X_j^{(s)}\right) - \sum_{j=1}^{q} \left|\frac{\partial f(\mathbf{X}^{(s)}, \mathbf{U}^c)}{\partial U_j}\right| U_j^w \\
\tilde{f}^R(\mathbf{X}) &= f(\mathbf{X}^{(s)}, \mathbf{U}^c) + \sum_{j=1}^{n} \frac{\partial f(\mathbf{X}^{(s)}, \mathbf{U}^c)}{\partial X_j}\left(X_j - X_j^{(s)}\right) + \sum_{j=1}^{q} \left|\frac{\partial f(\mathbf{X}^{(s)}, \mathbf{U}^c)}{\partial U_j}\right| U_j^w
\end{aligned} \tag{6.3}$$

Similarly, the intervals of the approximate constraints at any **X** in Eq. (6.1), can also be explicitly obtained:

$$\begin{aligned}
\tilde{g}_i^L(\mathbf{X}) &= g_i(\mathbf{X}^{(s)}, \mathbf{U}^c) + \sum_{j=1}^{n} \frac{\partial g_i(\mathbf{X}^{(s)}, \mathbf{U}^c)}{\partial X_j}\left(X_j - X_j^{(s)}\right) - \sum_{j=1}^{q} \left|\frac{\partial g_i(\mathbf{X}^{(s)}, \mathbf{U}^c)}{\partial U_j}\right| U_j^w,\ i = 1, 2, \ldots, l \\
\tilde{g}_i^R(\mathbf{X}) &= g_i(\mathbf{X}^{(s)}, \mathbf{U}^c) + \sum_{j=1}^{n} \frac{\partial g_i(\mathbf{X}^{(s)}, \mathbf{U}^c)}{\partial X_j}\left(X_j - X_j^{(s)}\right) + \sum_{j=1}^{q} \left|\frac{\partial g_i(\mathbf{X}^{(s)}, \mathbf{U}^c)}{\partial U_j}\right| U_j^w,\ i = 1, 2, \ldots, l
\end{aligned} \tag{6.4}$$

With the transformation model based on order relation of interval number, Eq. (6.1), can be converted into the following deterministic optimization problem:

$$\begin{cases}
\min\limits_{\mathbf{X}} \left(\tilde{f}^c(\mathbf{X}), \tilde{f}^w(\mathbf{X})\right) \\
\text{s.t. } P\left(\tilde{g}_i^I(\mathbf{X}) \le b_i^I\right) \ge \lambda_i,\ i = 1, 2, \ldots, l \\
\qquad \max\left[\mathbf{X}_l, \mathbf{X}^{(s)} - \boldsymbol{\delta}^{(s)}\right] \le \mathbf{X} \le \min\left[\mathbf{X}_r, \mathbf{X}^{(s)} + \boldsymbol{\delta}^{(s)}\right]
\end{cases} \tag{6.5}$$

It can be further transformed into:

$$\begin{cases}
\min\limits_{\mathbf{X}} \tilde{f}_p(\mathbf{X}) = \tilde{f}_d(\mathbf{X}) + \sigma \sum\limits_{i=1}^{l} \varphi\left(P\left(\tilde{g}_i^I(\mathbf{X}) \le b_i^I\right) - \lambda_i\right) \\
= (1-\beta)\left(\tilde{f}^c(\mathbf{X}) + \xi\right)/\phi + \beta\left(\tilde{f}^w(\mathbf{X}) + \xi\right)/\psi \\
\qquad + \sigma \sum\limits_{i=1}^{l} \varphi\left(P\left(\tilde{g}_i^I(\mathbf{X}) \le b_i^I\right) - \lambda_i\right) \\
\text{s.t. } \max\left[\mathbf{X}_l, \mathbf{X}^{(s)} - \boldsymbol{\delta}^{(s)}\right] \le \mathbf{X} \le \min\left[\mathbf{X}_r, \mathbf{X}^{(s)} + \boldsymbol{\delta}^{(s)}\right]
\end{cases} \tag{6.6}$$

where $\tilde{f}_d$ and $\tilde{f}_p$ are the multi-objective evaluation function and penalty function of the approximate interval optimization problem, respectively.

Although the approximated objective function and constraints in Eq. (6.1), are continuous and differentiable the continuity and differentiability of Eq. (6.6), cannot be guaranteed. Therefore, here IP-GA is also employed as the solver of Eq. (6.6). The flowchart of the optimization procedure is shown in Fig. 6.1. From the flowchart, it can be found that for arbitrary design vector, the intervals of the approximate objective function and constraints can be readily obtained through the explicit functions in Eqs. (6.3) and (6.4), hence the inner optimization is not necessary anymore. Therefore, via the linearization of the original optimization model, the approximated interval optimization problem in each iteration can be converted into a single-layer optimization problem, thereby promoting the convenience of the solution. In addition, the optimization process is based on the cheap explicit functions, which ensures a very high optimization efficiency.

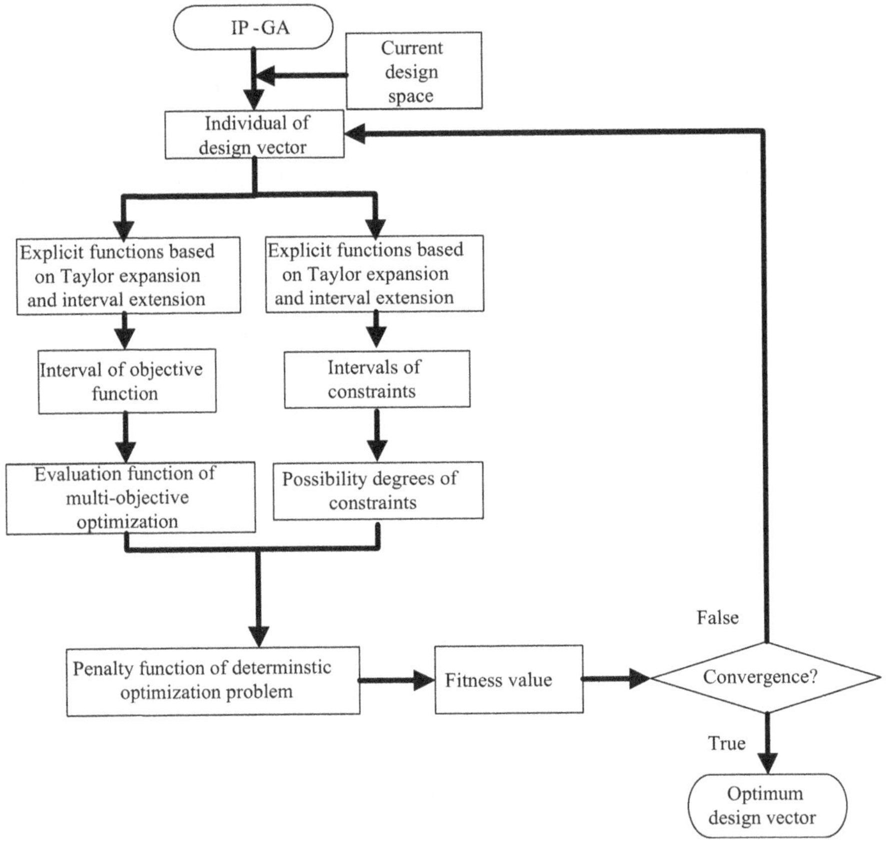


**Fig. 6.1** Solving procedure of the approximated interval optimization problem in each iteration [5]

### 6.1.2 Iteration Mechanism

The iteration process of the proposed method is as follows:

(1) Set the initial design vector $\mathbf{X}^{(1)}$, the initial step size vector $\boldsymbol{\delta}^{(1)}$, the scaling factor $\alpha \in (0, 1)$, the allowable errors $\varepsilon_1 > 0$, $\varepsilon_2 > 0$ and $\varepsilon_3 > 0$, and $s = 1$. $\mathbf{X}^{(1)}$ should be a feasible solution of the transformed deterministic optimization problem expressed in Eq. (4.2), namely $P\left(g_i^I\left(\mathbf{X}^{(1)}\right) \le b_i^I\right) \ge \lambda_i$, $i = 1, 2, \ldots, l$ and $\mathbf{X}^{(1)} \in \Omega^n$.

(2) Construct and solve the linear interval optimization problem expressed in Eq. (6.1), to obtain the optimal solution $\overline{\mathbf{X}}$.

(3) Compute the interval of the objective function at $\overline{\mathbf{X}}$, $\left[f^L\left(\overline{\mathbf{X}}\right), f^R\left(\overline{\mathbf{X}}\right)\right]$ and the corresponding multi-objective evaluation function $f_d\left(\overline{\mathbf{X}}\right)$; compute the intervals of the constraints at $\overline{\mathbf{X}}$, $\left[g_i^L\left(\overline{\mathbf{X}}\right), g_i^R\left(\overline{\mathbf{X}}\right)\right]$, $i = 1, 2, \ldots, l$ and corresponding possibility degrees $P\left(g_i^I\left(\overline{\mathbf{X}}\right) \le b_i^I\right)$, $i = 1, 2, \ldots, l$.

(4) If $\min\left\{\left(P\left(g_i^I\left(\overline{\mathbf{X}}\right) \le b_i^I\right) - \lambda_i\right), i = 1, 2, \ldots, l\right\} > -\varepsilon_1$ and $f_d\left(\overline{\mathbf{X}}\right) < f_d\left(\mathbf{X}^{(s)}\right)$, set $\mathbf{X}^{(s+1)} = \overline{\mathbf{X}}$ and go to Step (6), otherwise set $\boldsymbol{\delta}^{(s)} := \alpha\boldsymbol{\delta}^{(s)}$.

(5) If $\min\left\{\delta_i^{(s)}, i = 1, 2, \ldots, n\right\} < \varepsilon_2$, take $\mathbf{X}^{(s)}$ as the optimum design vector and stop the optimization, otherwise turn to Step (2).

(6) If $\left\|\mathbf{X}^{(s+1)} - \mathbf{X}^{(s)}\right\| < \varepsilon_3$, take $\mathbf{X}^{(s+1)}$ as the optimum design vector, and stop the optimization, otherwise set $\boldsymbol{\delta}^{(s+1)} = \boldsymbol{\delta}^{(s)}$ and $s := s + 1$, and go to Step (2).

The flowchart of the optimization process is shown in Fig. 6.2. In Step (4), we use $\min\left\{\left(P\left(g_i^I\left(\overline{\mathbf{X}}\right) \le b_i^I\right) - \lambda_i\right), i = 1, 2, \ldots, l\right\} > -\varepsilon_1$ and $f_d\left(\overline{\mathbf{X}}\right) < f_d\left(\mathbf{X}^{(s)}\right)$ as the criteria to ensure that $\overline{\mathbf{X}}$ can be taken as the better design into next iteration only when it is a descending feasible solution for Eq. (4.2). In each iteration, the linear approximate models of uncertain objective function and constraints are based on the hybrid space consisting of current design space and uncertainty domain, and they are explicit functions with respect to both design variables and uncertain variables. However, only the design space is updated during iteration, while the uncertainty domain remains unchanged. The reason is the uncertainty levels are assumed to be relatively small in this chapter, and thus the uncertainty domain is also small. Therefore, the approximation error of linear models in each iteration is mainly caused by relatively large design space. We only need to constantly shrink the design space to promote the accuracy of the approximate models to obtain better design vector. Generally, the algorithm converges fast in the early stage of optimization because the accuracy of the linear approximate model is improved with the shrink and moving of design space, which will move the design space towards the optimum to improve the design vector. Ideally, the optimization reaches convergence by the criterion $\left\|\mathbf{X}^{(s+1)} - \mathbf{X}^{(s)}\right\| < \varepsilon_3$. However, for some complex problems, when the local optimum is close and the step size of design vector is relatively small and even comparable to the uncertainty levels of variables, it may lead to the difficulty for searching a better design vector. In such a situation, the approximation error caused by uncertainty domain in the linear approximate models starts to dominate, and it cannot

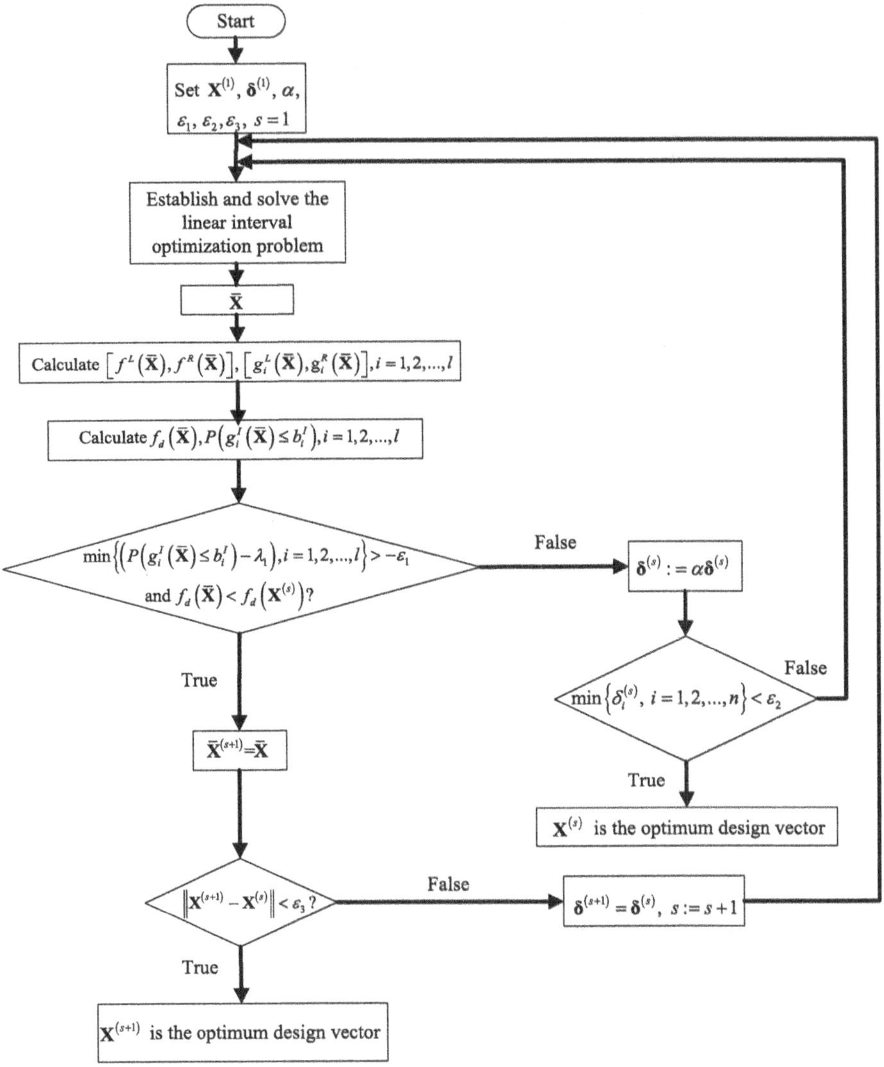


**Fig. 6.2** Solving procedure of the interval optimization method based on sequential linear programming [5]

be eliminated by iteration even though it is still relatively small. The reason is the unchanged uncertainty domain during iteration. At this point, improvement of design vector becomes very slow even stagnant. If we only use $\left\|\mathbf{X}^{(s+1)} - \mathbf{X}^{(s)}\right\| < \varepsilon_3$ as the termination criterion, the convergence will be quite slow even impossible. Therefore, another termination criterion $\min\left\{\delta_i^{(s)}, i = 1, 2, \ldots, n\right\} < \varepsilon_2$ is introduced, which means the update of design space is meaningless when the step size of the design vector is too small, thus the optimization will be stopped forcibly.

From the above analysis, it can be found that the computational cost for this method mainly comes from three parts. The first part is the construction of the approximate interval optimization problem where the actual models are called to calculate the first-order gradients of the uncertain objective function and constraints with respect to the design variables and the uncertain variables, and to compute the values of the objective function and constraints at middle points of the current design vector and the uncertain variables. The second part is in solving the approximate interval optimization problem, while the computational cost of this part can be neglected because the optimization process is based on the simple linear functions and the computational cost is usually much less compared with numerical analysis. The last part is the calculation of the intervals of the actual objective function and constraints at $\overline{\mathbf{X}}$, $\left[f^L(\overline{\mathbf{X}}), f^R(\overline{\mathbf{X}})\right]$ and $\left[g_i^L(\overline{\mathbf{X}}), g_i^R(\overline{\mathbf{X}})\right]$, $i = 1, 2, \ldots, l$, to judge whether $\overline{\mathbf{X}}$ is a descending feasible solution in each iteration. However, conventionally the computation of these intervals requires several optimizations in which the actual models of the objective function and constraints need to be called many times, thus affecting the efficiency of the whole uncertain optimization. To further improve the efficiency, the interval structural analysis method introduced in Chap. 5, will be employed.

### *6.1.3 Calculation of the Intervals of the Actual Objective Function and Constraints in Each Iteration*

The uncertainty levels of the variables in this chapter are assumed to be relatively small. Hence, the first-order Taylor expansion is employed to linearize the uncertain objective function with respect to $\overline{\mathbf{X}}$ in each iteration [6, 7]:

$$f(\overline{\mathbf{X}}, \mathbf{U}) \approx f(\overline{\mathbf{X}}, \mathbf{U}^c) + \sum_{j=1}^{q} \frac{\partial f(\overline{\mathbf{X}}, \mathbf{U}^c)}{\partial U_j}\left(U_j - U_j^c\right) \tag{6.7}$$

According to Eq. (6.2), and the natural interval extension, the bounds of the uncertain objective function at $\overline{\mathbf{X}}$ can be explicitly obtained:

$$\begin{aligned} f^L(\overline{\mathbf{X}}) &= \min_{\mathbf{U}\in\Gamma} f(\overline{\mathbf{X}}, \mathbf{U}) = f(\overline{\mathbf{X}}, \mathbf{U}^c) - \sum_{j=1}^{q} \left|\frac{\partial f(\overline{\mathbf{X}}, \mathbf{U}^c)}{\partial U_j}\right| U_j^w \\ f^R(\overline{\mathbf{X}}) &= \max_{\mathbf{U}\in\Gamma} f(\overline{\mathbf{X}}, \mathbf{U}) = f(\overline{\mathbf{X}}, \mathbf{U}^c) + \sum_{j=1}^{q} \left|\frac{\partial f(\overline{\mathbf{X}}, \mathbf{U}^c)}{\partial U_j}\right| U_j^w \end{aligned} \tag{6.8}$$

Similarly, the bounds of the uncertain constraint at $\overline{\mathbf{X}}$ can be explicitly obtained:

$$g_i^L(\overline{\mathbf{X}}) = \min_{\mathbf{U}\in\Gamma} g_i(\overline{\mathbf{X}}, \mathbf{U}) = g_i(\overline{\mathbf{X}}, \mathbf{U}^c) - \sum_{j=1}^{q} \left| \frac{\partial g_i(\overline{\mathbf{X}}, \mathbf{U}^c)}{\partial U_j} \right| U_j^w, i = 1, 2, ..., l$$
$$g_i^R(\overline{\mathbf{X}}) = \max_{\mathbf{U}\in\Gamma} g_i(\overline{\mathbf{X}}, \mathbf{U}) = g_i(\overline{\mathbf{X}}, \mathbf{U}^c) + \sum_{j=1}^{q} \left| \frac{\partial g_i(\overline{\mathbf{X}}, \mathbf{U}^c)}{\partial U_j} \right| U_j^w, \ i = 1, 2, ..., l \tag{6.9}$$

With Eqs. (6.8) and (6.9), the intervals of the objective function and constraints at $\overline{\mathbf{X}}$ can be obtained with only several evaluations of the actual models, and the main computational cost is caused by the calculation of the gradients of actual objective function and constraints with respect to the uncertain variables. The study conducted in Chap. 5, has indicated that for problems with variables of relatively small uncertainties, the above method is quite accurate. Therefore, the optimizations based on the actual models can be avoided, thus further improving the efficiency of the uncertain optimization.

It can be found that during the whole solution process of the algorithm, the optimizations based on the actual models are not involved. The only optimization to be solved is the linear interval optimization problem in each iteration. However, as mentioned above, the computational cost of this linear interval optimization is negligible. Therefore, the previous deterministic optimization problem obtained by the transformation model is a two-layer nested optimization problem based on the actual models, while in the proposed algorithm, both the inner and outer optimizations are eliminated and the primary task in each iteration is to obtain the gradients of the objective function and constraints, which only needs a few evaluations of the actual models.

Theoretically, when the uncertainties are relatively small, the computational accuracy of the above method is relatively high, but errors still exist due to the fact that the uncertainties can never be infinitely small. Those errors will affect the computational accuracy of $f_d(\overline{\mathbf{X}})$ and $P\left(g_i^I(\overline{\mathbf{X}}) \le b_i^I\right)$, $i = 1, 2, ..., l$, and the subsequent judgment on descending feasible solutions. It may even further influence the whole iteration process, leading to worse convergence and lower accuracy. To verify the influence of those errors and illustrate the effectiveness of the above approximate method, two different methods are used to deal with the first test function in the next section. In the first method, $\left[f^L(\overline{\mathbf{X}}), f^R(\overline{\mathbf{X}})\right]$ and $\left[g_i^L(\overline{\mathbf{X}}), g_i^R(\overline{\mathbf{X}})\right]$, $i = 1, 2, ..., l$ are calculated through the optimizations calling for the actual models. In the second method, the above approximate method is employed. For simplicity, the iteration algorithms with these two methods are called the optimal bound method and the approximate bound method, respectively. It is worth mentioning that the latter is recommended in this chapter and the former is only a reference aiming at showing the effectiveness of the latter.

## 6.2 Testing of the Proposed Method

### 6.2.1 Test Function 1

The following test function is investigated [4]:

$$\begin{cases} \min_{\mathbf{X}} f(\mathbf{X}, \mathbf{U}) = U_1(X_1 - 2)^2 + U_2(X_2 - 1)^2 + U_3^2 X_3 \\ \text{s.t. } U_1 X_1^2 - U_2^2 X_2 + U_3^2 X_3 \geq [6.5, 7.0] \\ \quad U_1^2 X_1 + U_2 X_2^2 + U_3^2 X_3 + 1 \geq [10.0, 15.0] \\ \quad 2.0 \leq X_1 \leq 12.0, 2.0 \leq X_2 \leq 12.0,\ 2.0 \leq X_3 \leq 12.0 \\ \quad U_1 \in [0.9, 1.1],\ U_2 \in [0.9, 1.1],\ U_3 \in [0.9, 1.1] \end{cases} \tag{6.10}$$

where the uncertainty levels of $U_1$, $U_2$, and $U_3$ are all 10%.

The factors $\xi$, $\phi$, and $\psi$ are specified as 0.0, 3.0 and 0.4, respectively. The multi-objective weighting factor $\beta$ is 0.5, the penalty factor $\sigma$ is 1000, and the possibility degree levels of the two constraints are both given as 0.9. The scaling factor $\alpha$ is set to be 0.5. The allowable errors $\varepsilon_1$, $\varepsilon_2$, and $\varepsilon_3$ are all set to be 0.01. In the solution process of the approximate linear interval problem, the maximum generation of IP-GA is 200. When using the optimal bound method, IP-GA is also employed as the solver to obtain the bounds of the uncertain objective function and constraints at $\overline{\mathbf{X}}$, and the maximum generation is also 200. In the optimization process, all of the gradients are computed analytically.

Firstly, with an initial design vector $\mathbf{X}^{(1)} = (7.0, 7.0, 7.0)^{\mathrm{T}}$ and initial step size vector $\boldsymbol{\delta}^{(1)} = (1.0, 1.0, 1.0)^{\mathrm{T}}$, the results of the two methods are listed in Table 6.1. From the table, it can be found both of these two methods converge after 12 iterations and obtain the same optimum design vector $(3.12, 2.98, 2.00)^{\mathrm{T}}$ and corresponding multi-objective evaluation function value of 2.34. In each iteration, the two methods share the same design vector and obtain very close multi-objective evaluation function values, the constraint intervals and the possibility degrees of constraints. It indicates that the approximate method given in Sect. 6.1.3, is quite accurate when the uncertainties of the variables are relatively small (10% in this numerical example), and hence the approximate bound method has almost the same convergence and accuracy in this example. In addition, during the optimization, the multi-objective function shows a monotone decreasing trend, and all the design vectors can meet the possibility degree requirements of constraints. The reason is that the original design is a feasible solution and only a descending feasible design will be kept to the next iteration in this method.

Secondly, five different cases are considered, in which the initial design vector $\mathbf{X}^{(1)}$ is kept the same, while the initial step size vector $\boldsymbol{\delta}^{(1)}$ is set to be $(0.5, 0.5, 0.5)^{\mathrm{T}}$, $(2.0, 2.0, 2.0)^{\mathrm{T}}$, $(3.0, 3.0, 3.0)^{\mathrm{T}}$, $(4.0, 4.0, 4.0)^{\mathrm{T}}$, and $(5.0, 5.0, 5.0)^{\mathrm{T}}$, respectively. The results obtained by the two methods are listed in Table 6.2. It can be seen from the table that both methods result in the same optimum design vector in each case and the optimum design vectors obtained by different initial step size vectors are

**Table 6.1** Iteration history of optimal bound method and approximate bound method [5]

| Iterative step $s$ | | Design vector $\mathbf{X}^{(s)}$ | Interval of constraint 1 | Interval of constraint 2 | Possibility degrees of constraints | Multi-objective evaluation function $f_d(\mathbf{X}^{(s)})$ |
|---|---|---|---|---|---|---|
| 1 | Optimal bound method | (7.00, 7.00, 7.00) | [41.31, 56.67] | [56.46, 71.84] | 1.00, 1.00 | 20.68 |
| | Approximate bound method | (7.00, 7.00, 7.00) | [41.30, 56.70] | [56.30, 71.70] | 1.00, 1.00 | 20.71 |
| 2 | Optimal bound method | (6.00, 6.00, 6.00) | [30.01, 41.95] | [43.14, 55.10] | 1.00, 1.00 | 14.30 |
| | Approximate bound method | (6.00, 6.00, 6.00) | [30.00, 42.00] | [43.00, 55.00] | 1.00, 1.00 | 14.46 |
| 3 | Optimal bound method | (5.00, 5.00, 5.00) | [20.51, 29.47] | [31.60, 40.58] | 1.00, 1.00 | 9.34 |
| | Approximate bound method | (5.00, 5.00, 5.00) | [20.50, 29.50] | [31.50, 40.50] | 1.00, 1.00 | 9.38 |
| 4 | Optimal bound method | (4.00, 4.00, 4.00) | [12.81, 19.19] | [21.90, 28.27] | 1.00, 1.00 | 5.46 |
| | Approximate bound method | (4.00, 4.00, 4.00) | [12.80, 19.20] | [21.80, 28.20] | 1.00, 1.00 | 5.46 |
| 5 | Optimal bound method | (3.03, 3.00, 4.05) | [8.02, 12.58] | [14.84, 19.47] | 1.00, 1.00 | 3.17 |
| | Approximate bound method | (3.03, 3.00, 4.05) | [7.92, 12.58] | [14.77, 19.40] | 1.00, 1.00 | 3.16 |
| 6 | Optimal bound method | (2.55, 2.61, 4.98) | [6.73, 11.05] | [13.24, 17.60] | 0.98, 0.93 | 2.92 |
| | Approximate bound method | (2.55, 2.61, 4.98) | [6.70, 11.03] | [13.15, 17.52] | 0.98, 0.92 | 2.92 |
| 7 | Optimal bound method | (2.85, 2.69, 4.05) | [7.34, 11.65] | [13.12, 17.30] | 1.00, 0.92 | 2.73 |

(continued)

**Table 6.1** (continued)

| Iterative step $s$ | | Design vector $\mathbf{X}^{(s)}$ | Interval of constraint 1 | Interval of constraint 2 | Possibility degrees of constraints | Multi-objective evaluation function $f_d(\mathbf{X}^{(s)})$ |
|---|---|---|---|---|---|---|
| | Approximate bound method | (2.85, 2.69, 4.05) | [7.32, 11.64] | [13.03, 17.23] | 1.00, 0.91 | 2.73 |
| 8 | Optimal bound method | (2.78, 2.78, 3.55) | [6.48, 10.54] | [13.10, 17.15] | 0.93, 0.91 | 2.58 |
| | Approximate bound method | (2.78, 2.78, 3.55) | [6.46, 10.54] | [13.03, 17.11] | 0.93, 0.91 | 2.58 |
| 9 | Optimal bound method | (3.11, 2.89, 2.55) | [7.30, 11.41] | [13.12, 17.06] | 1.00, 0.91 | 2.47 |
| | Approximate bound method | (3.11, 2.89, 2.55) | [7.29, 11.41] | [13.07, 17.01] | 1.00, 0.91 | 2.47 |
| 10 | Optimal bound method | (3.12, 2.98, 2.00) | [6.75, 10.68] | [13.14, 16.96] | 0.98, 0.91 | 2.34 |
| | Approximate bound method | (3.12, 2.98, 2.00) | [6.76, 10.70] | [13.08, 16.91] | 0.99, 0.90 | 2.34 |
| 11 | Optimal bound method | (3.12, 2.98, 2.00) | [6.75, 10.68] | [13.14, 16.96] | 0.98, 0.91 | 2.34 |
| | Approximate bound method | (3.12, 2.98, 2.00) | [6.76, 10.70] | [13.08, 16.91] | 0.99, 0.90 | 2.34 |
| 12 | Optimal bound method | (3.12, 2.98, 2.00) | [6.74, 10.68] | [13.14, 16.96] | 0.98, 0.91 | 2.34 |
| | Approximate bound method | (3.12, 2.98, 2.00) | [6.76, 10.69] | [13.08, 16.91] | 0.99, 0.90 | 2.34 |

very close. Besides, for the same initial step size vector, both methods need the same number of iterations, while different initial step sizes will lead to different numbers of iterations. For example, when $\boldsymbol{\delta}^{(1)} = (1.0, 1.0, 1.0)^{\mathrm{T}}$, 12 iterations are needed for convergence, but when $\boldsymbol{\delta}^{(1)} = (3.0, 3.0, 3.0)^{\mathrm{T}}$, 26 iterations are needed. Shown in Fig. 6.3, are the convergence curves of the optimization processes by means of the two methods with different initial step size vectors. It can be found that the convergence curves of the two methods are almost anastomotic for the same initial

**Table 6.2** Optimization results for different initial step size vectors with the same initial design vector $(7.0, 7.0, 7.0)^T$ [5]

| Initial step size vector | | Optimum design vector | Number of iterative steps | Possibility degrees of constraints | Multi-objective evaluation function $f_d$ |
|---|---|---|---|---|---|
| (0.5, 0.5, 0.5) | Optimal bound method | (3.05, 2.98, 2.00) | 21 | 0.90, 0.91 | 2.30 |
| | Approximate bound method | (3.05, 2.98, 2.00) | 21 | 0.90, 0.90 | 2.30 |
| (1.0, 1.0, 1.0) | Optimal bound method | (3.12, 2.98, 2.00) | 12 | 0.98, 0.91 | 2.34 |
| | Approximate bound method | (3.12, 2.98, 2.00) | 12 | 0.99, 0.90 | 2.34 |
| (2.0, 2.0, 2.0) | Optimal bound method | (3.03, 2.97, 2.09) | 15 | 0.90, 0.90 | 2.32 |
| | Approximate bound method | (3.03, 2.97, 2.09) | 15 | 0.90, 0.90 | 2.32 |
| (3.0, 3.0, 3.0) | Optimal bound method | (3.05, 2.98, 2.00) | 26 | 0.90, 0.91 | 2.30 |
| | Approximate bound method | (3.05, 2.98, 2.00) | 26 | 0.90, 0.90 | 2.30 |
| (4.0, 4.0, 4.0) | Optimal bound method | (3.07, 2.98, 2.00) | 17 | 0.94, 0.91 | 2.31 |
| | Approximate bound method | (3.07, 2.98, 2.00) | 17 | 0.94, 0.90 | 2.31 |
| (5.0, 5.0, 5.0) | Optimal bound method | (3.06, 2.98, 2.00) | 15 | 0.92, 0.91 | 2.31 |
| | Approximate bound method | (3.06, 2.98, 2.00) | 15 | 0.92, 0.90 | 2.31 |

step size vector, which further indicates that the approximate bound method has almost the same convergence performance and optimization results with the optimal bound method when the uncertainties of the variables are relatively small.

Thirdly, other two initial design vectors $\mathbf{X}^{(1)} = (3.0, 3.0, 3.0)^T$ and $\mathbf{X}^{(1)} = (10.0, 10.0, 10.0)^T$ are considered in combination with the above 6 initial step size vectors, which lead to totally 12 combinations of initial conditions. The optimization results are listed in Tables 6.3 and 6.4. The results show that the optimum design vectors and the corresponding multi-objective evaluation functions obtained by the two methods for different initial conditions are very close to the results listed in Table 6.2. Besides, both methods share the same number of iterations for most initial conditions. Figures 6.4 and 6.5 show the convergence curves of all optimization processes. It is obvious that for most initial conditions, the convergence curves of the two methods almost coincide. Only for a few initial conditions, such as the first

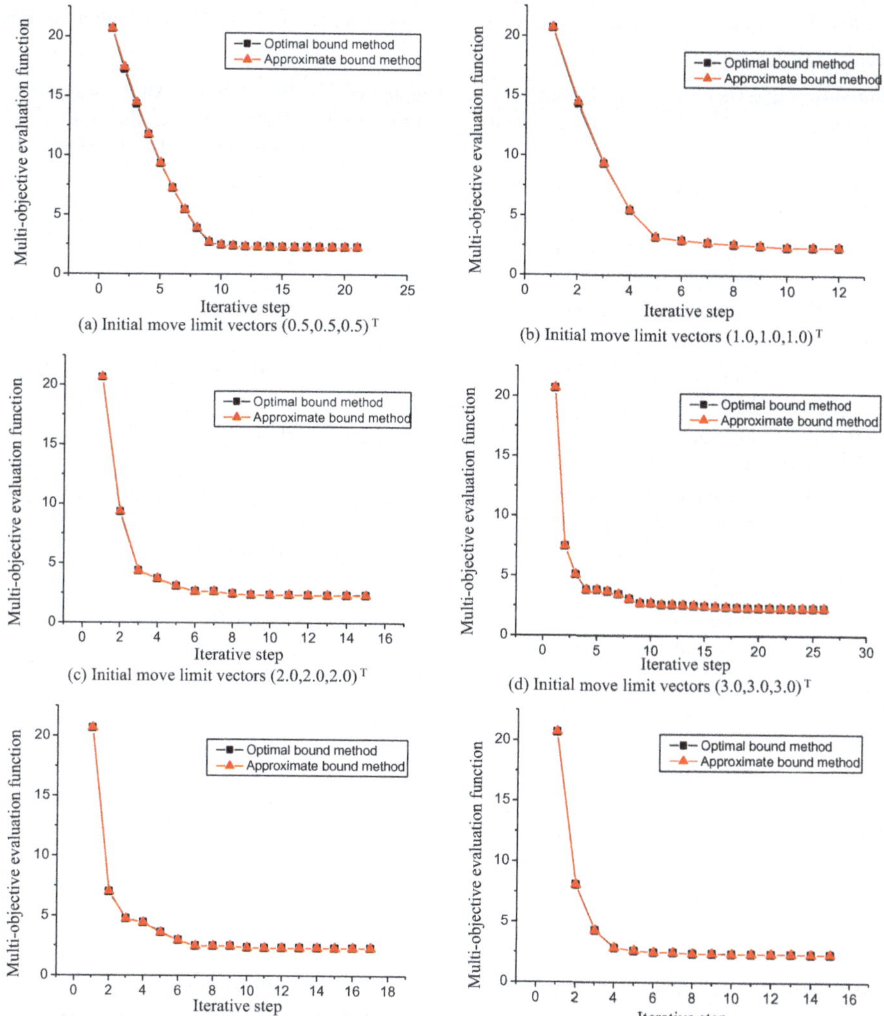


(a) Initial move limit vectors $(0.5,0.5,0.5)^T$
(b) Initial move limit vectors $(1.0,1.0,1.0)^T$
(c) Initial move limit vectors $(2.0,2.0,2.0)^T$
(d) Initial move limit vectors $(3.0,3.0,3.0)^T$
(e) Initial move limit vectors $(4.0,4.0,4.0)^T$
(f) Initial move limit vectors $(5.0,5.0,5.0)^T$

**Fig. 6.3** Convergence curves of the two methods for different initial step size vectors with the same initial design vector $(7.0, 7.0, 7.0)^T$ [5]

and second ones in Table 6.4, although both the iteration numbers and the convergence curves of the two methods are different, almost the same design vectors and multi-objective evaluation function values are obtained.

In the above analysis, the two methods almost converge to the same optimum design vector in different initial conditions, which indicates that both methods have relatively robust convergence. In addition, for most initial conditions, the convergence curves of the two methods are almost the same. Although the convergence curves are

**Table 6.3** Optimization results for different initial step size vectors with the same initial design vector $(3.0, 3.0, 3.0)^T$ [5]

| Initial step size vector | | Optimum design vector | Number of iterative step | Possibility degrees of constraints | Multi-objective evaluation function $f_d$ |
|---|---|---|---|---|---|
| (0.5, 0.5, 0.5) | Optimal bound method | (3.04, 2.98, 2.01) | 20 | 0.90, 0.91 | 2.30 |
| | Approximate bound method | (3.06, 2.98, 2.00) | 9 | 0.93, 0.90 | 2.31 |
| (1.0, 1.0, 1.0) | Optimal bound method | (3.05, 2.98, 2.00) | 13 | 0.90, 0.91 | 2.30 |
| | Approximate bound method | (3.05, 2.98, 2.00) | 10 | 0.90, 0.90 | 2.30 |
| (2.0, 2.0, 2.0) | Optimal bound method | (3.06, 2.98, 2.00) | 11 | 0.91, 0.91 | 2.31 |
| | Approximate bound method | (3.06, 2.98, 2.00) | 11 | 0.91, 0.90 | 2.30 |
| (3.0, 3.0, 3.0) | Optimal bound method | (3.12, 2.98, 2.00) | 8 | 0.98, 0.91 | 2.34 |
| | Approximate bound method | (3.12, 2.98, 2.00) | 8 | 0.99, 0.90 | 2.34 |
| (4.0, 4.0, 4.0) | Optimal bound method | (3.06, 2.98, 2.00) | 12 | 0.92, 0.91 | 2.31 |
| | Approximate bound method | (3.06, 2.98, 2.00) | 12 | 0.92, 0.90 | 2.31 |
| (5.0, 5.0, 5.0) | Optimal bound method | (3.07, 2.98, 2.00) | 10 | 0.93, 0.91 | 2.31 |
| | Approximate bound method | (3.07, 2.98, 2.00) | 10 | 0.93, 0.90 | 2.31 |

quite different in a few cases, the two methods can also obtain very close optimization results. It indicates that even the approximate bound method adopts the approximate method to compute the bounds of the actual objective function and constraints in each iteration, its convergence and accuracy are almost the same as the optimal bound method when the uncertainties of the variables in the problem are relatively small. Besides, it can be predicted that the convergence of the approximate bound method will approach that of the optimal bound method with the decrease of variable uncertainties. In conclusion, when the uncertainties of variables are relatively small, replacing the optimal bound method with the approximate bound method is feasible and the computational efficiency will be further promoted.

**Table 6.4** Optimization results for different initial step size vectors with the same initial design vector $(10.0, 10.0, 10.0)^T$ [5]

| Initial step size vector | | Optimum design vector | Number of iterative step | Possibility degrees of constraints | Multi-objective evaluation function $f_d$ |
|---|---|---|---|---|---|
| (0.5, 0.5, 0.5) | Optimal bound method | (3.05, 2.98, 2.00) | 27 | 0.90, 0.91 | 2.30 |
| | Approximate bound method | (3.05, 2.98, 2.00) | 27 | 0.90, 0.90 | 2.30 |
| (1.0, 1.0, 1.0) | Optimal bound method | (3.12, 2.98, 2.00) | 15 | 0.98, 0.91 | 2.34 |
| | Approximate bound method | (3.05, 2.98, 2.00) | 15 | 0.99, 0.90 | 2.34 |
| (2.0, 2.0, 2.0) | Optimal bound method | (3.05, 2.98, 2.00) | 20 | 0.90, 0.91 | 2.30 |
| | Approximate bound method | (3.05, 2.98, 2.00) | 20 | 0.90, 0.90 | 2.30 |
| (3.0, 3.0, 3.0) | Optimal bound method | (3.05, 2.98, 2.00) | 27 | 0.90, 0.91 | 2.30 |
| | Approximate bound method | (3.05, 2.98, 2.00) | 27 | 0.90, 0.90 | 2.30 |
| (4.0, 4.0, 4.0) | Optimal bound method | (3.05, 2.98, 2.00) | 15 | 0.90, 0.91 | 2.30 |
| | Approximate bound method | (3.05, 2.99, 2.00) | 17 | 0.90, 0.90 | 2.30 |
| (5.0, 5.0, 5.0) | Optimal bound method | (3.05, 2.99, 2.00) | 21 | 0.90, 0.91 | 2.30 |
| | Approximate bound method | (3.05, 2.99, 2.00) | 20 | 0.90, 0.90 | 2.30 |

### 6.2.2 Test Function 2

Shown in Fig. 6.6, is a cantilever beam whose cross-sectional dimensions $X_1$ and $X_2$ need to be optimized to obtain a minimum vertical deflection under the constraints of the cross-section area and the maximum stress. This example is modified from references [8, 9]. The parameters are given as follows: Young's modulus $E = 2 \times 10^4$ kN/cm$^2$, the loads $F_1 = 600$ kN and $F_2 = 50$ kN, the length of the beam $L = 200$ cm, and the cross-sectional dimensions $U_1 = 1.0$ cm and $U_2 = 2.0$ cm. Due to the manufacturing error, $U_1$ and $U_2$ are uncertain variables with the same uncertainty level of 10%. It is required that the cross-sectional area and the maximum stress of the beam are no more than 300 cm$^2$ and 10 kN/cm$^2$, respectively. Therefore, an uncertain optimization problem can be modeled as follows [4]:

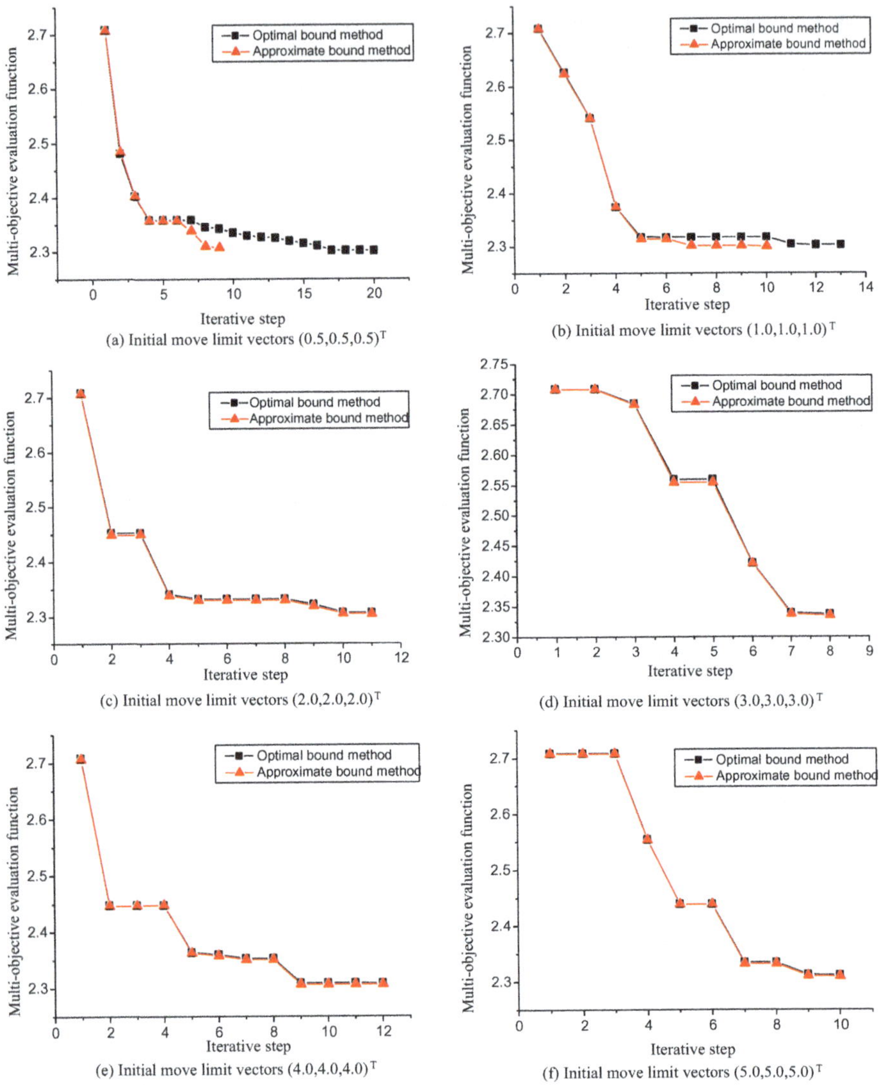


(a) Initial move limit vectors $(0.5,0.5,0.5)^T$

(b) Initial move limit vectors $(1.0,1.0,1.0)^T$

(c) Initial move limit vectors $(2.0,2.0,2.0)^T$

(d) Initial move limit vectors $(3.0,3.0,3.0)^T$

(e) Initial move limit vectors $(4.0,4.0,4.0)^T$

(f) Initial move limit vectors $(5.0,5.0,5.0)^T$

**Fig. 6.4** Convergence curves of the two methods for different initial step size vectors with the same initial design vector $(3.0, 3.0, 3.0)^T$ [5]

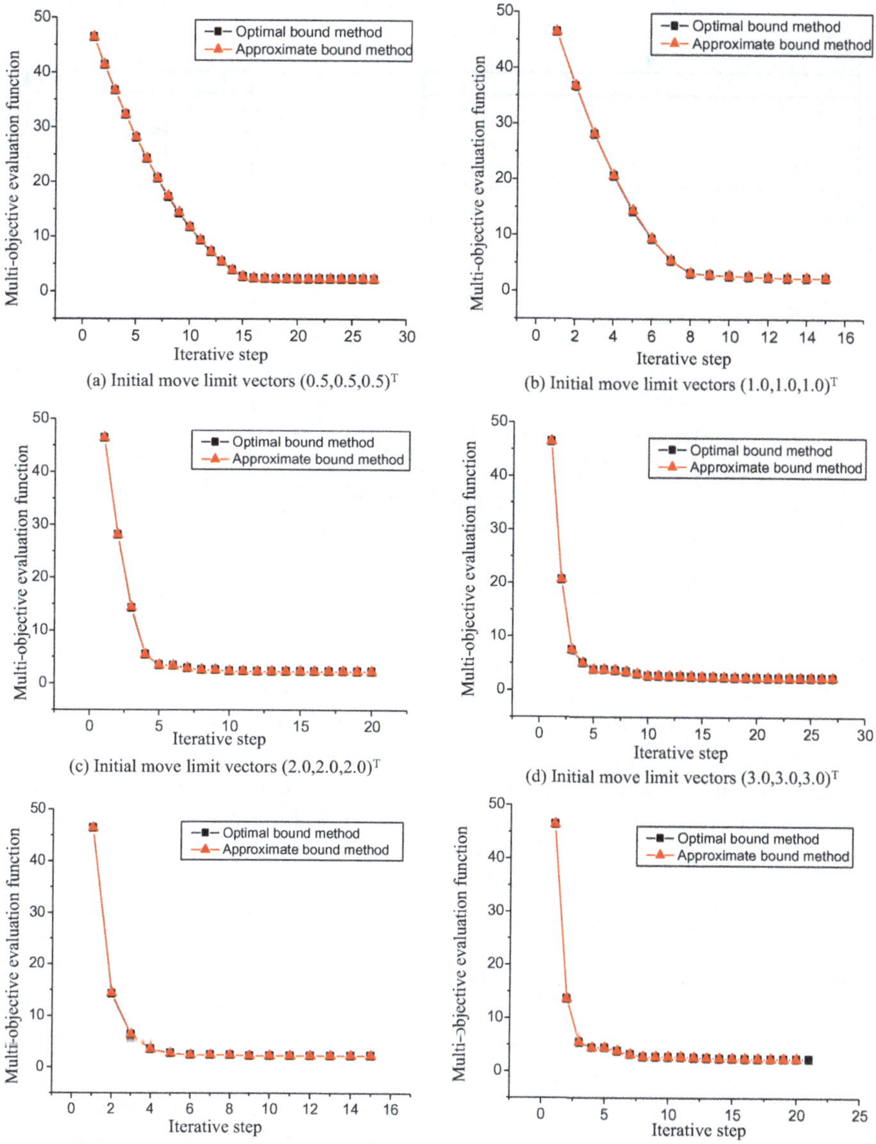


(a) Initial move limit vectors $(0.5,0.5,0.5)^T$

(b) Initial move limit vectors $(1.0,1.0,1.0)^T$

(c) Initial move limit vectors $(2.0,2.0,2.0)^T$

(d) Initial move limit vectors $(3.0,3.0,3.0)^T$

(e) Initial move limit vectors $(4.0,4.0,4.0)^T$

(f) Initial move limit vectors $(5.0,5.0,5.0)^T$

**Fig. 6.5** Convergence curves of the two methods for different initial step size vectors with the same initial design vector $(10.0, 10.0, 10.0)^T$ [5]

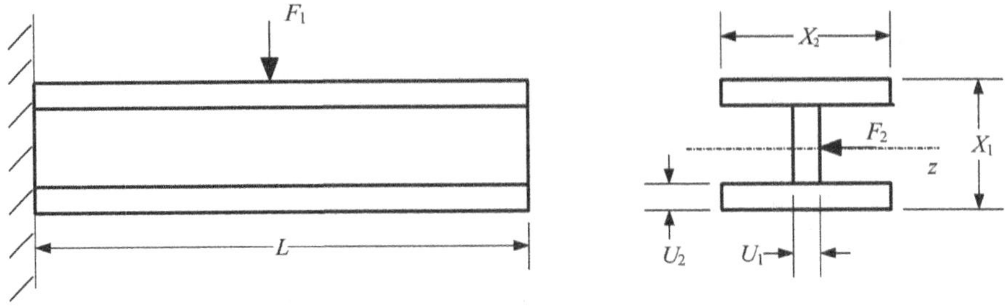


**Fig. 6.6** A cantilever beam [8]

$$\begin{cases} \min\limits_{\mathbf{X}} f(\mathbf{X},\mathbf{U}) = \frac{F_1 L^3}{48EI_z} = \frac{5000}{\frac{1}{12}U_1(X_1-2U_2)^3+\frac{1}{6}X_2U_2^3+2X_2U_2\left(\frac{X_1-U_2}{2}\right)^2} \\ \text{s.t.}\quad g_1(\mathbf{X},\mathbf{U}) = 2X_2U_2 + U_1(X_1-2U_2) \le 300\,\text{cm}^2 \\ \qquad g_2(\mathbf{X},\mathbf{U}) = \frac{180000X_1}{U_1(X_1-2U_2)^3+2X_2U_2\left[4U_2^2+3X_1(X_1-2U_2)\right]} \\ \qquad\quad + \frac{15000X_2}{(X_1-2U_2)U_1^3+2U_2X_2^3} \le 10\,\text{KN/cm}^2 \\ \qquad 10.0\,\text{cm} \le X_1 \le 120.0\,\text{cm},\ 10.0\,\text{cm} \le X_2 \le 120.0\,\text{cm} \\ \qquad U_1 \in \left[U_1^L, U_1^R\right] = [0.9\,\text{cm}, 1.1\,\text{cm}] \\ \qquad U_2 \in \left[U_2^L, U_2^R\right] = [1.8\,\text{cm}, 2.2\,\text{cm}] \end{cases} \tag{6.11}$$

where the objective function $f$ represents the beam's vertical deflection at the free end, the constraints $g_1$ and $g_2$ represent the cross-section area and the maximum stress, and $I_z$ represents the inertia moment of the beam to the neutral axis $z$.

The factors $\xi$, $\phi$, $\psi$, $\beta$, and $\sigma$ are given by 0.0, 0.007, 0.0007, 0.5, and 1000, respectively. The possibility degree levels of the two uncertain constraints are set to be 0.9, the scaling factor $\alpha$ is set to be 0.5, and the allowable errors $\varepsilon_1$, $\varepsilon_2$, and $\varepsilon_3$ are set to be 0.1, 0.1, and 0.001, respectively. The maximum generation number of IP-GA for the approximate linear interval optimization problem is set to be 100. In this example, only the approximate bound method is used to solve the optimization problem, and all gradients are obtained by the central difference method in the optimization process. In addition, the objective function in Eq. (6.11), is assumed to be a computationally expensive numerical analysis model and thus the number of the evaluations of the objective function is used to assess the optimization efficiency.

Considering three different initial design vectors $\mathbf{X}^{(1)} = (50.0, 50.0)^{\mathrm{T}}$, $(100.0, 40.0)^{\mathrm{T}}$, and $(35.0, 60.0)^{\mathrm{T}}$, and 4 different initial step size vectors $\boldsymbol{\delta}^{(1)} = (5.0, 5.0)^{\mathrm{T}}$, $(10.0, 10.0)^{\mathrm{T}}$, $(30.0, 30.0)^{\mathrm{T}}$, and $(50.0, 50.0)^{\mathrm{T}}$, and 12 different combinations of the initial condition can be obtained. The optimization results for different initial conditions are listed in Tables 6.5, 6.6 and 6.7. From the tables, it can be found that the proposed method almost obtains the same optimum design vector, the same possibility degrees of the constraints and the same multi-objective evaluation function value for all the initial conditions. It shows the robust convergence performance of the method since it stably converges to the optimal design despite the different initial conditions. However, the optimization processes need different numbers of evaluations of the objective function for different cases. For examples,

**Table 6.5** Optimization results for different initial step size vectors with the same initial design vector $(50.0, 50.0)^T$ [4]

| Initial step size vector (cm) | Optimum design vector (cm) | Number of iteration steps | Possibility degrees of constraints | Multi-objective evaluation function $f_d$ | Number of evaluation |
|---|---|---|---|---|---|
| (5.0, 5.0) | (120.00, 40.52) | 19 | 0.90, 1.00 | 1.00 | 239 |
| (10.0, 10.0) | (120.00, 40.52) | 20 | 0.90, 1.00 | 1.00 | 244 |
| (30.0, 30.0) | (120.00, 40.51) | 8 | 0.90, 1.00 | 1.00 | 94 |
| (50.0, 50.0) | (120.00, 40.52) | 14 | 0.90, 1.00 | 1.00 | 133 |

**Table 6.6** Optimization results for different initial step size vectors with the same initial design vector $(100.0, 40.0)^T$ [4]

| Initial step size vector (cm) | Optimum design vector (cm) | Number of iterative steps | Possibility degrees of constraints | Multi-objective evaluation function $f_d$ | Number of evaluations |
|---|---|---|---|---|---|
| (5.0, 5.0) | (120.00, 40.52) | 19 | 0.90, 1.00 | 1.00 | 248 |
| (10.0, 10.0) | (120.00, 40.52) | 9 | 0.90, 1.00 | 1.00 | 81 |
| (30.0, 30.0) | (120.00, 40.52) | 9 | 0.90, 1.00 | 1.00 | 81 |
| (50.0, 50.0) | (120.00, 40.56) | 12 | 0.90, 1.00 | 1.00 | 87 |

**Table 6.7** Optimization results for different initial step size vectors with the same initial design vector $(35.0, 60.0)^T$ [4]

| Initial step size vector (cm) | Optimum design vector (cm) | Number of iterative steps | Possibility degrees of constraints | Multi-objective evaluation function $f_d$ | Number of evaluations |
|---|---|---|---|---|---|
| (5.0, 5.0) | (120.00, 40.52) | 22 | 0.90, 1.00 | 1.00 | 290 |
| (10.0, 10.0) | (120.00, 40.52) | 17 | 0.90, 1.00 | 1.00 | 175 |
| (30.0, 30.0) | (120.00, 40.52) | 12 | 0.90, 1.00 | 1.00 | 114 |
| (50.0, 50.0) | (120.00, 40.52) | 13 | 0.90, 1.00 | 1.00 | 101 |

when $\mathbf{X}^{(1)} = (35.0, 60.0)^T$ and $\boldsymbol{\delta}^{(1)} = (5.0, 5.0)^T$, 290 evaluations are needed, while for the initial condition that $\mathbf{X}^{(1)} = (100.0, 40.0)^T$ and $\boldsymbol{\delta}^{(1)} = (10.0, 10.0)^T$, only 81 evaluations are required. The convergence curves of the optimization processes for different initial conditions are shown in Figs. 6.7, 6.8 and 6.9. It can be seen from the figures that the convergence rate and the numbers of iterations for different initial conditions are different but the same multi-objective evaluation function value, 1.0, is achieved for all the initial conditions.

Besides, the interval optimization method based on interval structural analysis proposed in Chap. 5, is also used to solve this problem, in which IP-GA is taken as the optimization solver and the maximum generation number is 100. The results are

**Fig. 6.7** Convergence curves for four different initial step size vectors with the same initial design vector $(50.0, 50.0)^T$ [4]

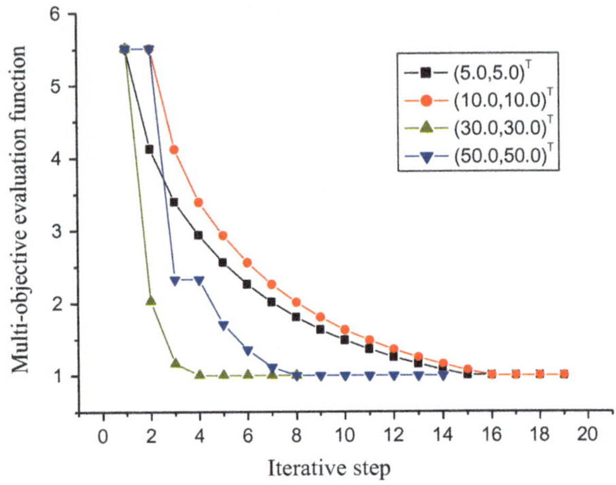


**Fig. 6.8** Convergence curves for four different initial step size vectors with the same initial design vector $(100.0, 40.0)^T$ [4]

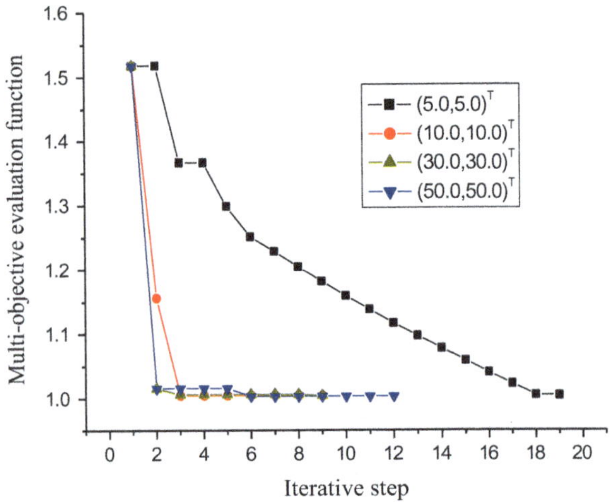


listed in Table 6.8. Obviously, the results are almost the same as the results obtained by the proposed method in this chapter. However, the computational efficiency varies greatly. The interval optimization method based on interval structural analysis needs 2500 evaluations of the actual objective function, while the present method needs no more than 290 evaluations under all the different initial conditions due to the elimination of both inner and outer optimizations that are based on the actual model. Therefore, the computational efficiency is greatly promoted.

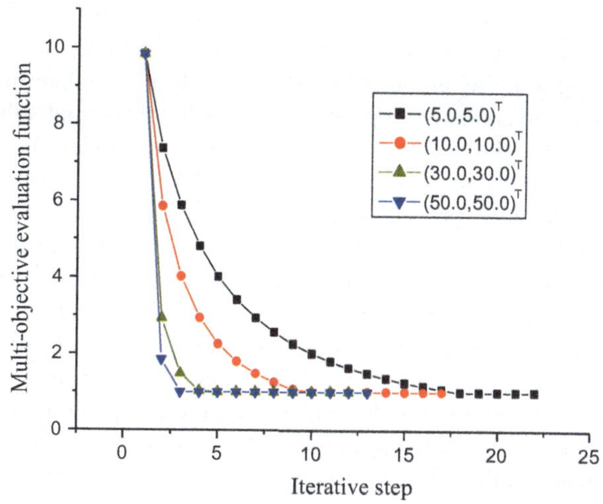


**Fig. 6.9** Convergence curves for four different initial step size vectors with the same initial design vector $(35.0, 60.0)^T$ [4]

**Table 6.8** Optimization results from the interval optimization method based on interval structural analysis [5]

| Optimum design vector (cm) | Possibility degrees of constraints | Multi-objective evaluation function $f_d$ | Number of evaluations |
|---|---|---|---|
| (120.00, 40.52) | 0.90, 1.00 | 1.00 | 2500 |

## 6.3 Discussions on Convergence of the Proposed Method

In the above analyses on the test functions, the method almost converges to the same optimum for different initial conditions, which proves the robustness of the proposed method. However, it does not mean that the method is capable of global optimization. If the corresponding deterministic optimization problem after transformation has multiple local optimums, the method may lead to different optimization results by different initial conditions. In addition, the convergence performance and optimization accuracy of the method is guaranteed only when the uncertainties of variables are relatively small. On the contrary, uncertainties may lead to greater errors during the modeling of approximate interval optimization and computation of intervals of the actual objective function and constraints at $\overline{\mathbf{X}}$ in each iteration. Therefore, the optimization accuracy is affected. Besides, the initial design vector $\mathbf{X}^{(1)}$ should be a feasible solution to the transformed optimization problem. In practice, the designer usually understands the problem to some extent. Hence, the feasible initial design can usually be set by choosing a conservative one.

Like many other engineering optimization methods, it is quite hard to prove the local convergence property or offer general convergence conditions through strict mathematical deductions, while the method can usually converge to a relatively good design. In the iteration process, only the descending feasible solutions are kept to the next iteration, so it can be guaranteed that a better design vector is updated, which is very significant to engineering designs. For the conventional sequential linear programming algorithms aiming at solving the deterministic optimization problem, the convergence of optimization to a local optimal solution can usually be ensured, because the accuracy of the approximate model after linearization can always be continually improved with the shrink of the design domain, thus the optimization results will approach the local optimal solution. However, in the proposed method, the linearized model is with respect to both the design variables and the uncertain variables, while only the design domain is updated while the uncertainty domain keeps unchanged in the iteration process. Therefore, theoretically, the accuracy of the linear approximate model cannot be infinitely improved since the approximation error caused by uncertainty domain that cannot be eliminated by updating the design domain will dominate when the design domain is small enough. That is why in the above two examples the proposed method converges relatively fast in the early stage, while as the optimization proceeds, the convergence becomes more and more slow. For some more complicated problems, the proposed method may not be able to converge through updating the approximate interval optimization problem. Therefore, another convergence criterion $\min\left\{\delta_i^{(s)},\ i = 1, 2, \ldots, n\right\} < \varepsilon_2$ is added into the method to avoid unnecessary computation cost in the late stage of the optimization, which is helpful in solving practical engineering problems.

## 6.4 Application to the Design of a Vehicle Occupant Restraint System

With the popularization of vehicles, more attention is paid to the vehicle safety. The vehicle occupant restraint system, a safety device used to help reduce or even avoid the secondary collision in vehicle crashes [10], is the key part of vehicle passive safety. Shown in Fig. 6.10, is the MADYMO simulation model of a mini car's passenger-side restraint system composed of the vehicle body, a dummy, a safety belt, a seat, a steering wheel and other models. The belted Hybrid III 50th Percentile Male Crash Test Dummy model [11], from the MADYMO dummy database is used as a typical occupant. The simulation combined with intuitively visual adjustments is used to locate the dummy. The safety belt is simulated by the hybrid method: the joint part of the safety belt and the vehicle body is treated as a multi-rigid-body belt, and the contact site of the safety belt and the dummy adopts the FEM safety belt.

To verify the validity and correctness of the simulation model, a real vehicle collision test is conducted according to the prescriptive test methods and procedures in GB11551-2014 "The Protection of the Occupant in the Event of a Frontal Collision for Motor Vehicle". The vehicle body's deceleration, the triaxial accelerations of the dummy's head and thorax, the axial force of the dummy's femurs, and the dummy's injury index can be obtained via the test. Inputting the vehicle body's acceleration curve (as shown in Fig. 6.11) and the intrusion of the vehicle body from the test to the

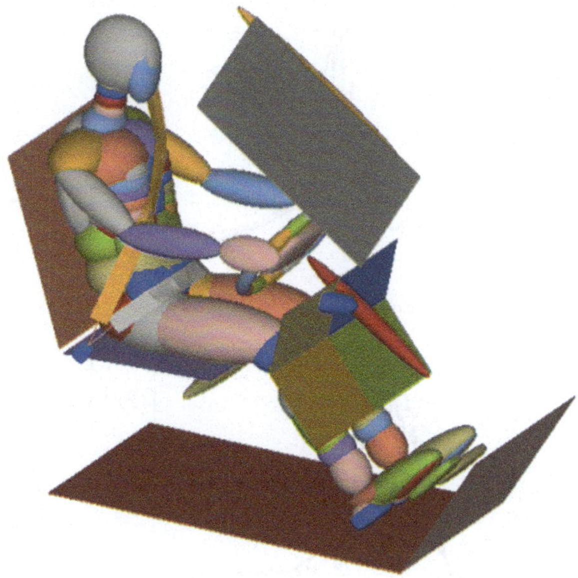

**Fig. 6.10** A passenger-side restraint system model [12, 13]

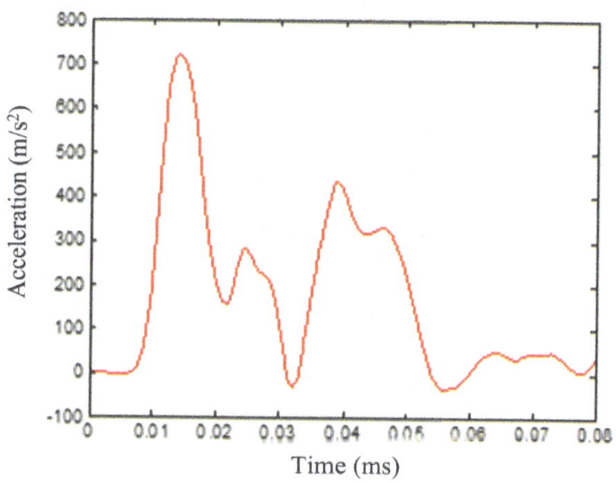


**Fig. 6.11** The vehicle body acceleration curve in frontal collision [12, 13]

restraint system model, the corresponding indexes and curves can then be obtained through the model to be compared with experimental data. Comparisons between the simulation results and the test data are shown in Fig. 6.12. The numerical results agree well with the experimental data, which indicates the validity of the simulation model.

The injury indexes of the passengers mainly include the head injury criterion ($HIC$), the thorax 3 ms criteria and the femur force criteria ($FFC$). Because there are several injury criterion, the weighted injury criterion ($WIC$) is usually used to evaluate the restraint system

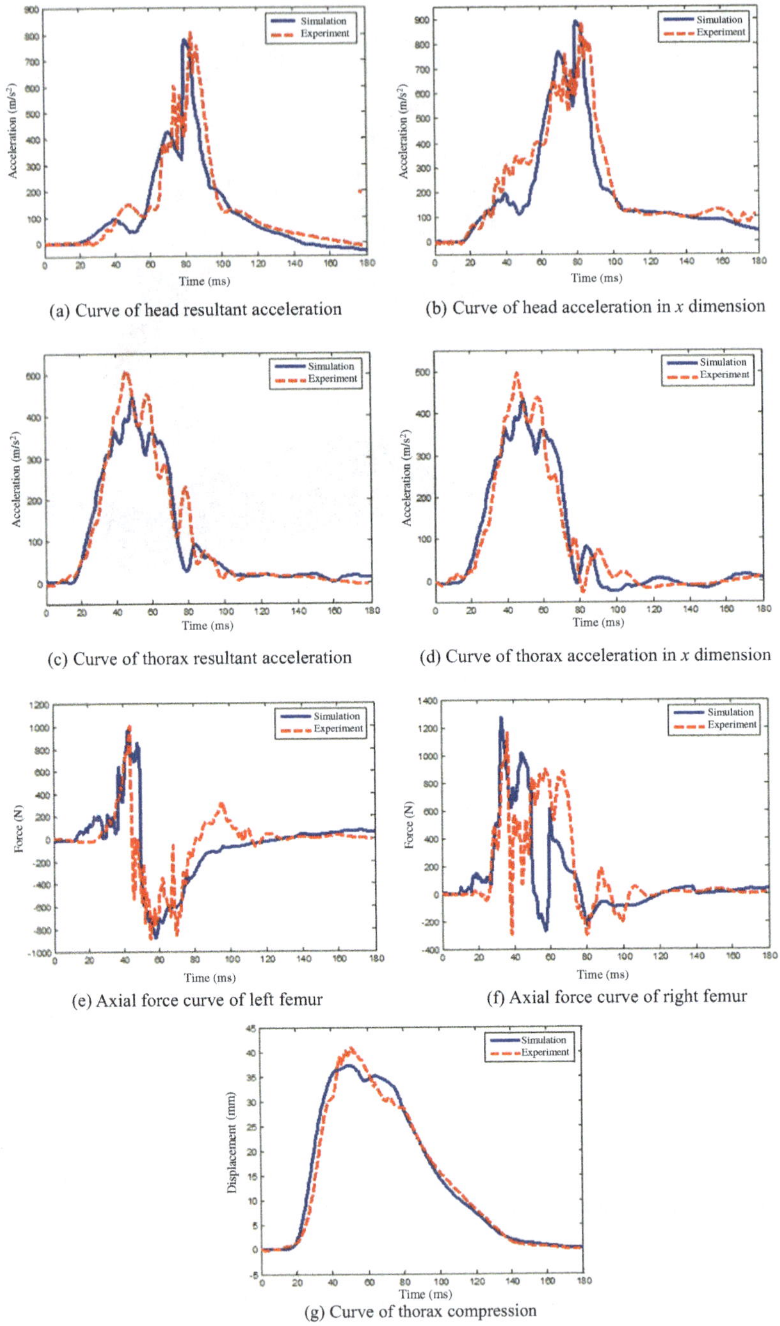


(a) Curve of head resultant acceleration

(b) Curve of head acceleration in $x$ dimension

(c) Curve of thorax resultant acceleration

(d) Curve of thorax acceleration in $x$ dimension

(e) Axial force curve of left femur

(f) Axial force curve of right femur

(g) Curve of thorax compression

**Fig. 6.12** Results comparison between simulation and experiment for the restraint system's key features [12]

$$WIC = 0.6\left(\frac{HIC_{36\text{ms}}}{1000}\right) + 0.35\left(\frac{C_{3\text{ms}}}{60} + \frac{D}{75}\right)/2 + 0.05\left(\frac{F_{FL} + F_{FR}}{20.0}\right) \tag{6.12}$$

where $HIC_{36\text{ms}}$ is the value of head injury 36 ms criterion; $C_{3\text{ms}}$ is the value of thorax 3 ms criterion (g); $D$ is the thorax compression (m); $F_{FL}$ and $F_{FR}$ are the maximum axial forces of the left and right femurs (kN), respectively. In the premise that cost is not increased, the upper hanging (D-Ring) location of the safety belt $h$, the stiffness $e$ of the safety belt (elongation of the belt), and the initial strain $s$ of the safety belt are optimized to reduce the head injury index of the dummy ($HIC_{36\text{ms}}$), thus improving the passive safety. The frictions between the seat and the dummy, and between the safety belt and the dummy in different collisions are not the same, and the lock-up time of the retractor cannot be precisely obtained through tests. Therefore, the friction coefficient $\mu_1$ between the safety belt and the dummy, the friction coefficient $\mu_2$ between the seat and the dummy, and the lock-up time $t$ of retractor are regarded to be uncertain parameters. The dummy's injury criteria are selected as the constraints and $WIC$ is the objective function, then the corresponding uncertain optimization model can be created as follows [12]:

$$\begin{cases} \min\limits_{\mathbf{X}} \; WIC(\mathbf{X}, \mathbf{U}) \\ \text{s.t.} HIC_{36\,\text{ms}} \le 1000,\; D \le 75\,\text{mm},\; C_{3\,\text{ms}} \le 60\,\text{g},\; F_{FL} \le 10\,\text{kN},\; F_{FR} \le 10\,\text{kN} \\ 0.82\,\text{m} \le h \le 0.92\,\text{m}, 5\% \le e \le 15\%, -0.04 \le s \le 0 \\ \quad \mu_1 \in \left[\mu_1^L, \mu_1^R\right] = [0.2, 0.4], \mu_2 \in \left[\mu_2^L, \mu_2^R\right] = [0.2, 0.4] \\ \quad t \in \left[t^L, t^R\right] = [0.6\,\text{ms}, 1.4\,\text{ms}] \end{cases} \tag{6.13}$$

where $\mathbf{X}=(h, e, s)^{\mathrm{T}}$ is the design vector; $\mathbf{U}=(\mu_1, t, \mu_2)^{\mathrm{T}}$ is the uncertain vector.

The above interval optimization method is used to solve the problem in Eq. (6.13). The injury indexes before and after the optimization are listed in Table 6.9. The

**Table 6.9** Optimization results for the vehicle occupant restraint system [12, 13]

| Design variables and injury indexes | Before optimization | After optimization |
|---|---|---|
| $h$ (m) | 0.87 | 0.8876 |
| $e$ (%) | 9 | 6.85 |
| $s$ | -0.002 | −0.0058 |
| $HIC_{36\,\text{ms}}$ | 1071.40 | [812.97, 954.31] |
| $D$ (mm) | 37.4 | [36.5, 39.8] |
| $C_{3\,\text{ms}}$ (g) | 41.920 | [41.088, 46.463] |
| $F_{FL}$ (kN) | 0.9675 | [0.9180, 0.9970] |
| $F_{FR}$ (kN) | 1.2783 | [1.2554, 1.2588] |
| $WIC$ | 0.8579 | [0.7059, 0.7958] |

value of $HIC_{36\text{ms}}$ decreases from 1071.4 to [812.97, 954.31] after the optimization. The upper bound of the interval [812.97, 954.31] is less than 1000, satisfying the corresponding regulation. Besides, the values of $D$, $C_{3\text{ms}}$ and the upper bounds of $F_{FL}$ and $F_{FR}$ are far less than the corresponding regulatory requirements. The value of $WIC$ decreases from 0.8579 to [0.7059, 0.7958] after optimization. Therefore, the restraint system's protection performance is improved, and under the same parameter disturbances, all injury indexes do not exceed the specified thresholds, improving the system's reliability. Besides, only 9 iterations in the interval optimization process are used to converge to the optimal design, and the MADYMO simulation model was called only 117 times.

## 6.5 Summary

In this chapter, the sequential linear programming technique and the mathematical transformation model based on order relation of interval number are combined to establish an efficient nonlinear interval optimization method. The whole optimization process is composed of a series of approximate uncertain optimization problems, and an iteration mechanism is proposed to guarantee the convergence. In each iteration, the approximate uncertain optimization base on the first-order Taylor expansion is actually a linear interval optimization problem whose solution can be obtained by a conventional single-layer optimization, which simplifies the computation process to a large extent. To improve the computational efficiency, the interval structural analysis method is used to approximately calculate the bounds of the actual objective function and constraints at the solution of the current iteration, thus avoiding the optimizations based on actual models. The analysis of two test functions indicates that the method has robust convergence and high efficiency. Besides, comparisons between the optimal bound method and the approximate bound method with different initial conditions are given. The comparisons show that when the uncertainties of the variables are relatively small, the approximate bound method almost has the same convergence performance and accuracy with the optimal bound method. Therefore, it can be used to replace the optimal bound method to solve practical engineering problems more efficiently. Finally, the method is applied to the design of a vehicle occupant restraint system to illustrate its capability for solving practical engineering problems.

## References

1. Marcotte P, Dussault J-P (1989) A sequential linear programming algorithm for solving monotone variational inequalities. SIAM J Control Optim 27(6):1260–1278
2. Yang RJ, Chuang CH (1994) Optimal topology design using linear programming. Comput Struct 52(2):265–275

3. Lamberti L, Pappalettere C (2000) Comparison of the numerical efficiency of different sequential linear programming based algorithms for structural optimisation problems. Comput Struct 76(6):713–728
4. Li D, Jiang C, Han X, Zhang Z (2011) An efficient optimization method for uncertain problems based on non-probabilistic interval model. Int J Comput Methods 8(4):837–850
5. Jiang C (2008) Theories and algorithms of uncertain optimizationbased on interval. PhD thesis, Hunan University, Changsha
6. Qiu ZP (2003) Comparison of static response of structures using convex models and interval analysis method. Int J Numer Meth Eng 56(12):1735–1753
7. Zhou YT, Jiang C, Han X (2006) Interval and subinterval analysis methods of the structural analysis and their error estimations. Int J Comput Methods 3(2):229–244
8. Wang GG (2003) Adaptive response surface method using inherited latin hypercube design points. Trans-Am Soc Mech Eng J Mech Design 125(2):210–220
9. Jiang C, Han X, Liu GP (2008) A sequential nonlinear interval number programming method for uncertain structures. Comput Methods Appl Mech Eng 197(49–50):4250–4265
10. Zhong ZH, Zhang WG, Cao LB, He W (2003) Automotive crash safety technology. Machinery Industry Press, Beijing
11. TNO company (2004) MADYMO theory manual version621 model manual. Netherlands
12. Ning HM, Jiang C, Liu J, Bai YC (2012) uncertainty optimization of vehicle occupant restraint system based on interval method. Automot Eng 34(12):1085–1089
13. Han X (2015) Numerical simulation-based design theory and methods. Science Press, Beijing

# Chapter 7
# Interval Optimization Based on Approximation Models

**Abstract** This chapter creates two nonlinear interval optimization methods based on the approximation model management strategy and the local-densifying approximation technique, respectively. Both the methods can greatly improve the computational efficiency of interval optimization by using approximation models, but different strategies are selected to guarantee accuracy during the iteration.

The engineering optimization problems are becoming more and more complicated and they usually involve computationally expensive numerical analysis models, making conventional optimization methods incapable of meeting computational efficiency demands. Currently, simple explicit functions are usually constructed as the approximation models to replace the original models. Then the approximate optimization problem can be established and efficiently solved. With the approximation model, the optimization efficiency can be greatly improved, making the solution of many complex practical engineering problems possible. In addition, what is significant is that the approximation model can smooth or denoise the actual function, which will achieve better results when solving many optimization problems based on complex analysis models. So far, the approximation optimization method has become a research hotspot in the field of engineering optimization, and a large number of methods have been proposed [1–9].

In this chapter, the approximation model is introduced to the nonlinear interval optimization, and two types of efficient uncertain optimization methods are proposed. The first one is a nonlinear interval optimization method based on the approximation model management strategy. The whole optimization process is composed of a series of approximate uncertain optimization problems. In the iteration process, the design space and the approximation models are updated by a model management tool. This method applies to optimization problems with variables of small uncertainties. The second one is a nonlinear interval optimization method based on the local-densifying approximation technique. The optimization process of this method is also composed of a series of uncertain optimization problems. However, the approximation space is not updated during the iteration. Instead, a local-densifying technique is used to improve the accuracy of the approximation model in key areas, thus improving the



C. Jiang et al., *Nonlinear Interval Optimization for Uncertain Problems*, Springer Tracts in Mechanical Engineering, https://doi.org/10.1007/978-981-15-8546-3_7

optimization accuracy. This method is not restricted by the uncertainty levels of the problem.

## 7.1 Nonlinear Interval Optimization Based on the Approximation Model Management Strategy

The optimization based on the approximation model management strategy has received more and more attention in recent years. In the strategy, the approximation model is employed to improve the optimization efficiency and the model management tool such as the trust region method is used to ensure optimization accuracy. As for the approximation model management, so far there have been a series of research achievements. It has been theoretically proven in references [10, 11] that the trust-region-based approximation optimization can ensure the convergence to the Karuch–Kuhn–Tucker solution. In reference [12], the trust region method was used to manage low-fidelity approximation models in a type of approximate constrained optimization problems. In reference [13], applications and developments of the trust-region-based model management strategy in the multi-disciplinary design optimization (MDO) were reviewed and discussed. In reference [14], three approximation model management strategies based on different nonlinear optimization algorithms were discussed and applied to the design of a practical aeronautic component. In reference [15], the approximation model management strategy is used to optimize the blank holder force in sheet metal forming. Above researches show that the approximation model management strategy is highly effective in ensuring the optimization convergence and accuracy. However, at present, the approximation model management strategy is mainly aimed at deterministic optimization problems. In order to apply the approximation model management strategy to interval optimizations, it must be modified and improved according to the interval optimization's characteristics considering the parametric uncertainties.

In this section, a brief introduction will be first given to the adopted approximation model and design of experiments (DOE), namely, the quadratic polynomial response surface and the Latin hypercube design (LHD). Afterward, by using the different transformation models, we will create two interval optimization methods based on the approximation management strategy, and two test functions are used to test the accuracy, efficiency, and other performance of the methods. Finally, the methods are applied to dealing with two practical engineering problems.

### 7.1.1 *Quadratic Polynomial Response Surface*

In the quadratic polynomial response surface [16], the approximation function is modeled by the regression analysis and variance analysis. Consider a function $h(\mathbf{x})$

with $n_d$ input variables and $n_s$ design samples, its approximation function can be expressed as follows:

$$\tilde{h}(\mathbf{x}) = c_0 + \sum_{i=1}^{n_d} c_i x_i + \sum_{i=1}^{n_d} \sum_{j=1}^{n_d} c_{ij} x_i x_j \tag{7.1}$$

The following equation holds at all design samples:

$$h^{(k)} = c_0 + \sum_{i=1}^{n_d} c_i x_i^{(k)} + \sum_{i=1}^{n_d} \sum_{j=1}^{n_d} c_{ij} x_i^{(k)} x_j^{(k)}, \ k = 1, 2, \ldots, n_s \tag{7.2}$$

where $h^{(k)}$ is the realization of $h(\mathbf{x})$ at the $k$th sample; $x_i^{(k)}$ and $x_j^{(k)}$ represent the realizations of the $i$th and of the $j$th input variables at the $k$th sample, respectively; $c_0$, $c_i$, and $c_{ij}$ are coefficients of the constant term, the linear term, and the quadratic term, respectively. If $c_{ij} = c_{ji}$, the total number of the unknown coefficients is $n_t = (n_d + 1)(n_d + 2)/2$, so $n_s \geq n_t$ must be guaranteed. Equation (7.2) can be rewritten in a matrix form as follows:

$$\mathbf{h} = \mathbf{B}\mathbf{c} \tag{7.3}$$

where $\boldsymbol{h}$ is an $n_s$-dimensional vector of the actual function values at the samples, and $\mathbf{B}$ is an $n_s \times n_t$ matrix expressed as follows:

$$\mathbf{B} = \begin{bmatrix} 1 & x_1^{(1)} & x_2^{(1)} & \ldots & \left(x_{n_d}^{(1)}\right)^2 \\ \vdots & \vdots & \vdots & \ddots & \vdots \\ 1 & x_1^{(n_s)} & x_2^{(n_s)} & \ldots & \left(x_{n_d}^{(n_s)}\right)^2 \end{bmatrix} \tag{7.4}$$

A least-square estimation $\tilde{\mathbf{c}}$ of $\mathbf{c}$ can be obtained:

$$\tilde{\mathbf{c}} = \left(\mathbf{B}^{\mathrm{T}}\mathbf{B}\right)^{-1}\mathbf{B}^{\mathrm{T}}\mathbf{h} \tag{7.5}$$

Substituting $\tilde{\mathbf{c}}$ into Eq. (7.1), the approximation function $\tilde{h}(\mathbf{x})$ of $h(\mathbf{x})$ can be obtained.

When constructing the response surface model, there exist some contradictions in the selection of the optimal regression model: To better fit the actual function, the approximation function should contain more terms of basis functions. However, too many terms of basis functions may lead to overfitting and hence the approximation accuracy will unexpectedly decrease. Therefore, in the construction of an approximation function, it is necessary to keep important terms of basis functions and remove insensitive terms. Here the $F$-statistics-test-based back elimination procedure [16] is employed to determine the form of the approximation model, and its flowchart is given in Fig. 7.1. Firstly, the process starts from a full model containing all the basis functions. Then the significance testing on coefficients is conducted after using the least-square method to estimate the coefficients. If the minimal $F$-value is smaller than a given threshold, its corresponding term of basis function will be deleted until no basis function can be deleted and the final optimal approximation model is constructed. In the based back elimination procedure, if there are more than one insignificant terms, they cannot be deleted one time. Only the most insignificant term will be deleted.

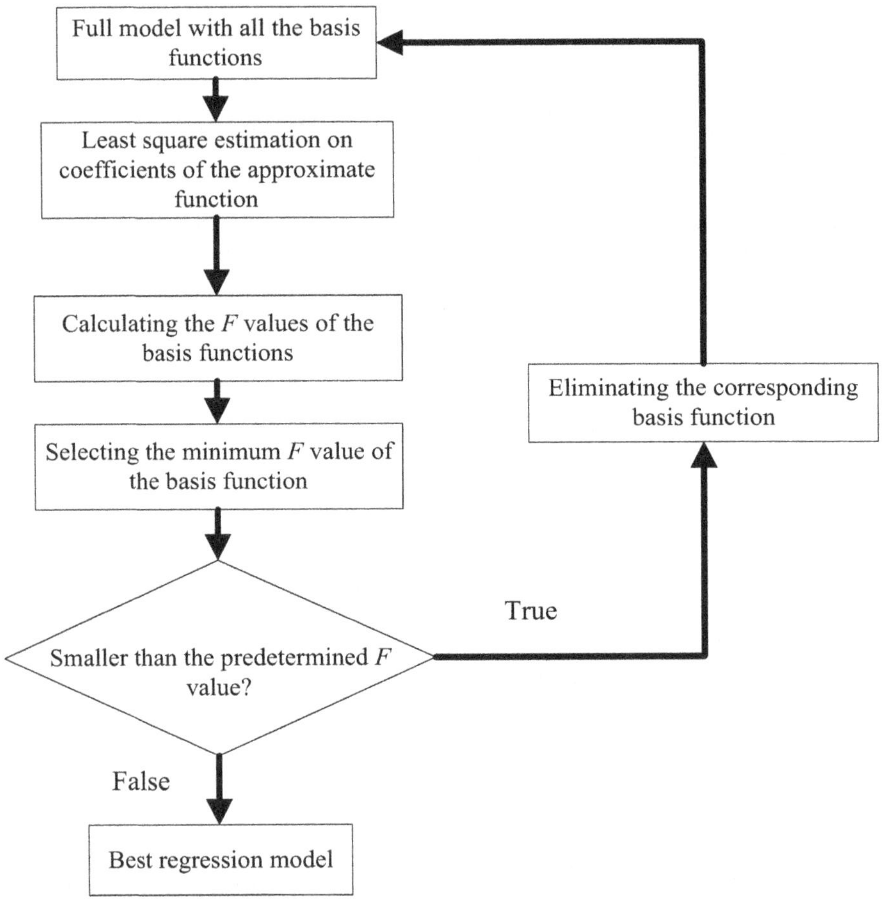


**Fig. 7.1** The back elimination procedure based on $F$-statistics test [17]

### 7.1.2 Design of Experiment Method

DOE is used to obtain the experimental samples required in the construction of the approximation model. The choice of DOE has a great influence on the approximation accuracy and the cost of the approximate model. In this paper, the Latin hypercube design (LHD) is selected as the DOE method. LHD is relatively flexible that can casually control the size of sample, thus it can generate saturated design samples. The samples obtained by LHD is uniformly distributed in the sample space. Besides, LHD can realize the inheritance of sample points between two consecutive generations to greatly save computational cost. In LHD, each variable is equally divided into $s$ intervals ($s$ is the variable level), and $s$ samples will be generated by repeating the following two steps $s$ times:

(1) Randomly select a non-repetitive interval from $s$ intervals for each variable.
(2) Randomly generate one sample in each interval following the uniform distribution.
Because the above process is based on random sampling, there are infinite sampling schemes. To make sure the final sample is uniformly distributed, an optimization method is employed to select the best scheme. In this chapter, the optimal LHD [18] proposed by Morris and Mitchell is adopted. In this method, the simulation annealing algorithm is used as the optimization solver to search for the optimal scheme.

### 7.1.3 The Method by Using the Transformation Model Based on Order Relation of Interval Number

The optimization problem to be investigated is expressed in Eq. (4.1), where the uncertainty levels of all uncertain variables in $\mathbf{U}$ are assumed to be small, so this method is only suitable to problems with small uncertainties. The optimization process is composed of a series of approximation uncertain optimizations and the trust region method is employed to manage the approximation models in order to update the design space and design vector and to guarantee the adaptive improvement of the optimization results.

1. Construction and solution of the approximate interval optimization problem

In the $s$th iteration, for the optimization problem expressed in Eq. (4.1), an approximate interval optimization problem as shown in follows can be constructed:

$$\begin{cases} \min\limits_{\mathbf{X}} \tilde{f}(\mathbf{X},\mathbf{U}) \\ \text{s.t. } \tilde{g}_i(\mathbf{X},\mathbf{U}) \le b_i^I = \left[b_i^L, b_i^R\right],\ i = 1,\ldots,l \\ \qquad \max\left[\mathbf{X}_l, \mathbf{X}^{(s)} - \boldsymbol{\Delta}^{(s)}\right] \le \mathbf{X} \le \min\left[\mathbf{X}_r, \mathbf{X}^{(s)} + \boldsymbol{\Delta}^{(s)}\right] \\ \qquad \mathbf{U} \in \mathbf{U}^I = \left[\mathbf{U}^L, \mathbf{U}^R\right],\ U_i \in U_i^I = \left[U_i^L, U_i^R\right],\ i = 1,2,\ldots,q \end{cases} \tag{7.6}$$

where $\mathbf{\Delta}^{(s)}$ is the trust region radius vector in the $s$-th iteration, which constitutes the current design space, or the current trust region, with the current design vector $\mathbf{X}^{(s)}$. $\mathbf{\Delta}^{(s)}$ will be updated in each iteration. $\tilde{f}(\mathbf{X}, \mathbf{U})$ and $\tilde{g}_i(\mathbf{X}, \mathbf{U})$ are the approximation models constructed by quadratic polynomial response surface for the objective function and the $i$-th constraint, respectively.

For deterministic optimization problems, when constructing the approximation models of the objective function and constraints, the sampling is carried out only in the design space. However, for the uncertain optimization problem expressed in Eq. (4.1), both the design space and the uncertainty domain need to be considered when constructing the approximation models. Therefore, attention should be paid to the following problems when creating $\tilde{f}(\mathbf{X}, \mathbf{U})$ and $\tilde{g}_i(\mathbf{X}, \mathbf{U})$:

(1) Employ LHD to generate samples in the hybrid space composed of the current design space and the uncertainty domain, and use the samples generated by the actual models to construct the approximation models.
(2) In the construction of approximation models, both the design variables and the uncertain variables are treated as input variables, hence $\tilde{f}(\mathbf{X}, \mathbf{U})$ and $\tilde{g}_i(\mathbf{X}, \mathbf{U})$ are explicit quadratic functions with respect to both design variables and uncertain variables.

With the transformation model based on order relation of interval number, the above approximate interval optimization problem can be converted into the following deterministic optimization problem:

$$\begin{cases} \min\limits_{\mathbf{X}} \tilde{f}_p(X) = (1-\beta)\left(\tilde{f}^c(X)+\xi\right)/\phi + \beta\left(\tilde{f}^w(X)+\xi\right)/\psi \\ \qquad\qquad + \sigma \sum\limits_{i=1}^{l} \varphi\left(P\left(\tilde{g}_i^I(X) \le b_i^I\right) - \lambda_i\right) \\ \text{s.t.} \max\left[\mathbf{X}_l, \mathbf{X}^{(s)} - \mathbf{\Delta}^{(s)}\right] \le \mathbf{X} \le \min\left[\mathbf{X}_r, \mathbf{X}^{(s)} + \mathbf{\Delta}^{(s)}\right] \end{cases} \tag{7.7}$$

where $\tilde{f}_p$ is the penalty function based on the approximation models, namely, the approximate penalty function, and the corresponding penalty function $f_p$ based on the actual models is called actual penalty function.

The two-layer nested optimization method proposed in Chap. 3 is used to solve the above optimization problem. The difference is that in the optimization process, the computation is based on the approximation models instead of the actual models. Since the approximation models are computationally cheap explicit quadratic functions, the efficiency will be very high even IP-GA is used.

2. Accuracy test of the approximation models and update of the design space

After solving the current approximation optimization, a trust region test is used to assess the accuracy of current approximation models in order to further assess the current design vector. A reliability index $\rho^{(s)}$ is thus employed to monitor the approximation degree between the current approximate model and actual model:

$$\rho^{(s)} = \frac{f_p(\mathbf{X}^{(s)}) - f_p(\mathbf{X}^{(s)*})}{f_p(\mathbf{X}^{(s)}) - \tilde{f}_p(\mathbf{X}^{(s)*})} \tag{7.8}$$

where $\mathbf{X}^{(s)*}$ denotes the optimum solution of Eq. (7.7). In fact, $\rho^{(s)}$ represents the ratio of the actual change of the penalty function between $\mathbf{X}^{(s)}$ and $\mathbf{X}^{(s)*}$ to the predictive change of the approximate penalty function between $\mathbf{X}^{(s)}$ and $\mathbf{X}^{(s)*}$. The more $\rho^{(s)}$ approaches 1, the more accurate the approximation models are.

The penalty function here is selected as the criterion to construct the reliability index because the penalty function contains information of both the objective function and the constraints. To make the value of $\rho^{(s)}$ approach 1, the approximation models of both the objective function and constraints should be accurate.

According to the value of $\rho^{(s)}$, the trust region radius vector can be updated as follows:

(1) If $\rho^{(s)} \leq 0.0$, the accuracy of the approximation models is relatively bad and the trust region radius vector $\mathbf{\Delta}^{(s)}$ should be reduced to improve the approximation precision.
(2) If $\rho^{(s)} \approx 1.0$, the accuracy of the approximation models is good. When $\mathbf{X}^{(s)*}$ is just located on the boundary of the current trust region, the actual optimum design may be quite far from the current one, so $\mathbf{\Delta}^{(s)}$ should be amplified in order to accelerate the optimization process. When $\mathbf{X}^{(s)*}$ is located in the trust region, it indicates that the actual optimum design should be near the current one, so $\mathbf{\Delta}^{(s)}$ should be kept unchanged.
(3) If $\rho^{(s)} \geq 1.0$, it also implies that the approximation is not good, but a desirable research direction is achieved. When $\mathbf{X}^{(s)*}$ is just located on the boundary of the current trust region, $\mathbf{\Delta}^{(s)}$ should be amplified in order to accelerate the optimization process. When $\mathbf{X}^{(s)*}$ is in the trust region, $\mathbf{\Delta}^{(s)}$ should be kept unchanged.
(4) If $0.0 < \rho^{(s)} < 1.0$, the update of $\mathbf{\Delta}^{(s)}$ depends on the distance between $\rho^{(s)}$ and 0 or 1.

Generally, different coefficients to amplify or shrink the trust region radius vector $\mathbf{\Delta}^{(s)}$ can be selected. In this method the following update scheme is adopted [17]:

$$\begin{cases} \mathbf{\Delta}^{(s+1)} = 0.5\mathbf{\Delta}^{(s)}, \mathbf{X}^{(s+1)} = \mathbf{X}^{(s)} \text{ if } \rho^{(s)} \leq 0.0 \\ \mathbf{\Delta}^{(s+1)} = 0.5\mathbf{\Delta}^{(s)}, \mathbf{X}^{(s+1)} = \mathbf{X}^{(s)*} \text{ if } 0.0 < \rho^{(s)} \leq 0.25 \\ \mathbf{\Delta}^{(s+1)} = \mathbf{\Delta}^{(s)}, \mathbf{X}^{(s+1)} = \mathbf{X}^{(s)*} \text{ if } 0.25 < \rho^{(s)} \leq 0.75 \\ \mathbf{\Delta}^{(s+1)} = 2.0\mathbf{\Delta}^{(s)}, \mathbf{X}^{(s+1)} = \mathbf{X}^{(s)*} \text{ if } \rho^{(s)} > 0.75,\ \mathbf{X}^{(s)*} \text{ is on the boundary of} \\ \qquad \text{the curren trust region} \\ \mathbf{\Delta}^{(s+1)} = \mathbf{\Delta}^{(s)}, \mathbf{X}^{(s+1)} = \mathbf{X}^{(s)*} \text{ if } \rho^{(s)} > 0.75,\ \mathbf{X}^{(s)*} \text{ is inside the current trust} \\ \qquad \text{region} \end{cases} \tag{7.9}$$

3. Calculation of the actual penalty function

In each iteration, $f_p\left(\mathbf{X}^{(s)}\right)$, $f_p\left(\mathbf{X}^{(s)*}\right)$, and $\tilde{f}_p\left(\mathbf{X}^{(s)*}\right)$ should be obtained to calculate $\rho^{(s)}$ via Eq. (7.8). Obviously, $\tilde{f}_p\left(\mathbf{X}^{(s)*}\right)$ has been obtained by solving Eq. (7.7), while it would be difficult to calculate $f_p\left(\mathbf{X}^{(s)}\right)$ through the following equation:

$$\begin{aligned} f_p\left(\mathbf{X}^{(s)}\right) &= (1-\beta)\left(f^c\left(\mathbf{X}^{(s)}\right)+\xi\right)/\phi + \beta\left(f^w\left(\mathbf{X}^{(s)}\right)+\xi\right)/\psi \\ &\quad + \sigma\sum_{i=1}^{l}\varphi\left(P\left(g_i^I\left(\mathbf{X}^{(s)}\right)\leq b_i^I\right)-\lambda_i\right) \end{aligned} \tag{7.10}$$

The reason is that each calculation for the interval of actual objective function $f^I\left(\mathbf{X}^{(s)}\right)$ or interval of each constraint $g_i^I\left(\mathbf{X}^{(s)}\right)$, $i = 1, 2, \ldots, l$ requires twice optimizations, and the actual models are usually computationally expensive numerical analysis models in practical engineering, hence leading to a very low efficiency in each iteration.

Shown in Fig. 7.2 is an efficient method to calculate $f_p\left(\mathbf{X}^{(s)}\right)$, and its calculation process is as follows:

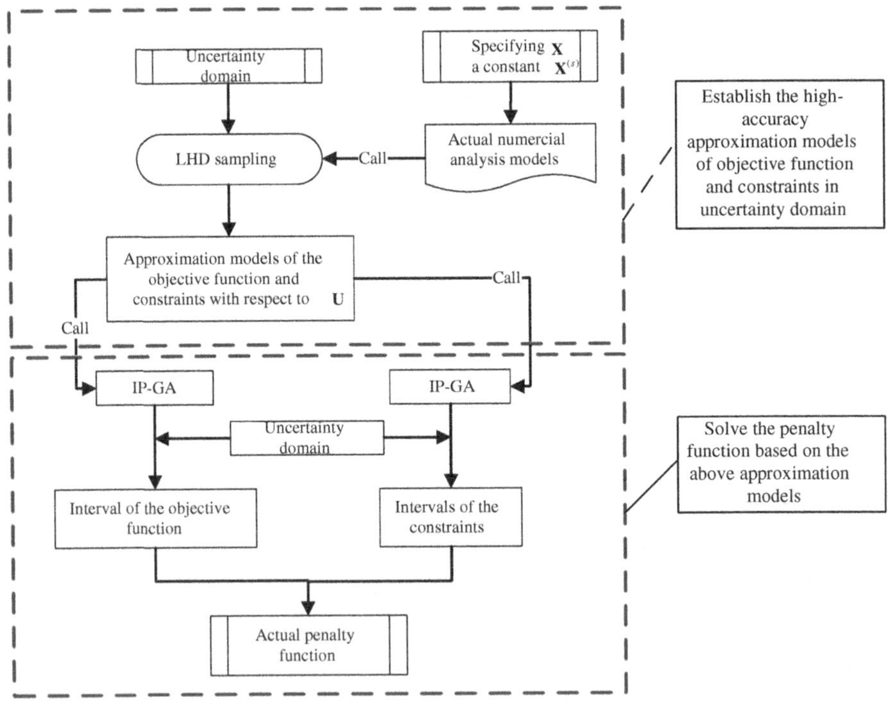


**Fig. 7.2** Calculation of the actual penalty function at $\mathbf{X}^{(s)}$ [17]

(1) Generate samples in the uncertainty domain $\Gamma = \{U | U_i^L \le U_i \le U_i^R,\ i = 1, 2, \ldots, q\}$ by means of LHD, and calculate the actual values of the objective function and constraints at these samples, with the given specific $\mathbf{X}^{(s)}$.
(2) Construct the approximation models $\tilde{f}(\mathbf{X}^{(s)}, \mathbf{U})$ and $\tilde{g}_i(\mathbf{X}^{(s)}, \mathbf{U}),\ i = 1, 2, \ldots, l$ by quadratic polynomial response surface approach. The approximation models are explicit functions with respect to the uncertain variables.
(3) Based on the approximation models, construct optimization problems to calculate the bounds of the objective function and constraints at $\mathbf{X}^{(s)}$:

$$f^L(\mathbf{X}^{(s)}) = \min_{U \in \Gamma} \tilde{f}(\mathbf{X}^{(s)}, \mathbf{U}),\ f^R(\mathbf{X}^{(s)}) = \max_{U \in \Gamma} \tilde{f}(\mathbf{X}^{(s)}, \mathbf{U}) \tag{7.11}$$

$$g_i^L(\mathbf{X}^{(s)}) = \min_{U \in \Gamma} \tilde{g}_i(\mathbf{X}^{(s)}, \mathbf{U}),\ g_i^R(\mathbf{X}^{(s)}) = \max_{U \in \Gamma} \tilde{g}_i(\mathbf{X}^{(s)}, \mathbf{U}),\ i = 1, 2, \ldots, l \tag{7.12}$$

(4) Calculate $f_p(\mathbf{X}^{(s)})$ based on the bounds of the objective function and constraints at $\mathbf{X}^{(s)}$.

Afterward, $f_p(\mathbf{X}^{(s)*})$ can be also obtained in a similar way, while the design vector should be given as $\mathbf{X}^{(s)*}$ when calling the actual models. In the calculations of $f_p(\mathbf{X}^{(s)})$ and $\tilde{f}_p(\mathbf{X}^{(s)*})$, approximation models are actually both employed. However, the difference is that the approximation models are constructed on the uncertainty domain (the design vector is given a constant) when calculating $f_p(\mathbf{X}^{(s)})$, while the approximation models are based on the hybrid space composed of the current design space and the uncertainty domain when calculating $\tilde{f}_p(\mathbf{X}^{(s)*})$. In the investigated optimization problem, since the uncertainties are assumed to be relatively small, the accuracy of using a quadratic polynomial to approximate the objective function and constraints in a relatively small uncertainty domain can be well guaranteed. Theoretically, we can construct an infinitely precise approximation model with an infinitely small uncertainty domain. In practical engineering problems, the disturbances of uncertain parameters are generally small compared with their nominal values, such as manufacture error and measurement error. Therefore, the approximation models $\tilde{f}(\mathbf{X}^{(s)}, \mathbf{U}), \tilde{g}_i(\mathbf{X}^{(s)}, \mathbf{U}),\ i = 1, 2, \ldots, l$ generally have high accuracies and the penalty function $f_P(\mathbf{X}^{(s)})$ obtained from these "accurate" approximation models will also have a high accuracy. That is why $f_P(\mathbf{X}^{(s)})$ and $f_P(\mathbf{X}^{(s)*})$ here are named the actual penalty function in order to be distinguished from the approximate penalty function obtained via Eq. (7.7), though they are still obtained through the approximation model technique.

4. Algorithm flow

The calculation process of the above interval optimization method is shown in Fig. 7.3. At the start of optimization, an initial design space is specified by giving an initial design vector and an initial trust region radius vector. In each iteration, approximation models of the uncertain objective function and constraints are constructed on

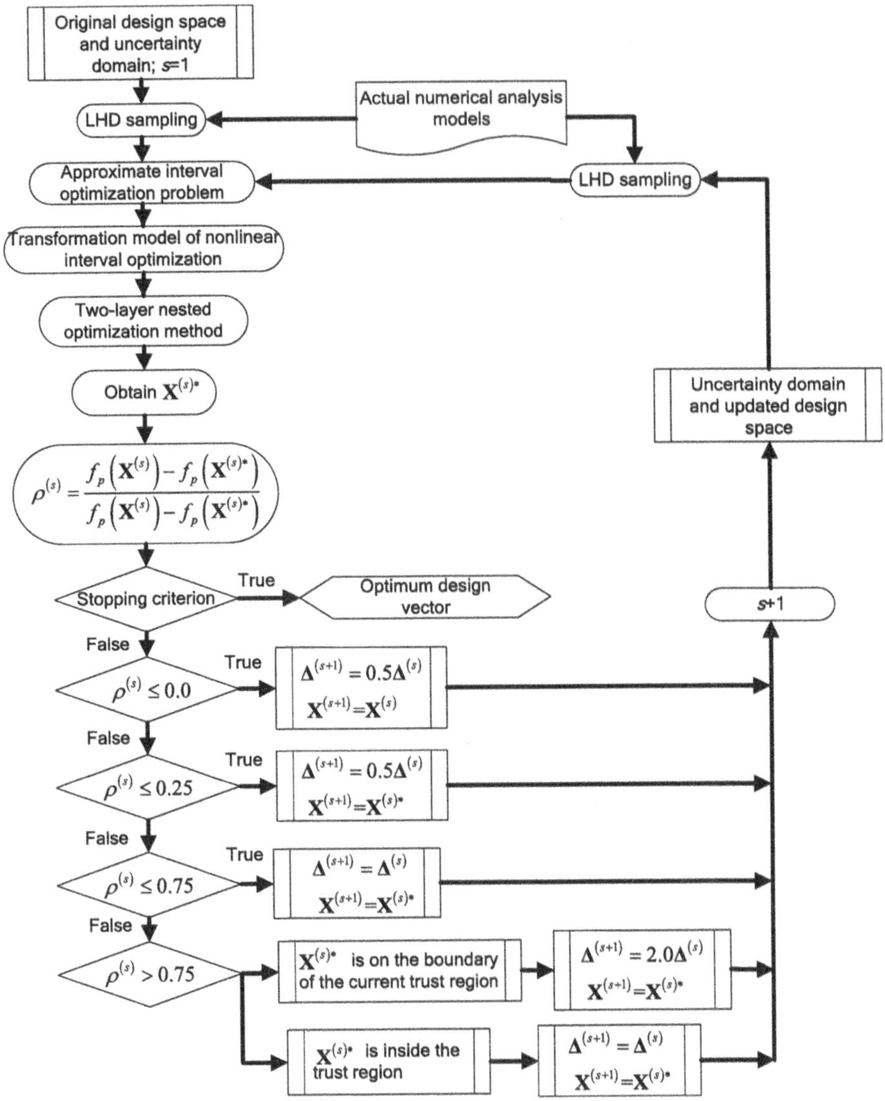


**Fig. 7.3** Algorithm flowchart of the interval optimization method based on the approximation model management strategy [17]

the hybrid space composed of the current design space and the uncertainty domain to constitute an approximate interval optimization problem. Afterward, the approximate interval optimization problem is converted into a deterministic optimization by using the order relation transformation model, which will be then solved by the two-layer nested optimization method. The reliability index is thus calculated to assess the accuracy of the current approximation models and thereby update the design space and the design vector. The above process is repeated until convergence is reached. In the method, the maximum iteration number is selected as the convergence criterion.

During the iterations, the design space is updated along with changes in design vector and trust region radius vector. By updating the design vector, a better design is kept and the design space will move to the optimum solution. By updating the trust region radius vector, the accuracy of approximation models can be guaranteed. In addition, the approximation models constructed in each iteration are functions with respect to the design variables and the uncertain variables, while the uncertainty domain is not updated in iterations. It is still assumed here that the variables' uncertainties are relatively small. Therefore, the quadratic polynomial response surface is sufficiently accurate on the uncertainty domain and hence only the design space needs to be updated. From the above analysis, we can see that the assumption of small uncertainties is taken in two parts in the proposed method. The first part is in the calculation of the actual penalty function and the second part is that the design space is updated while the uncertainty domain is kept unchanged. Theoretically, the method will get ideal optimization results only when the uncertainty level is small enough. Besides, the optimization performance will decrease with the increase of uncertainties of the variables. It seems difficult to mathematically and technically propose a general assessment index to quantify the influence of the uncertainty level on optimization accuracy and convergence performance. However, in the subsequent test functions, such influence will be numerically analyzed.

Besides, in optimization, the computation cost is mainly consumed in two parts. The first part is in the computation for samples when constructing the approximate interval optimization problem. The second part is in calculating the actual penalty function. Besides, using the IP-GA-based nested optimization algorithm to solve the approximate interval optimization problem will also result in a few computational costs. Nonetheless, such cost is generally negligible in practical engineering problem as mentioned above. Therefore, in the subsequent example analysis, only the computation cost spent in numerical analysis models will be considered in the assessment of computational efficiency.

### *7.1.4 The Method by Using the Transformation Model Based on Possibility Degree of Interval Number*

The method in the last section is based on the approximation model management strategy and the order relation transformation model, while the possibility degree transformation model will replace the order relation transformation model to construct another corresponding nonlinear interval optimization method in this section. The whole process of this method is similar to that in the last section, while there remain some differences as follows:

Firstly, in each iteration, the order relation transformation model is employed to transform the approximate uncertain optimization problem expressed in Eq. (7.6) into the deterministic optimization problem like Eq. (3.41) [19]:

$$\begin{cases} \max_{\mathbf{X}} \tilde{f}_P(\mathbf{X}) = P\left(\tilde{f}^I(\mathbf{X}) \le V^I\right) - \sigma \sum_{i=1}^{l} \varphi\left(P\left(\tilde{g}_i^I(\mathbf{X}) \le b_i^I\right) - \lambda_i\right) \\ \text{s.t. } \max\left[\mathbf{X}_l, \mathbf{X}^{(s)} - \mathbf{\Delta}^{(s)}\right] \le \mathbf{X} \le \min\left[\mathbf{X}_r, \mathbf{X}^{(s)} + \mathbf{\Delta}^{(s)}\right]. \end{cases} \tag{7.13}$$

Secondly, after an optimum design vector $\mathbf{X}^*$ is obtained by solving the optimization problem as shown in Fig. 7.3, the robustness criterion can be further introduced to perform a similar optimization, if $P\left(f^I(\mathbf{X}^*) \le V^I\right) = 1.0$. In each iteration of this process, the approximate interval optimization problem expressed in Eq. (7.6) is transformed into the deterministic optimization problem as shown in Eq. (3.42) to maximize the robustness of the uncertain objective function:

$$\begin{cases} \min_{\mathbf{X}} \tilde{f}_P(\mathbf{X}) = \tilde{f}^w(\mathbf{X}) + \sigma\left[\sum_{i=1}^{l} \varphi\left(P\left(\tilde{g}_i^I(\mathbf{X}) \le b_i^I\right) - \lambda_i\right) + \left(P\left(\tilde{f}^I(\mathbf{X}) \le V^I\right) - 1\right)^2\right] \\ \text{s.t. } \max\left[\mathbf{X}_l, \mathbf{X}^{(s)} - \mathbf{\Delta}^{(s)}\right] \le \mathbf{X} \le \min\left[\mathbf{X}_r, \mathbf{X}^{(s)} + \mathbf{\Delta}^{(s)}\right] \end{cases} \tag{7.14}$$

Therefore, two independent iterative solution processes may be performed in the above interval optimization method. In each iteration, the deterministic optimizations transformed from the approximate interval optimization problem have the different expressions in these two iterative solution processes.

### 7.1.5 Test Functions

In this section, two test functions are used to test the optimization accuracy, convergence performance, efficiency, and other properties of the proposed method. In test function 1, the interval optimization method based on the order relation transformation model described in Sect. 7.1.3 is employed to solve the optimization problem, while the method based on the possibility degree transformation model in Sect. 7.1.4 is adopted for test function 2.

1. Test function 1

Consider the cantilever beam structure in Chap. 6. The cross-sectional dimensions are still the design variables to be optimized to minimize the vertical deflection under two constraints: the cross-sectional area and maximum stress are not allowed to exceed the given thresholds. The structure, materials, and loads are the same and uncertain variables are still $U_1$ and $U_2$, while the allowable maximum stress and design space are different. The corresponding interval optimization problem is modeled as follows [17]:

$$\begin{cases} \min_{\mathbf{X}} f(\mathbf{X}, \mathbf{U}) = \dfrac{F_1 L^3}{48 E I_z} = \dfrac{5000}{\frac{1}{12}U_1(X_1 - 2U_2)^3 + \frac{1}{6}X_2 U_2^3 + 2X_2 U_2\left(\frac{X_1 - U_2}{2}\right)^2} \\ \text{s.t. } g_1(\mathbf{X}, \mathbf{U}) = 2X_2 U_2 + U_1(X_1 - 2U_2) \le 300 \text{ cm}^2 \\ \quad g_2(\mathbf{X}, \mathbf{U}) = \dfrac{180,000 X_1}{U_1(X_1 - 2U_2)^3 + 2X_2 U_2\left[4U_2^2 + 3X_1(X_1 - 2U_2)\right]} \\ \quad + \dfrac{15,000 X_2}{(X_1 - 2U_2)U_1^3 + 2U_2 X_2^3} \le 8 \text{ kN/cm}^2 \\ \quad 10.0 \text{ cm} \le X_1 \le 80.0 \text{ cm},\ 10.0 \text{ cm} \le X_2 \le 50.0 \text{ cm} \\ \quad U_1 \in \left[U_1^L, U_1^R\right] = [0.9 \text{ cm}, 1.1 \text{ cm}],\ U_2 \in \left[U_2^L, U_2^R\right] = [1.8 \text{ cm}, 2.2 \text{ cm}] \end{cases} \tag{7.15}$$

where the uncertainty levels of $U_1$ and $U_2$ are both 10%.

In the optimization process, the factors $\xi$, $\phi$, $\psi$, $\beta$, and $\sigma$ are specified as 0.0, 0.015, 0.0029, 0.5, and 1000, respectively. The possibility degree levels of the two uncertain constraints are both set to be 0.8. The maximum generation of IP-GA is set to be 100 for both outer and inner layers. Besides, the maximum number of iterations of the whole optimization process is set to be 8. In each iteration, LHD is employed to generate 17 samples (for 4 input variables) to model the approximate interval optimization problem and 7 samples (for 2 input variables) to calculate the actual penalty function value at the single design vector.

1) Iteration process and analysis of the method's convergence performance

Three different initial conditions are taken into consideration. For the first initial condition, the initial design vector is $\mathbf{X}^{(1)} = (30.0, 20.0)^{\mathrm{T}}$ and the initial trust region radius vector is $\mathbf{\Delta}^{(1)} = (17.5, 10.0)^{\mathrm{T}}$. To clearly describe the algorithm procedure, the details of the first three iterations are given as follows:

(1) Iteration 1
The current design vector is $\mathbf{X}^{(1)} = (30.0, 20.0)^{\mathrm{T}}$, and the current trust region radius vector is $\mathbf{\Delta}^{(1)} = (17.5, 10.0)^{\mathrm{T}}$. The current trust region (or current design space) is $12.5 \le X_1 \le 47.5$, $10.0 \le X_2 \le 30.0$. The approximation models of the objective function and constraints are:

$$\tilde{f}(\mathbf{X}, \mathbf{U}) = 1.27 - 0.71X_1 - 0.22X_2 + 0.34X_1X_2 + 0.13X_2U_2 - 0.49X_2^2 - 0.92U_1^2 - 0.87U_2^2$$

$$\tilde{g}_1(\mathbf{X}, \mathbf{U}) = 106.0 + 17.5X_1 + 40.0X_2 + 2.6U_1 + 7.6U_2 + 1.75X_1U_1 + 4.0X_2U_2 - 0.04U_1U_2$$

$$\tilde{g}_2(\mathbf{X}, \mathbf{U}) = 64.0 - 27.2X_1 - 28.7X_2 - 2.9U_2 + 12.7X_1X_2 - 26.1U_1^2 - 25.5U_2^2$$

In the above approximation models, the values of $X_1$, $X_2$, $U_1$, and $U_2$ are all in the interval $[-1.0, 1.0]$, so the actual variables should be mapped into this interval first when using these approximation models.
Solve the approximate interval optimization problem and obtain:

$$\mathbf{X}^{(1)*} = (40.6, 29.9)^{\mathrm{T}},\ \tilde{f}_P\left(\mathbf{X}^{(1)*}\right) = 457.5$$

The reliability index thus can be calculated:

$$\rho^{(1)} = \frac{653.9 - 645.0}{653.9 - 457.5} = 0.045$$

Since $0 < \rho^{(1)} < 0.25$, the trust region radius vector should be shrunk and $\mathbf{X}^{(1)*}$ is kept to the next iteration.

(2) Iteration 2.
The current design vector is $\mathbf{X}^{(2)} = (40.6, 29.9)^{\mathrm{T}}$, and the current trust region radius vector is $\mathbf{\Delta}^{(2)} = (8.75, 5.0)^{\mathrm{T}}$. The current trust region is $31.8 \le X_1 \le 49.3$, $24.9 \le X_2 \le 34.9$. The approximation models of the objective function and the constraints are:

$$\tilde{f}(\mathbf{X}, \mathbf{U}) = 0.11 - 0.05X_1 - 0.02X_2 - 0.0008U_1 - 0.009U_2 + 0.009X_1X_2 + 0.003X_1U_2 + 0.002X_2U_2 + 0.01X_1^2 - 0.006U_1^2 - 0.004U_2^2$$

$$\tilde{g}_1(\mathbf{X}, \mathbf{U}) = 156.1 + 8.75X_1 + 20.0X_2 + 3.66U_1 + 11.6U_2 + 0.88X_1U_1 + 2.0X_2U_2 - 0.04U_1U_2$$

$$\tilde{g}_2(\mathbf{X}, \mathbf{U}) = 16.7 - 3.4X_1 - 3.4X_2 - 0.1U_1 - 1.4U_2 + 0.6X_1X_2 + 0.2X_1U_2 + 0.3X_2U_2 + 0.9X_1^2 + 0.8X_2^2 + 0.2U_1^2$$

Solve the approximate interval optimization problem and obtain:

$$\mathbf{X}^{(2)*} = (49.3, 34.9)^{\mathrm{T}},\ \tilde{f}_P\left(\mathbf{X}^{(2)*}\right) = 644.0$$

The reliability index thus can be calculated:

$$\rho^{(2)} = \frac{645.0 - 642.8}{645.0 - 644.0} = 2.2$$

Since $\rho^{(2)} > 1.0$ and $\mathbf{X}^{(2)*}$ is located on the boundary of the current trust region, the trust region radius vector should be amplified and $\mathbf{X}^{(2)*}$ is kept to the next iteration.

(3) Iteration 3

The current design vector is $\mathbf{X}^{(3)} = (49.3, 34.9)^{\mathrm{T}}$, and the current trust region radius vector is $\mathbf{\Delta}^{(3)} = (17.5, 10.0)^{\mathrm{T}}$. The current trust region is $31.8 \le X_1 \le 66.8$, $24.9 \le X_2 \le 44.9$. The approximation models of the objective function and constraints are:

$$\begin{aligned}\tilde{f}(\mathbf{X}, \mathbf{U}) = {} & 0.07 - 0.05X_1 - 0.02X_2 - 0.005U_2 + 0.02X_1X_2 + 0.002X_1U_2 \\ & + 0.003X_2U_2 + 0.02X_1^2 - 0.01U_1^2 - 0.01U_2^2\end{aligned}$$

$$\begin{aligned}\tilde{g}_1(\mathbf{X}, \mathbf{U}) = {} & 184.8 + 17.5X_1 + 40.0X_2 + 4.53U_1 + 13.6U_2 + 1.75X_1U_1 \\ & + 4.0X_2U_2 - 0.04U_1U_2\end{aligned}$$

$$\begin{aligned}\tilde{g}_2(\mathbf{X}, \mathbf{U}) = {} & 11.5 - 4.1X_1 - 4.4X_2 - 1.0U_2 + 1.2X_1X_2 + 0.3X_1U_2 \\ & + 0.3X_2U_2 + 1.9X_1^2 + 1.7X_2^2\end{aligned}$$

Solve the approximate interval optimization problem and obtain:

$$\mathbf{X}^{(3)*} = (65.8, 41.1)^{\mathrm{T}},\ \tilde{f}_P\left(\mathbf{X}^{(3)*}\right) = 6.5$$

The reliability index thus can be calculated:

$$\rho^{(3)} = \frac{642.8 - 1.32}{642.8 - 6.5} = 1.0$$

Since $\rho^{(3)} \ge 1.0$ and $\mathbf{X}^{(3)*}$ is located in the trust region, the trust region radius vector and $\mathbf{X}^{(3)*}$ should be kept to the next iteration.

After another 5 iterations similar to the above process, the calculation results of the whole optimization are obtained and listed in Table 7.1. Besides, the results obtained by the two-layer nested optimization method given in Chap. 3 (the maximum generation of IP-GA of both layers is set to be 100) that directly calls the actual objective function and constraints are also listed in Table 7.1. Since in general, it is very difficult to obtain the exact solution of a nonlinear interval optimization problem, the results from this two-layer nested optimization method will be used as "reference solution" or "exact solution" to measure the accuracy of the method proposed in this chapter. From Table 7.1, it can be found that the design vector $\mathbf{X}^{(s)}$ gradually approaches to the reference solution with the iterations and finally converges to the reference solution $(80.0\ 50.0)^{\mathrm{T}}$ in the 6-th iteration. The middle point and radius of the uncertain objective function are 0.015 and 0.0014, respectively. The possibility

**Table 7.1** Optimization results with the first initial condition [17]

| Iterative step $s$ | $\mathbf{X}^{(s)}$(cm) | $f_P(\mathbf{X}^{(s)})$ | Midpoint of objective function (cm) | Radius of objective function (cm) | Possibility degrees of constraints |
|---|---|---|---|---|---|
| 1 | (30.0, 20.0) | 653.9 | 0.29 | 0.024 | 1.00, 0.00 |
| 2 | (40.6, 29.9) | 645.0 | 0.10 | 0.0091 | 1.00, 0.00 |
| 3 | (49.3, 34.9) | 642.8 | 0.059 | 0.0053 | 1.00, 0.00 |
| 4 | (65.8, 41.1) | 1.32 | 0.027 | 0.0025 | 1.00, 0.80 |
| 5 | (69.5, 49.8) | 0.74 | 0.015 | 0.0014 | 0.97, 1.00 |
| 6 | (80.0, 50.0) | 0.73 | 0.015 | 0.0014 | 0.94, 1.00 |
| 7 | (80.0, 50.0) | 0.73 | 0.015 | 0.0014 | 0.94, 1.00 |
| 8 | (80.0, 50.0) | 0.73 | 0.015 | 0.0014 | 0.94, 1.00 |

The obtained solution of design vector after 8 iterations is (80.0, 50.0); the reference solution is (80.0, 50.0)
The deviations between the calculated result and the reference solution are (0.0%, 0.0%)

degrees of two constraints are 0.94 and 1.00, respectively, and the penalty function value is 0.73. In the first several iterations, the penalty function value at the current design vector is far larger than that in the last several iterations. It is because the design vector in the trust region cannot satisfy the possibility degree levels of constraints and hence gets penalized in the first several iterations, while a design vector satisfying the possibility degree levels of constraints is obtained via the update of trust region in the last several iterations, thus the penalty is avoided. The reason may also be that the approximation models are too rough in the early stage of iterations so that the obtained design vector violates the possibility-degree-level requirements. Therefore, the design vector gets a penalty.

Then the other two initial conditions are considered. In the second initial condition, $\mathbf{X}^{(1)} = (45.0, 30.0)^{\mathrm{T}}$, $\boldsymbol{\Delta}^{(1)} = (35.0, 20.0)^{\mathrm{T}}$, and in the third initial condition, $\mathbf{X}^{(1)} = (60.0, 40.0)^{\mathrm{T}}$, $\boldsymbol{\Delta}^{(1)} = (4.4, 2.5)^{\mathrm{T}}$. Tables 7.2 and 7.3 list the optimization results with these two initial conditions. From the tables, it can be found that the design vector $\mathbf{X}^{(s)}$ still approaches to the reference solution with the iterations and finally reaches it both in the 6-th iteration. Besides, the convergence curves corresponding to the three initial conditions are given in Fig. 7.4. It shows that the proposed method has a relatively high convergence rate. In the 6-th iteration, a stable solution is obtained for all the three initial conditions. Additionally, the convergence rates in the first several iterations for the last two initial conditions are higher than that in the first initial condition because the initial design vector in the first initial condition is father from the real optimum than that in the other two initial conditions.

Actually, the above three initial conditions are typical. Their initial design vectors are located on the left, the center, and the right of the design space, respectively, and their initial trust region radiuses are 1/4, 1, and 1/64 of the initial design space radius, respectively. For all the three typical initial conditions, the proposed method can obtain the same optimum as the reference solution and has relatively

**Table 7.2** Optimization results with the second initial condition [17]

| Iterative step $s$ | $\mathbf{X}^{(s)}$(cm) | $f_P(\mathbf{X}^{(s)})$ | Midpoint of objective function (cm) | Radius of objective function (cm) | Possibility degrees of constraints |
|---|---|---|---|---|---|
| 1 | (45.0, 30.0) | 644.0 | 0.082 | 0.0073 | 1.00, 0.00 |
| 2 | (63.1, 48.8) | 1.23 | 0.025 | 0.0023 | 1.00, 1.00 |
| 3 | (79.9, 44.8) | 0.80 | 0.016 | 0.0015 | 1.00, 1.00 |
| 4 | (79.9, 44.8) | 0.80 | 0.016 | 0.0015 | 1.00, 1.00 |
| 5 | (80.0, 49.8) | 0.73 | 0.015 | 0.0014 | 0.96, 1.00 |
| 6 | (80.0, 50.0) | 0.73 | 0.015 | 0.0014 | 0.94, 1.00 |
| 7 | (80.0, 50.0) | 0.73 | 0.015 | 0.0014 | 0.94, 1.00 |
| 8 | (80.0, 50.0) | 0.73 | 0.015 | 0.0014 | 0.94, 1.00 |

The obtained solution of design vector after 8 iterations is (80.0, 50.0); the reference solution is (80.0, 50.0)
The deviations between the calculated result and the reference solution are (0.0%, 0.0%)

**Table 7.3** Optimization results with the third initial condition [17]

| Iterative step | $\mathbf{X}^{(s)}$(cm) | $f_P(\mathbf{X}^{(s)})$ | Midpoint of objective function (cm) | Radius of objective function (cm) | Possibility degrees of constraints |
|---|---|---|---|---|---|
| 1 | (60.0, 40.0) | 345.1 | 0.034 | 0.0031 | 1.00, 0.21 |
| 2 | (64.4, 42.5) | 1.34 | 0.027 | 0.0025 | 1.00, 0.93 |
| 3 | (63.7, 44.1) | 1.33 | 0.027 | 0.0025 | 1.00, 1.00 |
| 4 | (72.5, 49.1) | 0.91 | 0.019 | 0.0017 | 1.00, 1.00 |
| 5 | (78.4, 49.8) | 0.76 | 0.016 | 0.0015 | 0.99, 1.00 |
| 6 | (80.0, 50.0) | 0.73 | 0.015 | 0.0014 | 0.94, 1.00 |
| 7 | (80.0, 50.0) | 0.73 | 0.015 | 0.0014 | 0.94, 1.00 |
| 8 | (80.0, 50.0) | 0.73 | 0.015 | 0.0014 | 0.94, 1.00 |

The obtained solution of design vector after 8 iterations is (80.0, 50.0); the reference solution is (80.0, 50.0)
The deviations between the calculated result and the reference solution are (0.0%, 0.0%)

high convergence rate, which indicates that the method has stable convergence performance.

2) Analysis to the optimization efficiency

In this test function, the objective function is assumed to be a computationally expensive numerical analysis model, and the number of evaluations of the objective function will be used to quantitatively assess the optimization efficiency. When using the actual function-based nested optimization method to obtain the reference solution, computing the interval of the objective function needs two times of inner IP-GA and the number of calculations will be $100 \times 5 \times 2 = 1000$. Therefore, the whole

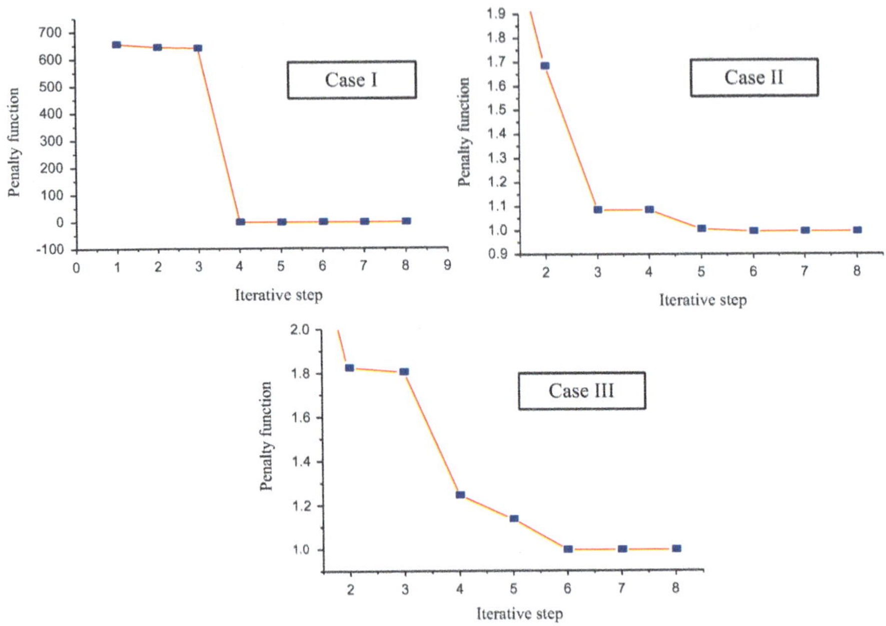


**Fig. 7.4** Convergence curves of the three cases with different initial conditions [17]

optimization needs $1000 \times 500 = 5 \times 10^5$ evaluations of the objective function. On the other hand, by the proposed method, only 17 calculations are needed to construct the approximation model of the objective function and 7 times of calculations are needed to compute the actual penalty function value at $\mathbf{X}^{(s)*}$ in each iteration (except the first one). Therefore, only 24 times of calculations are needed in one iteration. An exception is the first iteration where 31 times of calculations are required because the 7 times of calculations are used to evaluate the actual penalty function at the initial design vector. Totally, 199 times of evaluations of the objective function are needed for the proposed method, which indicates that the method is far more efficient than the original two-layer nested optimization method. On the other hand, in the two-layer nested optimization we can of courese use a gradient-based conventional optimization method as the solver of the inner optimization to improve the optimization efficiency. However, its computation cost is still far higher than the proposed method and what's worse, it takes the risk of violating the actual constraints due to the local optimum problem existing in the inner optimization.

3) Effects of the uncertainty levels

Other four uncertainty levels of the variables, 20%, 30%, 40%, and 50% are studied. The initial condition is $\mathbf{X}^{(1)} = (30.0, 20.0)^T$, $\mathbf{\Delta}^{(1)} = (17.5, 10.0)^T$, the maximum generation is 13, and other parameters are kept unchanged. The optimization results and the corresponding reference solutions for the 4 cases are listed in Table 7.4 and

**Table 7.4** Optimization results under four different uncertainty levels of the variable [17]

| Uncertainty level | Optimization result of **X** (cm) | Penalty function $f_P$ | Midpoint of objective function (cm) | Radius of objective function (cm) | Reference solution of **X** (cm) | Deviations from the reference solution |
|---|---|---|---|---|---|---|
| 20% | (79.0, 48.3) | 1.02 | 0.016 | 0.0029 | (80.0, 47.5) | 1.3%, 1.7% |
| 30% | (76.8, 46.5) | 1.67 | 0.019 | 0.0052 | (79.3, 45.0) | 3.3%, 3.3% |
| 40% | (74.9, 45.8) | 12.4 | 0.021 | 0.0074 | (80.0, 45.4) | 6.4%, 0.9% |
| 50% | (56.6, 27.7) | 646.9 | 0.067 | 0.027 | (80.0, 49.3) | 29.3%, 43.8% |

the convergence curves of the optimization process are shown in Fig. 7.5. When the uncertainty level is 20%, the optimization result is relatively accurate since the deviation between the obtained result and the reference solution for the design variables is only 1.3% and 1.7%, respectively. With the increase of the uncertainty level, the deviation of the optimization result gradually increases, peaking at 29.3% and 43.8% when the uncertainty level is 50%. Shown in Fig. 7.6 is the relationship between the maximum deviation of optimization result and the uncertainty level of variables. It

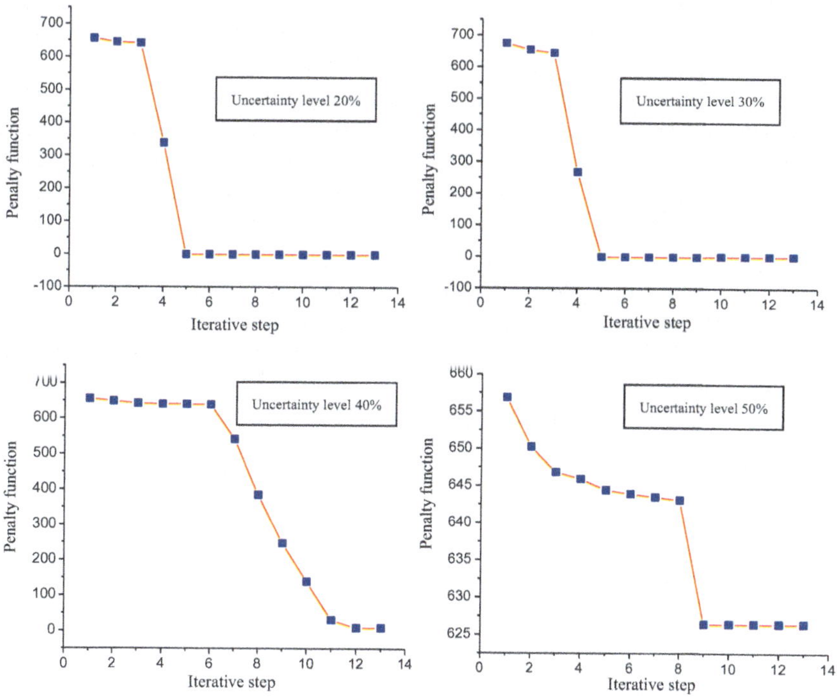


**Fig. 7.5** Convergence curves of four cases with different uncertainty levels of the variables [17]

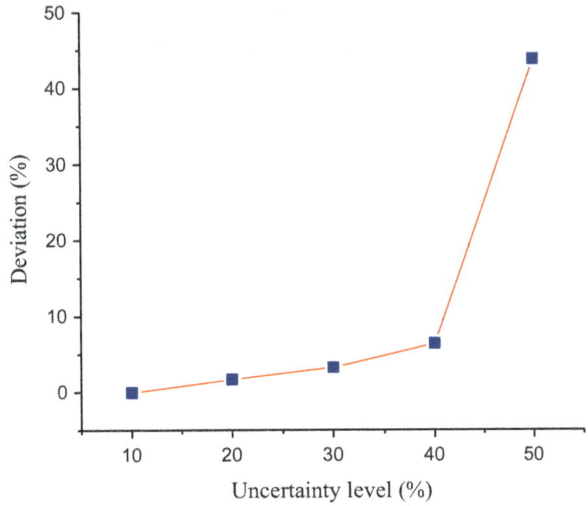


**Fig. 7.6** Relationship between the uncertainty level of variables and the maximum deviation of optimization result [17]

is obvious that when the uncertainty level is relatively small (less than 40%), the deviation of the optimization result is relatively small and increases slowly, while it increases dramatically once the uncertainty level is relatively large.

Above analysis indicates that the proposed method has a relatively high accuracy only when the uncertainty levels of the variables are relatively small. There are mainly two reasons. Firstly, in the approximate interval optimization problem, the approximation models are constructed in the hybrid space composed of the current design space and the uncertainty domain which is not updated in the iterations. Therefore, a relatively large error still exists in the approximation models for problems with large uncertainties even though the design space is shrunk to a tiny size, thus affecting the optimization accuracy. Secondly, when calculating the actual penalty function value in each iteration, the approximation model also needs to be constructed on the uncertainty domain. For problems with relatively large uncertainties of the variables, the accuracy of the approximation models could be relatively bad. Hence, the calculation of the reliability index will be inaccurate, further affecting the update of the design space and leading to a low optimization accuracy. In conclusion, a relatively small uncertainty level is vital to the proposed method in this chapter.

2. Test function 2

The following test function is investigated [19]:

$$\begin{cases} \max_{\mathbf{X}} f(\mathbf{X}, \mathbf{U}) = U_1(X_1 + 2)^2 + U_2^3(X_2 + 1) + X_3^2 \\ \text{s.t. } g(\mathbf{X}, \mathbf{U}) = U_1^2(X_1 + X_3) + U_2(X_2 - 4)^2 \leq b^I \\ \quad 2 \leq X_1 \leq 14,\ 2 \leq X_2 \leq 14,\ 2 \leq X_3 \leq 14 \\ \quad U_1 \in \left[U_1^L, U_1^R\right] = [0.9, 1.1],\ U_2 \in \left[U_2^L, U_2^R\right] = [0.9, 1.1] \end{cases} \tag{7.16}$$

where the uncertainty levels of $U_1$ and $U_2$ are both 10%.

The performance interval of the objective function is $V^I = [45, 68]$. According to the transformation model based on possibility degree of interval number, the possibility of the objective function, $\max_{\mathbf{X}} P\left(f^I(\mathbf{X}) \geq V^I\right)$, should be maximized. The possibility degree level $\lambda$ of the constraints and the penalty factor $\sigma$ are set to be 0.8 and 1000, respectively. When solving the approximate interval optimization, the maximum generation number is set to be 200 and 100 for the outer layer and the inner layer of IP-GA, respectively. In each iteration, 30 samples for 5 input variables are generated by LHD to construct the approximation interval optimization problem and 8 samples for 2 input variables are generated to compute the actual penalty function value at the design vector. In the following text, three cases are considered to analyze this test function.

(1) Case I

In this case, the maximum allowable interval of constraint is $b^I = [8, 9]$, the initial design vector $\mathbf{X}^{(1)}$, and the initial trust region radius vector are set to be (8.00, 8.00, 8.00) and (6.00, 6.00, 6.00), respectively. The optimization results obtained by the interval optimization method proposed in Sect. 7.1.4 are listed in Table 7.5. Additionally, the reference solution obtained by the actual function-based nested optimization method is also given in the table. From the table, it can be found that the penalty function value increases gradually as the iteration proceeds. After 10 iterations, an optimum design vector $\mathbf{X}^{(10)} = (5.44, 4.08, 2.07)$ is finally obtained. The convergence curve is shown in Fig. 7.7, from which it can be found that the proposed method has a high

**Table 7.5** Optimization results of case I [19]

| Iterative step | $\mathbf{X}^{(s)}$ | $f_P(\mathbf{X}^{(s)})$ |
|---|---|---|
| 1 | (8.00, 8.00, 8.00) | −639.00 |
| 2 | (3.49, 3.88, 2.36) | 0.00 |
| 3 | (3.87, 4.21, 3.29) | 0.24 |
| 4 | (4.79, 3.80, 2.39) | 0.51 |
| 5 | (4.79, 3.80, 2.39) | 0.51 |
| 6 | (4.89, 4.55, 2.24) | 0.58 |
| 7 | (4.94, 4.39, 2.20) | 0.59 |
| 8 | (5.44, 4.08, 2.05) | 0.84 |
| 9 | (5.44, 4.08, 2.05) | 0.84 |
| 10 | (5.44, 4.08, 2.07) | 0.84 |

The obtained solution of design vector after 10 iterations is (5.44, 4.08, 2.07)
The possibility degree of the uncertain objective function is 0.84
The possibility degree of the uncertain constraint is 0.8
The reference solution is (5.45, 4.06, 2.05)
The deviations between the calculated result and the reference solution is (0.2%, 0.5%, 1.0%)

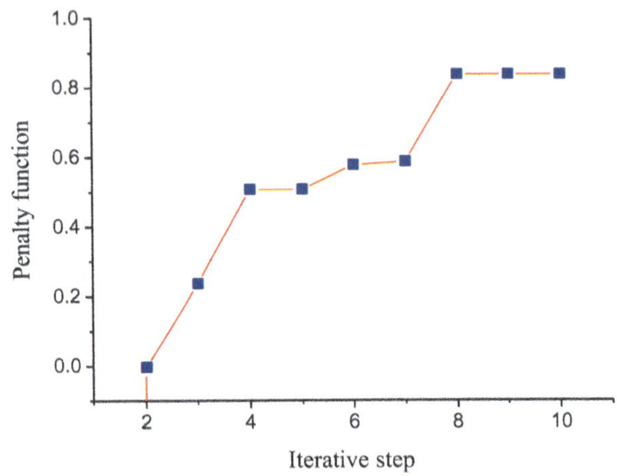


**Fig. 7.7** Convergence curve of Case I [19]

convergence rate and the design vector converges to a stable solution in the 8-th iteration. In addition, the deviations of the optimization results from the reference solution are 0.2%, 0.5%, and 1.0%, respectively, which indicates the relatively high computation accuracy of the proposed method.

In this test function, the objective function is also assumed to be a computationally expensive numerical analysis model, whose evaluation number thus can quantify the optimization efficiency. With the actual function-based nested optimization method, $1.0 \times 10^6$ evaluations are required, while only 388 evaluations (46 in the first iteration, 38 in every other iteration) are needed with the proposed method.

(2) Case II

In this case, $b^I = [8, 11]$ and other initial conditions are kept the same. The optimization results are listed in Table 7.6. The table shows the possibility degree of the constraint corresponding to the optimum design vector is 0.81 which can meet the requirement, and the possibility degree of the objective function reaches 1.0. The convergence curve is shown in Fig. 7.8, which shows the penalty function reaches its maximum value in the 9-th iteration.

Since the possibility degree of the objective function reaches 1.0 at the optimum, a robust optimization thus can be constructed to further optimize the design. The optimization results are listed in Table 7.7. The possibility degrees of the

**Table 7.6** Optimization results of case II (maximizing the possibility degree of the objective function) [19]

| Optimum of **X** | Radius of objective function | Penalty function $f_P$ | Possibility degree of objective function | Possibility degree of constraint |
|---|---|---|---|---|
| (6.20, 4.04, 2.00) | 8.24 | 1.0 | 1.0 | 0.81 |

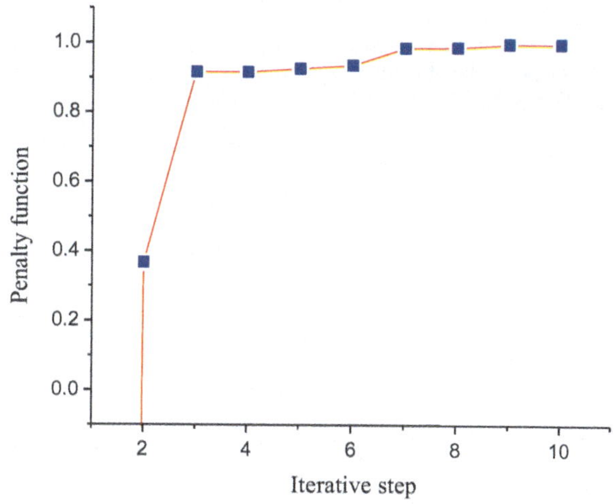


**Fig. 7.8** Convergence curve of case II (maximizing the possibility degree of the objective function) [19]

**Table 7.7** Optimization results of case II (considering the robustness of the objective function) [19]

| Optimum of **X** | Penalty function $f_P$ | Radius of objective function | Possibility degree of objective function | Possibility degree of constraint |
|---|---|---|---|---|
| (5.89, 3.92, 2.35) | 7.84 | 7.84 | 1.0 | 0.80 |

objective function and constraint at the optimum design vector are 1.0 and 0.80, respectively, which both meet the requirements. In addition, the radius of the uncertain objective function is 7.84, which is smaller than 8.24 in Table 7.6, indicating that the robustness of the design has been improved after the robustness optimization. Shown in Fig. 7.9 is the convergence curve.
In this case, the whole optimization process consists of two successive optimizations with respect to the objective function's possibility degree and the objective function's robustness respectively, both of which needs 388 evaluations of the objective function. Therefore, 766 evaluations are required for the whole optimization.

(3) Case III

In this case, $b^I = [8, 9]$, the uncertainty level of the two uncertain variables is increased to 30%, which means $U_1 \in [0.7, 1.3]$, $U_2 \in [0.7, 1.3]$, and the initial condition is kept unchanged. Optimization results obtained via the proposed method and the actual function-based nested optimization method are both listed in Table 7.8. As seen in the table, the deviations between the obtained optimum solution and the reference solution are 2.9%, 1.6%, and 8.5%, respectively, which are obviously larger than the deviations in Case I. The only difference between Case I and Case III is the uncertainty level of the variables,

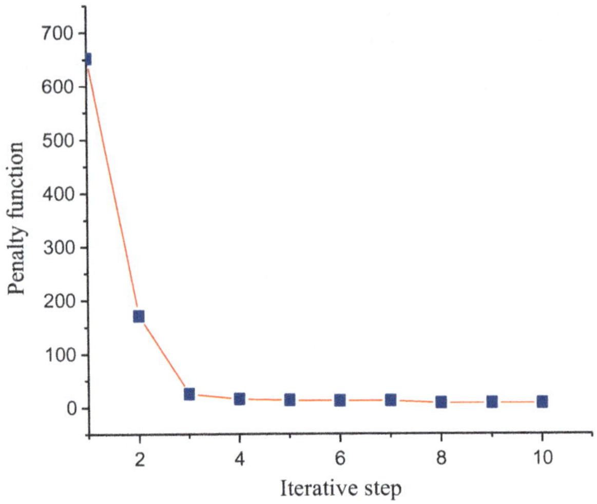


**Fig. 7.9** Convergence curve of Case II (considering the robustness of the objective function) [19]

**Table 7.8** Optimization results of case III [19]

| Optimum of **X** | Penalty function $f_p$ | Possibility degree of objective function | Possibility degree of constraint | Reference solution of **X** | Deviations from the reference solution |
|---|---|---|---|---|---|
| (3.57, 4.19, 2.14) | 0.10 | 0.10 | 0.83 | (3.47, 4.26, 2.34) | 2.9%, 1.6%, 8.5% |

which demonstrates again the influence of uncertainty level on the optimization results. The uncertainty level should be kept in a relatively small range to obtain a reliable optimization result by the proposed method. Shown in Fig. 7.10 is the convergence curve. We can see that the penalty function becomes stable and the design vector is no longer updated to a better one only after the 2nd iteration. The decrease of convergence performance is just caused by the larger uncertainty level of the variables.

### 7.1.6 *Discussions on the Convergence*

Above analyses show that the proposed method has a relatively high convergence rate in the early stage, while the current design vector will generally stay at a fixed point after a certain number of iterations. It is mainly because the accuracy of the approximation models can be improved continuously through the update of the design space in the early iterations. Therefore, a better design vector can be found. However, after a certain number of iterations, it becomes more and more difficult to improve the accuracy of the approximation models since only the design space is updated while

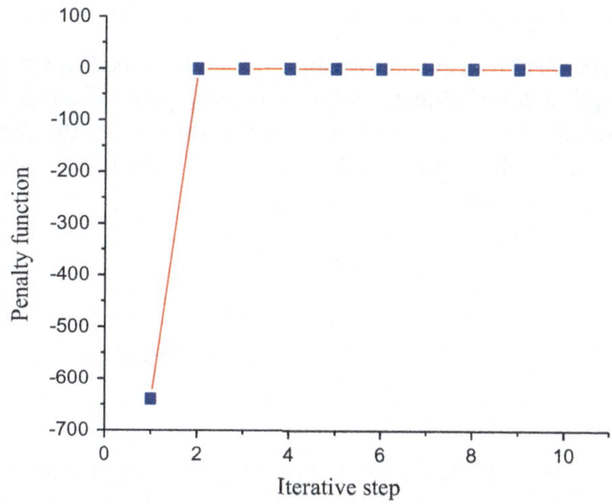


**Fig. 7.10** Convergence curve of case III [19]

the uncertainty domain remains unchanged. The approximation error caused by the unchanged uncertainty domain starts to dominate and cannot be eliminated by only updating the design space. Generally, the design vector will keep stable when the design space becomes small enough and thus further optimization will be difficult. From this perspective, the algorithm is similar to the nonlinear interval optimization algorithm proposed in Chap. 6.

In addition, for the two test functions with the same optimization results can be obtained despite the different initial conditions, which demonstrates that the proposed method has robust convergence when the uncertainties of the variables are relatively small. However, it does not mean that the proposed method has the global optimization performance. Conversely, it is a local optimization method. If the transformed deterministic optimization problem has multiple local optima, the proposed method may obtain a different optimum solution with different initial conditions. In addition, in the proposed method, the initial design vector is not required to be a feasible solution of the transformed deterministic optimization problem.

### *7.1.7 Engineering Applications*

In this section, two engineering application problems are investigated. In the first application, the interval optimization method based on the order relation transformation model described in Sect. 7.1.3 is employed, while the method based on the possibility degree transformation model in Sect. 7.1.4 is adopted for the second application.

1. Application to the design of an automobile frame

The design of the automobile frame described in Chap. 5 is investigated here. The optimization objectives are still the layout of cross beams to obtain a maximum stiffness in $y$-direction. The only difference is the maximum allowable equivalent stress of the frame which is set as 200 MPa here. An interval optimization problem can be modeled as follows [17]:

$$\begin{cases} \min\limits_{\mathbf{l}} d_{\max}(\mathbf{l}, E, \nu) \\ \text{s.t. } \sigma_{\max}(\mathbf{l}, E, \nu) \le 200 \text{ MPa} \\ \quad 200 \text{ mm} \le l_i \le 800 \text{ mm}, \ i = 1, 2, 3, 4 \\ \quad E \in \left[E^L, E^R\right], \ \nu \in \left[\nu^L, \nu^R\right] \end{cases} \tag{7.17}$$

The FEM is used to compute objective function and constraint. The four-node shell element combining 2-D solid element and plate element [20] is adopted for meshing and the number of elements is 1563. The factors $\xi$, $\phi$ and $\psi$ are set to be 0.0, 1.16, and 0.13, respectively. The weighting factor $\beta$, the penalty factor $\sigma$, and the possibility degree level of the uncertain constraint are given as 0.5, 1000, and 0.8, respectively. When solving the approximate interval optimization in each iteration the maximum generation numbers of both inner and outer layers of IP-GA are set to be 200, and the maximum iteration number is set to be 10 for the whole optimization process. In each iteration, 30 samples are used to construct the approximation optimization problem, and 7 samples are used to calculate the actual penalty function.

Two different uncertainty levels are investigated for the uncertain parameters. The first one is 10%, and the corresponding Young's modulus and the Poisson's ratio of the frame structure are $E \in \left[1.8 \times 10^5 \text{ MPa}, 2.2 \times 10^5 \text{ MPa}\right]$ and $\nu \in [0.27, 0.33]$, respectively. The second one is 20%, and the corresponding Young's modulus and the Poisson's ratio are $E \in \left[1.6 \times 10^5 \text{ MPa}, 2.4 \times 10^5 \text{ MPa}\right]$ and $\nu \in [0.24, 0.36]$, respectively. In both cases, the same initial condition, $\mathbf{X}^{(1)} = (500, 500, 500, 500)^{\mathrm{T}}$ and $\boldsymbol{\Delta}^{(1)} = (300, 300, 300, 300)^{\mathrm{T}}$ are given. The optimization results are listed in Table 7.9. As shown in the table, under these two cases the variation intervals of the maximum nodal displacement of the frame in $y$-direction caused by the uncertain

**Table 7.9** Optimization results of two cases with different uncertainty levels of the material properties [17]

| Uncertainty level | Optimum of $\mathbf{l}$ (mm) | Penalty function $f_p$ | Interval of objective function (mm) | Interval of constraint (MPa) | Possibility degree of constraint | Number of FEM evaluations |
|---|---|---|---|---|---|---|
| 10% | (486, 455, 424, 373) | 1.38 | [1.43,1.77] | [192,200] | 1.0 | 377 |
| 20% | (534, 663, 715, 373) | 1.69 | [1.04,1.60] | [181,185] | 1.0 | 377 |

material properties are [1.43 mm, 1.77 mm] and [1.04 mm, 1.60 mm], respectively. Besides, the possibility degree of the constraint at the optimum design vector is 1.0 in both cases. In other words, the maximum equivalent stress would never exceed the allowable value under uncertainty level of 10% or 20% for the material properties. Shown in Fig. 7.11 are the convergence curves for these two cases. From the figure, it can be found that even though the FEM model instead of the explicit function is employed, the proposed method shows similar convergence performance to that in test function examples. In early iterations, the convergence rate is relatively high, while the results converge to a stable value in later iterations. Besides, in the whole optimization process, the numbers of FEM evaluations for two cases are both 377.

2. Application to the crashworthiness design of a thin-walled beam of the vehicle body

Nowadays, the requirements on vehicle safety are becoming higher and higher and thus the vehicle safety design has become a significant research field. According to the basic requirements on crashworthiness of vehicle impact safety design, there are several indexes to assess the structural crashworthiness in practical engineering, such as energy absorption, average impact force, maximum impact force, etc. The vehicle body structure outside the driving cab should be designed to deform as greatly as possible in one impact to absorb the energy. Besides, the impact force should be decreased to reduce acceleration. Thin-walled beams connected by spot welding are major structures of a vehicle body serving for load support and energy absorption. Optimization on the thin-walled beam structures can improve their crashworthiness performance and it is very significant to vehicle safety. As shown in Fig. 7.12, the closed-hat beam, a typical thin-walled beam, is constructed by a hat beam and a web plate, connected by spot welding points along the two rims of the hat beam. The closed-hat beam impacts the rigid wall with an initial velocity of 10 m/s. Based on the work in reference [21], here the closed-hat beam will be optimized to maximize the energy absorption subjected to a constraint of axial impact force (average normal impact force on the rigid wall). Reference [22] indicated that the plate thickness $t$,

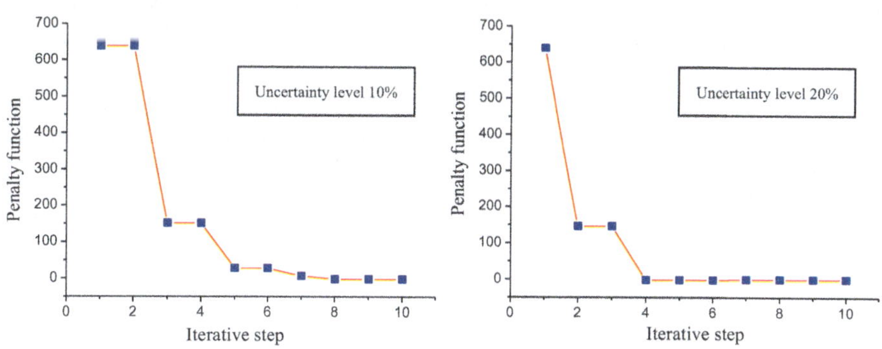


**Fig. 7.11** Convergence curves of two cases with different uncertainty levels of the material properties [17]

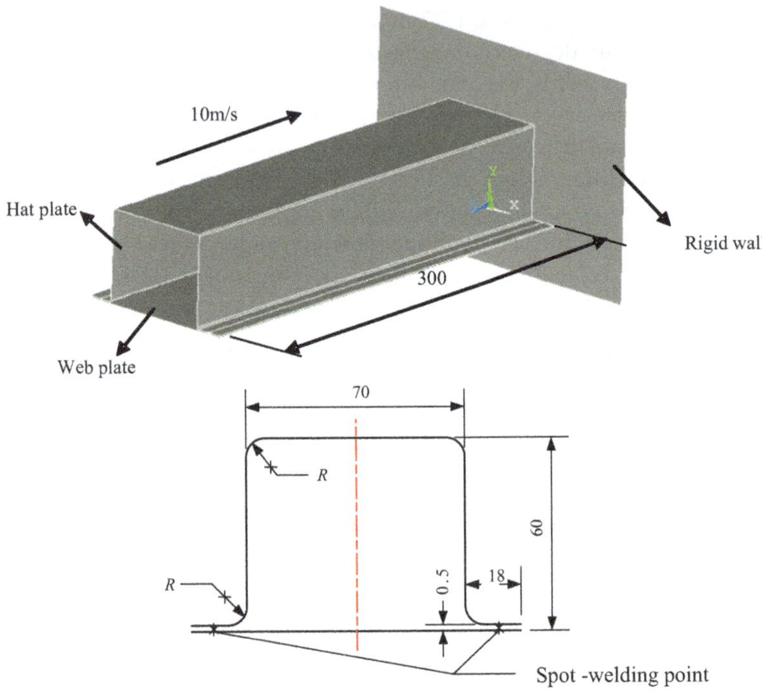


**Fig. 7.12** A closed-hat beam impacting the rigid wall and its cross-sectional dimensions (mm) [19]

**Table 7.10** Material properties of the closed-hat beam [19]

| Young's modulus $E$ (MPa) | Poisson's ratio $\nu$ | Density $\rho$ (kg/mm$^3$) | Yield stress $\sigma_s$ (MPa) | Tangent modulus $E_t$ (MPa) |
|---|---|---|---|---|
| $2.0 \times 10^5$ | 0.27 | $7.85 \times 10^{-6}$ | 310 | 763 |

round radius $R$ of the hat beam and distance $d$ between two neighboring spot welding points have dominant effects on the crashworthiness performance of the closed-hat beam, and hence these three parameters are selected as design variables in our study. The material is the normal low-carbon steel, whose parameters are listed in Table 7.10. Due to the measuring and manufacturing errors, the yield stress $\sigma_s$ and the tangent modulus $E_t$ are uncertain parameters whose uncertainty levels are both 5%. Therefore, an interval optimization problem can be created as follows [19]:

$$
\begin{cases}
\max\limits_{t,R,d} f_e(t, R, d, \sigma_s, E_t) \\
\text{s.t. } g_f(t, R, d, \sigma_s, E_t) \leq [65\text{ kN}, 70\text{ kN}] \\
\quad 0.5\text{ mm} \leq t \leq 2.5\text{ mm} \\
\quad 1\text{ mm} \leq R \leq 8\text{ mm} \\
\quad 10\text{ mm} \leq d \leq 60\text{ mm} \\
\quad \sigma_s \in [294.5\text{ MPa}, 325.5\text{ MPa}] \\
\quad E_t \in [724.85\text{ MPa}, 801.15\text{ MPa}]
\end{cases}
\tag{7.18}
$$

where the objective function $f_e$ and constraint $g_f$ represent the absorbed energy (internal energy) of the closed-hat beam and the axial impact force, respectively, and they are both obtained through FEM.

The Belytschko–Tsay shell element [23] is used for meshing in the FEM model and the number of elements is 4200. An elasto-plasticity material model of bilinear kinematic hardening is used. A 250 kg of concentrated mass is added to the tail of the closed-hat beam in order to supply enough impacting energy. The duration of the impacting process is 20 ms. The FEM simulation is carried out on the commercial software ANSYS/LS-DYNA. The FEM mesh and a possible deformation of the impacting system are shown in Fig. 7.13.

The possibility degree level $\lambda$ of the constraint $g_f$ and the penalty factor $\sigma$ are set to be 0.8 and 1000, respectively. For the approximation optimization solution, the maximum generation numbers of the inner and outer IP-GA are 100 and 200, respectively. The maximum iteration number of the whole interval optimization process is set to be 10. The original design vector is $\mathbf{X}^{(1)} = (1.5, 4.5, 35.0)^{\mathrm{T}}$, and the original trust region radius vector is $\boldsymbol{\Delta}^{(1)} = (1.0, 3.5, 25.0)^{\mathrm{T}}$. In each iteration, 30 samples are selected to create the approximation models for the uncertain objective function and constraint, and 8 samples are selected for calculating the actual penalty function.

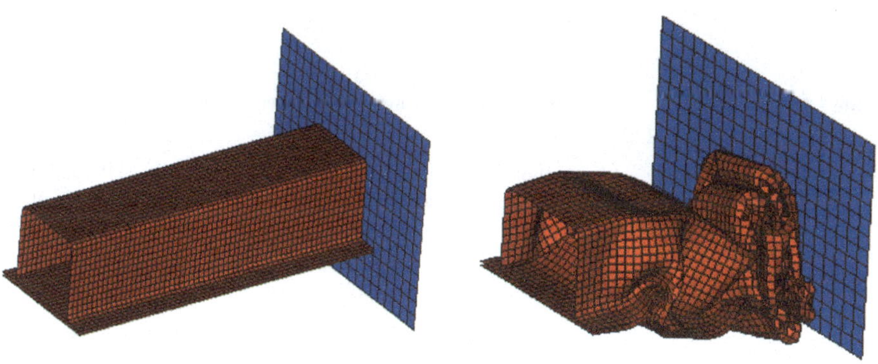

**Fig. 7.13** The FEM mesh and a possible deformation of the closed-hat beam impacting a rigid wall [19]

Firstly, the performance interval of the objective function is set to be $V^I$ = [8kJ, 10kJ]. Therefore, the possibility degree that the interval of the objective function is larger than [8 kJ, 10 kJ] should be maximized. The optimization results are listed in Table 7.11. As shown in the table, the interval of energy absorption for the optimum solution is [9.15 kJ, 9.8 kJ], the corresponding possibility degree of the uncertain objective function is 0.74, and the interval of the average impact force is [54.6 kN, 70.1 kN] with a possibility degree of 0.83, which meets the demand of the possibility degree level. Besides, in the whole optimization process, the number of FEM evaluations is 388.

Secondly, the performance interval $V^I$ is set to be [7 kJ, 8 kJ] which is lower than the first case. A new interval optimization based on this $V^I$ is performed to maximize the possibility degree of the uncertain objective function. After three iterations, the optimum solution $(1.7\text{ mm}, 3.6\text{ mm}, 28.75\text{ mm})^{\mathrm{T}}$ is obtained, and the corresponding possibility degrees of the objective function and constraint are both 1.0. Since the possibility degree of the objective function is 1.0, a robust optimization can be further taken into consideration and its optimization results are listed in Table 7.12. From the table, it can be found that the intervals of the absorbed energy and axial impact force are [8.46 kJ, 8.80 kJ] and [57.1 kN, 61.67 kN], respectively. The corresponding possibility degrees are both 1.0. The radius of the energy absorption interval is only 0.17 kJ, which reflects a good robustness of the optimum design with respect to the uncertain material properties. Besides, in the whole optimization process, the number of FEM evaluations is 510.

**Table 7.11** Optimization results of the closed-hat beam for $V^I$ = [8 kJ, 10 kJ] [19]

| Optimum of $(t, R, d)$ (mm) | Penalty function $f_p$ | Interval of objective function (kJ) | Interval of constraint (kN) | Possibility degree of objective function | Possibility degree of constraint |
|---|---|---|---|---|---|
| (2.10, 2.45, 35.41) | 0.74 | [9.15, 9.8] | [54.6, 70.1] | 0.74 | 0.83 |

**Table 7.12** Optimization results of the closed-hat beam considering the robustness of the objective function for $V^I$ = [7 kJ, 8 kJ] [19]

| Optimum of $(t, R, d)$ (mm) | Penalty function $f_p$ | Interval of objective function (kJ) | Radius of objective function (kJ) | Interval of constraint (kN) | Possibility degree of objective function | Possibility degree of constraint |
|---|---|---|---|---|---|---|
| (2.00, 4.50, 33.20) | 0.17 | [8.46, 8.80] | 0.17 | [57.1, 61.67] | 1.00 | 1.00 |

## 7.2 Nonlinear Interval Optimization Based on the Local-Densifying Approximation Technique

In the above nonlinear interval optimization method based on the approximation model management strategy, the approximation space is updated by the trust region method in optimization process to make sure that the optimization result is approaching the optimal solution. However, new samples have to be reselected and recalculated in each iteration, while the samples in the last iteration are not kept, which increases the computational cost. On the other hand, current approximation models generally adopt the orthogonal design, LHD or other design of experiment methods to generate samples. Such methods can ensure the uniformly distributed samples in the approximation space. However, for some complicated problems, especially for high-dimensional design problems, a great number of samples will be required to ensure the accuracy of the approximation model in such methods. In addition, the necessary sample size dramatically increases with increase of the number of dimensions, making the construction of the approximation models quite computationally expensive. What's worse, sometimes too many samples may lead to matrix singularity or ill-posed problems, which will unexpectedly reduce the approximation accuracy.

Therefore, according to characteristics of nonlinear interval optimizations, a local-densifying approximation model technique is proposed in this chapter to efficiently solve the transformed two-layer nested optimization problem. The main idea of the proposed method is to densify the samples according to the intermediate computation results during optimization to improve the accuracy of the approximation models in some important regions instead of in the whole approximation space. During optimization, the design vector is improved by solving the approximate uncertain optimization problem and the samples in early iterations are kept, which can save the computational cost.

The rest of this section is organized as follows. Firstly, the adopted approximation model, namely the radial basis functions (RBF) is briefly introduced. Secondly, based on the local-densifying approximation model technique and the order relation transformation model, a nonlinear interval optimization method is formulated. Finally, the method is applied to two test functions and a practical engineering problem.

### *7.2.1 Radial Basis Function*

The radial function is a category of functions whose input variables are the Euclidean distances between the estimation point and the sample points. The linear combination of radial functions yields the so-called radial basis function [24]. The radial basis function has a great ability to approximate nonlinear functions. Reference [25] used radial basis function and other three methods to systematically study and research 14 numerical examples that represent different types of problems and found the radial basis function was the most reliable one in terms of accuracy and robustness.

With the response of given sample point $\mathbf{x}^i$, $i = 1, 2, \ldots, n$, the response at estimation point $\mathbf{x}$ can be expressed by linear combination of radial functions as follows [24]:

$$\tilde{f}(\mathbf{x}) = \sum_{i=1}^{n_s} w_i \phi\left(r^i\right) \tag{7.19}$$

where $n_s$ is the number of samples, $w_i$, $i = 1, 2, \ldots, n_s$ are the coefficients of the linear combinations, $r^i = \left\|\mathbf{x} - \mathbf{x}^i\right\|$ represents the Euclidean distance between $\mathbf{x}$ and $\mathbf{x}^i$, and $\phi(r)$ is the radial function that is monotonous with respect to $r$. In this chapter, the following Gaussian function is taken as the radial function:

$$\phi(r) = \exp\left(-r^2/c^2\right) \tag{7.20}$$

where $c$ is a positive constant. According to the interpolation condition $\tilde{f}\left(\mathbf{x}^i\right) = f\left(\mathbf{x}^i\right)$, $i = 1, 2, \ldots, n_s$, the following equations can be obtained:

$$\mathbf{f} = \Phi\mathbf{w} \tag{7.21}$$

where $\mathbf{f}$ is the $n$-dimensional response vector of samples, $\mathbf{w}$ is the $n_s$-dimensional coefficient vector, and $\mathbf{\Phi} = \Phi_{ij} = \Phi\left(\left\|\mathbf{x}^i - \mathbf{x}^j\right\|\right)$, $i, j = 1, 2, \ldots, n_s$ is an $n_s \times n_s$ matrix. If the inverse matrix of $\mathbf{\Phi}$ exists, the coefficient vector can be obtained by:

$$\mathbf{w} = \mathbf{\Phi}^{-1}\mathbf{f} \tag{7.22}$$

By substituting the coefficient vector into Eq. (7.19), the radial basis function model can be obtained.

### 7.2.2 *Algorithm Flow*

The problem expressed in Eq. (4.1) is studied here, and there is no assumption of small uncertainty level. In other words, the proposed method is suitable for any uncertainty levels. Firstly, an approximation space $\Omega'$ is defined as follows:

$$\Omega' = \left\{(\mathbf{X},\mathbf{U})\middle|X_i^L \le X_i \le X_i^R,\ U_j^L \le U_j \le U_j^R,\ i = 1, 2, \ldots, n,\ j = 1, 2, \ldots, q\right\} \tag{7.23}$$

It is obvious that $\Omega'$ is $(n+q)$-dimensional and it consists of the design space and the uncertainty domain.

The optimization process of the proposed method is still composed of a series of approximate interval optimization problems, and in the $s$-th iteration, the corresponding approximate interval optimization problem of Eq. (4.1) is constructed as follows:

$$\begin{cases} \min\limits_{\mathbf{X}} \tilde{f}(\mathbf{X},\mathbf{U}) \\ \text{s.t. } \tilde{g}_i(\mathbf{X},\mathbf{U}) \le b_i^I = \left[b_i^L, b_i^R\right],\ i=1,2,\ldots,l \\ \quad \mathbf{X}_l \le \mathbf{X} \le \mathbf{X}_r \\ \quad \mathbf{U} \in \mathbf{U}^I = \left[\mathbf{U}^L, \mathbf{U}^R\right],\ U_i \in U_i^I = \left[U^L, U^R\right],\ i=1,2,\ldots,q \end{cases} \tag{7.24}$$

Being different from Eq. (7.6), $\tilde{f}(\mathbf{X},\mathbf{U})$ and $\tilde{g}_i(\mathbf{X},\mathbf{U}),\ i=1,2,\ldots,l$ in the above equation are constructed in $\Omega'$ which keeps unchanged during iterations. All approximation models are constructed by radial basis functions, and they are functions with respect to both the design variables and the uncertain variables.

Based on the order relation transformation model, Eq. (7.24) can be converted into the following deterministic optimization problem:

$$\begin{cases} \min\limits_{\mathbf{X}} \tilde{f}_p(\mathbf{X}) = \tilde{f}_d(\mathbf{X}) + \sigma \sum\limits_{i=1}^{l} \varphi\left(P\left(\tilde{g}_i^I(\mathbf{X}) \le b_i^I\right) - \lambda_i\right) \\ = (1-\beta)\left(\tilde{f}^c(\mathbf{X}) + \xi\right)/\phi + \beta\left(\tilde{f}^w(\mathbf{X}) + \xi\right)/\psi \\ + \sigma \sum\limits_{i=1}^{l} \varphi\left(P\left(\tilde{g}_i^I(\mathbf{X}) \le b_i^I\right) - \lambda_i\right) \\ \text{s.t. } \mathbf{X}_l \le \mathbf{X} \le \mathbf{X}_r \end{cases} \tag{7.25}$$

where $\tilde{f}_d$ is the approximate multi-objective evaluation function.

The two-layer nested optimization method given in Chap. 3 is employed to solve the above problem. Since the optimization is based on the simple and computationally cheap approximation models, it is therefore very efficient to solve even though it is still a nested optimization problem.

The iteration process of the proposed method thus can be summarized as follows [26]:

(1) Generate samples in $\Omega'$ using LHD and call the actual models to compute the initial objective function samples and constraint samples. Set the allowable errors $\varepsilon_1 > 0$ and $\varepsilon_2 > 0$, and set $s = 1$.
(2) Construct the radial-basis-function-based approximation models of the objective function and constraints using the samples obtained in the last step and then construct an approximate interval optimization problem as shown in Eq. (7.24). Obtain the optimum solution $\mathbf{X}^{(s)}$ of the approximate interval optimization

problem by using the order relation transformation model and the two-layer nested optimization method.

(3) The interval $\left[\tilde{f}^L(\mathbf{X}^{(s)}), \tilde{f}^R(\mathbf{X}^{(s)})\right]$ of the approximate objective function at $\mathbf{X}^{(s)}$ can be obtained by solving the approximate interval optimization problem. Besides, the realizations of uncertain variables corresponding to the upper and lower bounds, $\mathbf{U}_f^R$ and $\mathbf{U}_f^L$, can also be obtained. Their relationships are as follows:

$$\tilde{f}^L(\mathbf{X}^{(s)}) = \tilde{f}(\mathbf{X}^{(s)}, \mathbf{U}_f^L),\ \tilde{f}^R(\mathbf{X}^{(s)}) = \tilde{f}(\mathbf{X}^{(s)}, \mathbf{U}_f^R) \tag{7.26}$$

Therefore, two coordinate points $\left(\mathbf{X}^{(s)}, \mathbf{U}_f^L\right)$ and $\left(\mathbf{X}^{(s)}, \mathbf{U}_f^R\right)$ in $\Omega'$ can be obtained, in which the approximate objective function $\tilde{f}$ reaches its response bounds.

Similarly, intervals of the approximate constraints $\left[\tilde{g}_i^L(\mathbf{X}^{(s)}), \tilde{g}_i^R(\mathbf{X}^{(s)})\right]$, $i = 1, 2, \ldots, l$ at $\mathbf{X}^{(s)}$, and corresponding $\mathbf{U}_{g_i}^R$ and $\mathbf{U}_{g_i}^L$, $i = 1, 2, \ldots, l$ can be obtained, and their relationships are as follows:

$$\tilde{g}_i^L(\mathbf{X}^{(s)}) = \tilde{g}_i(\mathbf{X}^{(s)}, \mathbf{U}_{g_i}^L),\ \tilde{g}_i^R(\mathbf{X}^{(s)}) = \tilde{g}_i(\mathbf{X}^{(s)}, \mathbf{U}_{g_i}^R),\ i = 1, 2, \ldots, l \tag{7.27}$$

Therefore, two coordinate points $\left(\mathbf{X}^{(s)}, \mathbf{U}_{g_i}^L\right)$ and $\left(\mathbf{X}^{(s)}, \mathbf{U}_{g_i}^R\right)$ in $\Omega'$ can be obtained, in which the $i$-th approximate constraint $\tilde{g}_i$ reaches its response bounds.

(4) Calculate the actual objective function values $f^L(\mathbf{X}^{(s)}) = f\left(\mathbf{X}^{(s)}, \mathbf{U}_f^L\right)$ and $f^R(\mathbf{X}^{(s)}) = f\left(\mathbf{X}^{(s)}, \mathbf{U}_f^R\right)$, and also calculate the actual constraints at corresponding points:

$$g_i^L(\mathbf{X}^{(s)}) = g_i(\mathbf{X}^{(s)}, \mathbf{U}_{g_i}^L),\ g_i^R(\mathbf{X}^{(s)}) = g_i(\mathbf{X}^{(s)}, \mathbf{U}_{g_i}^R),\ i = 1, 2, \ldots, l \tag{7.28}$$

(5) Calculate the error $e_{\max}$:

$$e_{\max} = \max\left\{\left|\frac{f^L - \tilde{f}^L}{f^L}\right| + \left|\frac{f^R - \tilde{f}^R}{f^R}\right|, \left|\frac{g_i^L - \tilde{g}_i^L}{g_i^L}\right| + \left|\frac{g_i^R - \tilde{g}_i^R}{g_i^R}\right|, i = 1, 2, \ldots, l\right\} \tag{7.29}$$

If $e_{\max} < \varepsilon_1$, $\mathbf{X}^{(s)}$ is obtained as the optimum design vector, then stop the iteration, otherwise go to the next step.

(6) Calculate the Euclidean distances between $\left(\mathbf{X}^{(s)}, \mathbf{U}_f^L\right)$ and all samples of the current objective function and then denote the minimum distance by $d_{\min}$. If $d_{\min} > \varepsilon_2$, add $\left(\mathbf{X}^{(s)}, \mathbf{U}_f^L\right)$ into the sample set of the objective function. Calculate

the Euclidean distances between $\left(\mathbf{X}^{(s)}, \mathbf{U}_f^R\right)$ and all samples of the current objective function and then denote the minimum distance by $d_{\min}$. If $d_{\min} > \varepsilon_2$, add $\left(\mathbf{X}^{(s)}, \mathbf{U}_f^R\right)$ into the sample set of the objective function. Similarly, use the above method on all constraints so that the sample set of each constraint will be updated. If all the sample sets of objective function and constraints cannot be updated, $\mathbf{X}^{(s)}$ is selected as the final optimum design vector; otherwise, turn to Step (2).

The optimization process is shown in Fig. 7.14. In the construction of the above method, it is assumed that the objective function and all the constraints are calculated via computationally expensive numerical analysis models. However, usually, only one or a few functions need to be calculated by numerical analysis models in practical problems, and the others are usually explicit functions. Therefore, only those functions based on numerical analysis models need approximation models and local-densifying technique, while the computation of the other functions is based on original functions.

In Step (1), both design variables and uncertain variables are involved in the construction of approximation models for an interval optimization. The dimension of the problem thus could be relatively high, which may lead to high computational cost if we try to obtain accurate approximation models which generally need a large number of uniformly distributed samples. For the proposed method, the initial samples are only required to demonstrate the basic variation tendency of the original function and the solution can be improved in following local-densifying procedure. Therefore, it is actually unnecessary to generate too many initial samples by LHD to achieve high-accuracy approximation models.

In Step (5), the stop criterion $e_{\max} < \varepsilon_1$ is used. The satisfaction of this criterion indicates that the approximation models have a good approximation accuracy in the neighborhoods of the points where objective function and constraints achieve their lower and upper bounds. Therefore, it can ensure the accuracy of the optimal solution.

Step (6) is the process of sample densifying. In the nonlinear interval optimization, the uncertain optimization is transformed into a deterministic optimization problem, and the calculation of the deterministic optimization problem is based on the intervals of objective function and constraints. Therefore, the bounds of objective function and constraints are very important for the optimization. In the constructions of approximate objective function and constraints, the accuracy of approximation models can be improved by densifying the samples in regions near the upper and lower bounds of objective function and constraints in approximation space, and thus the optimization accuracy can be ensured. Besides, in the sample densifying process, an allowable error $\varepsilon_2$ is used to judge whether the new samples should be added into the current sample sets of objective function and constraints. If a new sample is very close to one or some current samples, it would contribute little to the improvement of the approximation accuracy. What's worse, too close samples may cause the matrix singularity problem of approximation models, which will unexpectedly reduce the approximation accuracy. Therefore, it is necessary to use $\varepsilon_2$ to avoid those samples.

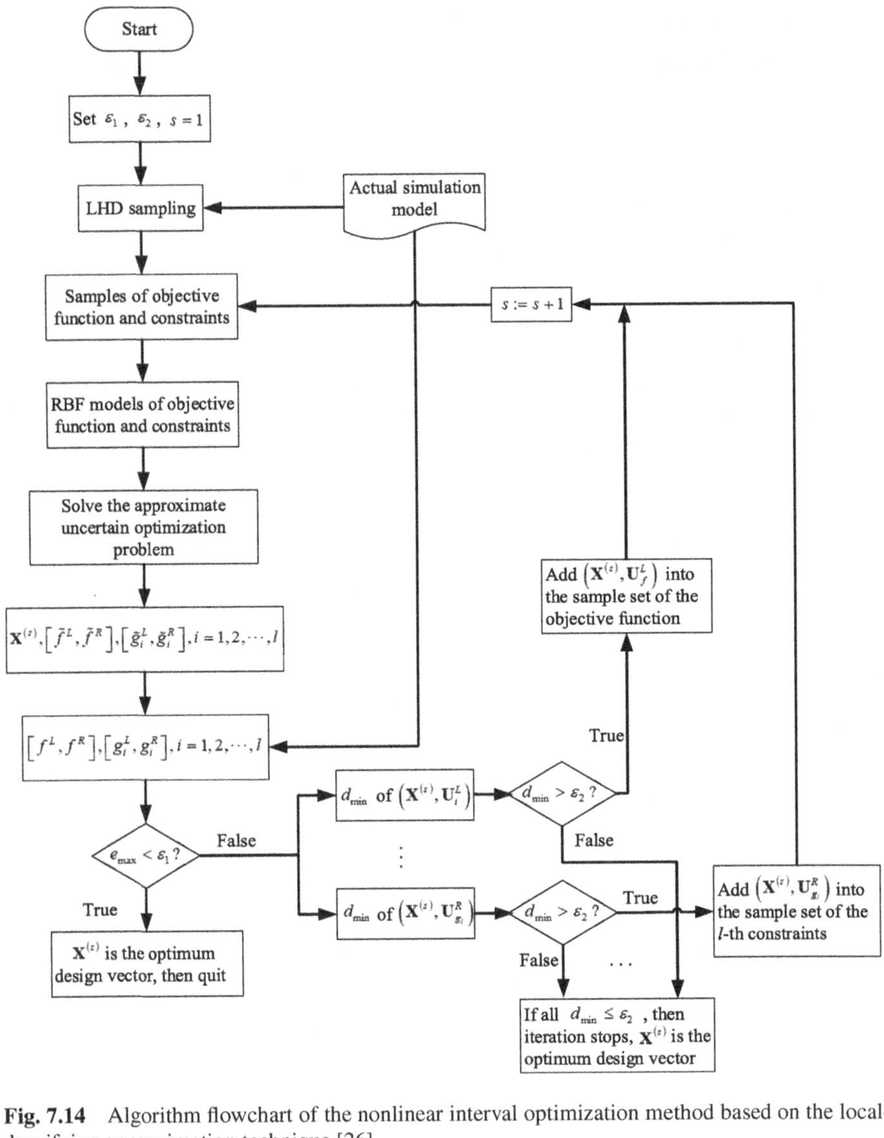


**Fig. 7.14** Algorithm flowchart of the nonlinear interval optimization method based on the local-densifying approximation technique [26]

If no sample can be added into the current sample sets, the approximation accuracy in the key regions is sufficiently high. That is why this stop criterion is adopted.

It should be mentioned that the initial samples of objective function and constraints can either be the same or not. However, after iterations, the sample sets of objective function and constraints will be different even the same initial samples are selected, because different samples are added into the sample sets in each iteration and the numbers of added samples may be different due to $\varepsilon_2$.

In the optimization process, the computational cost is consumed in three parts. The first part is in the computation of initial samples in Step (1). The second part is in solving the approximate interval optimization problem in Step (2). As is mentioned above, generally the computational cost of this part can be negligible as it is based on simple approximation models. The third part is in calling the actual models to compute the objective function and constraints at boundary coordinate points in Step (4).

### 7.2.3 *Test Functions*

In this section, two test functions are investigated. In order to intuitively describe the iteration process of the method, there are only one design variable and one uncertain variable in the first test function. Besides, no uncertain constraint is involved. The second test function is relatively more complicated. There are multiple design variables, uncertain variables and uncertain constraints.

1. Test function 1

The following test function is considered [27]:

$$\begin{cases} \min\limits_{X} f(X,U) = -24\sin\sqrt{\left(\frac{X-40}{2}\right)^2 + (U-5)^2} \Big/ \sqrt{\left(\frac{X-40}{2}\right)^2 + (U-5)^2} + 46 \\ \text{s.t. } 24 \le X \le 43 \\ \qquad U \in [3.5, 16.5] \end{cases} \tag{7.30}$$

where the variable has a relatively large uncertainty level, 65%.

The parameters $\xi$,$\phi$,$\psi$, and $\beta$ are set to 0.0, 5.0, 5.0, and 0.4, respectively. The allowable errors $\varepsilon_1$ and $\varepsilon_2$ are set to be 0.003 and 0.15, respectively. For the approximate interval optimization in each iteration, the maximum generation numbers of IP-GA of inner and outer layers are both 100. In this test function, the objective function is also assumed to be a computationally expensive numerical analysis model, and the number of evaluations of the objective function will be used to quantitatively assess the optimization efficiency. Firstly, the actual function-based nested optimization method (the maximum generation numbers of IP-GA of both layers are 100) is employed to solve this problem. The optimal solution is $X = 39.99$, the optimal interval of the objective function is [22.00, 51.21], and the values of uncertain variable $U$ corresponding to the lower and upper bounds of this interval are 5.00 and 9.5, respectively. This result is treated as the reference solution to evaluate the accuracy of the proposed method. Four cases with 4 different initial sample sizes, 5, 10, 20, 30, are investigated, and the results obtained by the proposed method are listed in Table 7.13. As shown in the table, the proposed method almost achieves the same results as the reference solution for the latter three cases. We can also see that the total

**Table 7.13** Optimization results with different initial sample sizes for test function 1 [27]

| Initial sample size | Optimum of $X$ | Interval of objective function $[f^L, f^R]$ | Values of uncertain variables at lower and upper objective function bounds | Multi-objective evaluation function $f_d$ | Number of iterations | Number of evaluations |
|---|---|---|---|---|---|---|
| 5 | 24.00 | [43.03, 48.19] | 4.96, 12.41 | 5.67 | 8 | 21 |
| 10 | 39.99 | [22.00, 51.20] | 4.99, 9.57 | 5.56 | 13 | 36 |
| 20 | 40.03 | [22.01, 51.21] | 4.95, 9.50 | 5.56 | 8 | 36 |
| 30 | 39.92 | [22.01, 51.21] | 5.03, 9.49 | 5.56 | 2 | 34 |

number of iterations decreases as the initial sample size increases. It needs only two iterations when the initial sample size is 30, which implies that 30 initial samples are enough to construct a high-quality approximation model for this simple problem and there is no need to add too many densifying samples. In the fourth case, the number of objective function evaluations is less than the second and third cases even it has the largest initial sample size. Besides, for the first case, the algorithm converges after 8 iterations but achieves an inaccurate solution, which indicates that too small initial sample size will prevent the initial approximation model from describing the variation tendency of actual model in approximation space. Therefore, the algorithm may fall into a local optimum.

In order to better show the details of the optimization process, Table 7.14 lists each iteration history of the third case. From the table, we can see the approximate

**Table 7.14** Iteration history of the optimization with the initial sample size of 20 [27]

| Iterative step | Design variable | $U^L, U^R$ | $[\tilde{f}^L, \tilde{f}^R]$ | $[f^L, f^R]$ |
|---|---|---|---|---|
| 1 | 39.14 | 3.50, 16.50 | [20.63, 50.28] | [30.63, 47.82] |
| 2 | 38.06 | 4.62, 9.59 | [26.49, 50.20] | [26.12, 51.12] |
| 3 | 43.00 | 5.61, 11.46 | [21.33, 49.25] | [31.19, 44.78] |
| 4 | 42.65 | 5.00, 16.50 | [29.08, 47.69] | [28.42, 47.73] |
| 5 | 42.89 | 4.94, 9.61 | [29.49, 48.74] | [29.53, 50.93] |
| 6 | 40.86 | 4.97, 9.85 | [23.23, 50.94] | [22.74, 50.86] |
| 7 | 40.33 | 4.91, 9.61 | [22.13, 51.07] | [22.14, 51.18] |
| 8 | 40.03 | 4.95, 9.50 | [22.04, 51.21] | [22.01, 51.21] |

interval $\left[\tilde{f}^L, \tilde{f}^R\right]$ is quite far from the actual function-based interval $\left[f^L, f^R\right]$ in the early iterations, while these two intervals become very close in the last two iterations. It indicates that approximations in the two local regions corresponding to the lower and upper bounds of objective function achieve a high accuracy after several local-densifying operations. Figure 7.15 shows the updates of the approximation model. Obviously, the actual model and the approximation model get closer in two key regions (upper and lower bounds of the function) as the optimization proceeds. In the 8-th iteration, these two curves almost coincide in the two key regions. Even the two curves are separate in the rectangle area in Fig. 7.15a, the optimization accuracy is not hampered since only the upper and lower bounds are used in the transformed deterministic optimization problem. This just shows the characteristic of the local-densifying method, which only focuses on the approximation accuracy in key regions and properly ignores approximation accuracy in other regions to improve the efficiency of constructing the approximation model and the whole interval optimization.

The initial and final distributions of samples are shown in Fig. 7.16. As shown in the figure, the samples are densified in two regions marked by circles in right part of the figure. In these two regions, the objective function achieves its upper bound and lower bound, respectively. Additionally, the samples are much denser than in other regions. If a uniformly distributed sampling method is employed, the same sample density should be adopted in the entire space to get the same optimization results, which will dramatically increase the sample size required, hence leading to very low optimization efficiency.

2. Test function 2

The following interval optimization problem is investigated [27]:

$$\begin{cases} \min\limits_{\mathbf{X}} f(\mathbf{X}, \mathbf{U}) = U_1(X_1 - 2)^2 + U_2(X_2 - 1)^2 + U_3X_3 \\ \text{s.t. } U_1X_1^2 - U_2^2X_2 + U_3X_3 = [6.5, 7.0] \\ \quad U_1X_1 + U_2X_2 + U_3^2X_3^2 + 1 \geq [10.0, 15.0] \\ \quad\quad 2 \leq X_1 \leq 6,\ -4 \leq X_2 \leq 7,\ -3 \leq X_3 \leq 8 \\ \quad U_1 \in [0.6, 1.8],\ U_2 \in [0.5, 1.5],\ U_3 \in [0.6, 2.0] \end{cases} \tag{7.31}$$

where the uncertainty levels of $U_1, U_2$, and $U_3$ are 50.0%, 50.0%, and 53.8%, respectively, which are relatively large.

In the optimization, $\xi, \phi, \psi, \beta$, and $\sigma$ are set to be 4.0, 1.4, 2.0, 0.5, and 100,000, respectively. The possibility degree levels of two uncertain constraints are both 0.8. The allowable errors $\varepsilon_1$ and $\varepsilon_2$ are set to be 0.05 and 0.15, respectively. Besides, for the approximate interval optimization in each iteration, the maximum generation numbers of IP-GA in inner and outer layers are both 300. Here the number of objective function evaluations is also used to quantify the optimization efficiency.

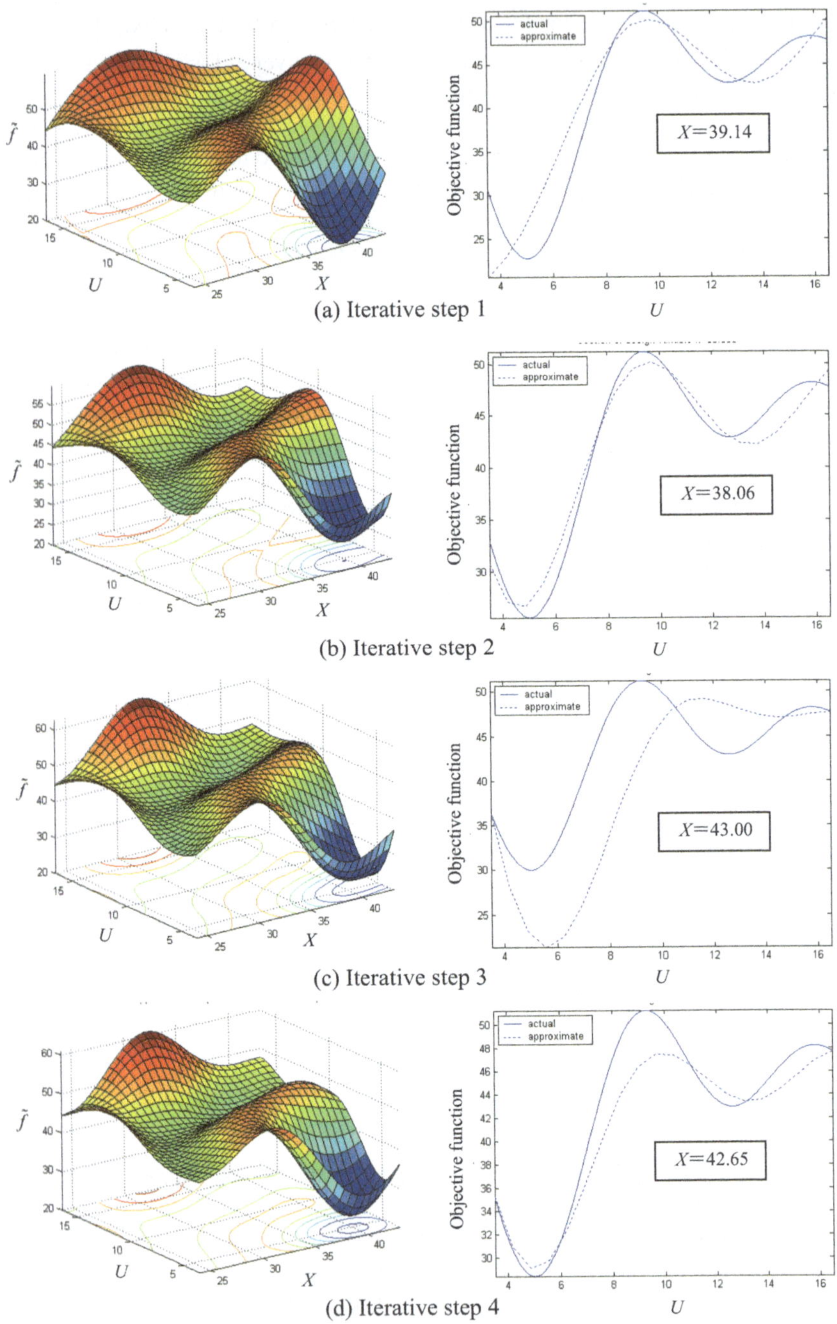


(a) Iterative step 1

(b) Iterative step 2

(c) Iterative step 3

(d) Iterative step 4

**Fig. 7.15** Comparison between approximation model and actual model during optimization with the initial sample size of 20 [26]

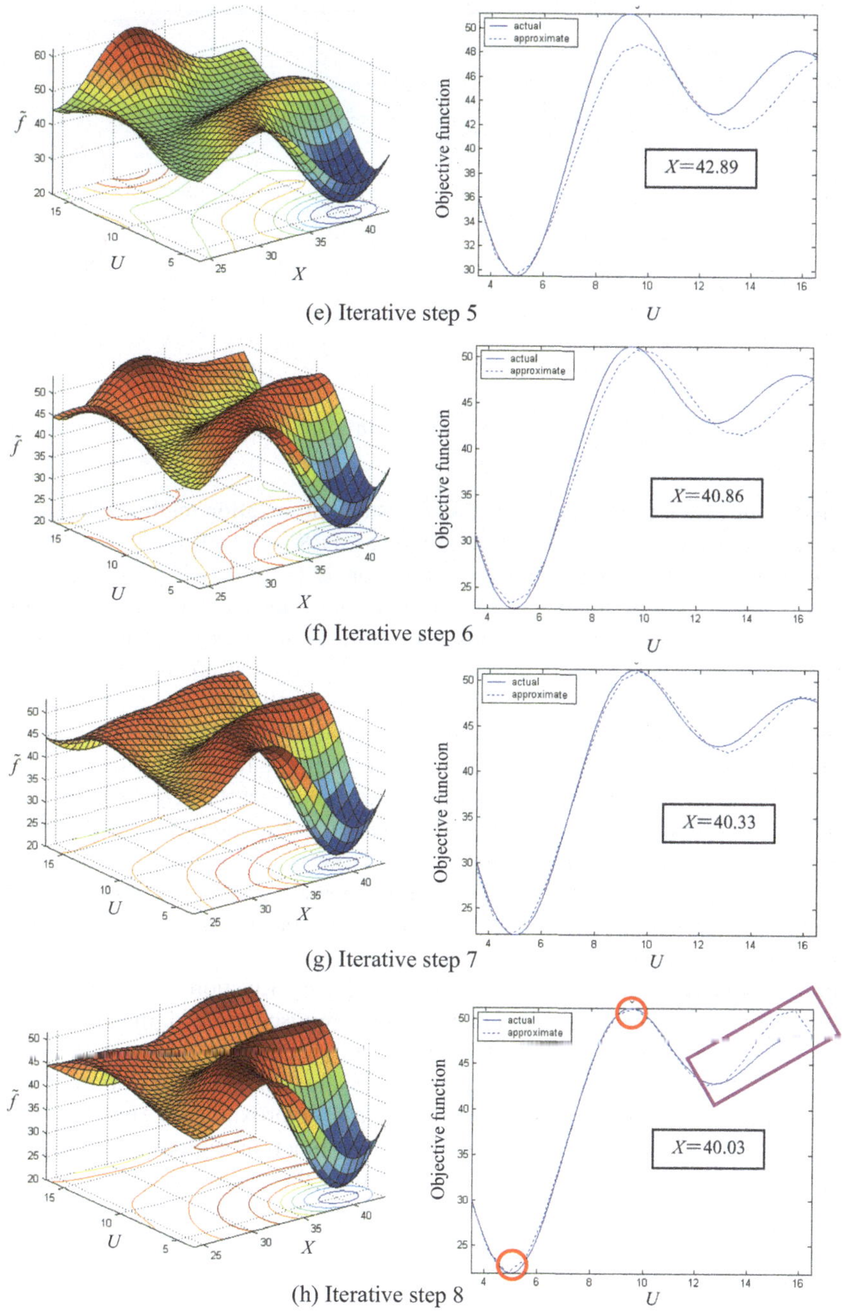


(e) Iterative step 5

(f) Iterative step 6

(g) Iterative step 7

(h) Iterative step 8

**Fig. 7.15** (continued)

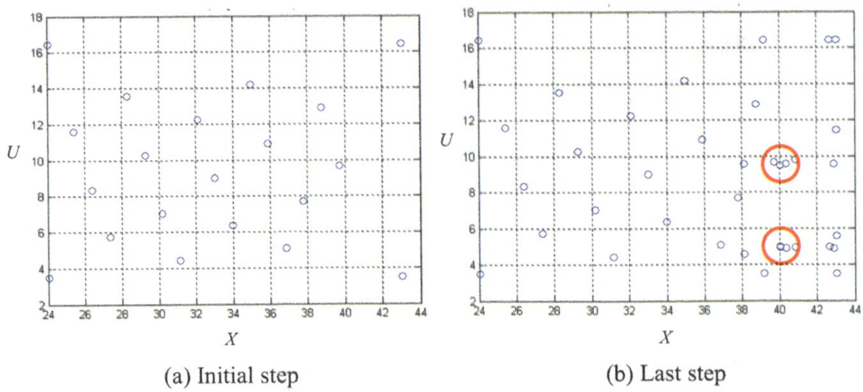


**Fig. 7.16** Sample distributions of initial and last steps in approximate space [27]

Firstly, the actual function-based IP-GA nested optimization method is employed to solve this problem. The optimal design vector is $\mathbf{X} = (4.01, 0.87, -3.00)^{\mathrm{T}}$, the corresponding interval of the objective function is $[-3.49, 5.49]$, and the corresponding penalty function value is 3.91. This result is viewed as a reference solution to evaluate the accuracy of the proposed method. Four cases with different initial sample sizes (30, 40, 50, and 60, respectively) are investigated, and the corresponding results are listed in Table 7.15. As shown in the table, the required number of iterations decreases with the increase of initial sample size, while the evaluations of the objective function show an increasing trend. In all the four cases, the possibility degrees of the actual constraints at the optimum can meet the requirement of 0.8. Besides, the optimum design vectors in the four cases slightly differ from the reference one, especially the second design variable. However, the penalty function values in the four cases are very close to the reference value $f_P = 3.91$, which indicates that relatively good design vectors are achieved by the proposed method even though they are not very accurate.

In the above four cases, the evaluations of objective function are 72, 80, 88, and 92, respectively. To verify the advantages of the proposed method, the approximate optimization based on uniformly distributed samples is performed for these four cases (with 72, 80, 88, and 92 samples, respectively). In the optimization process, the samples are uniformly generated in the entire approximation space by LHD. These samples are then used to obtain objective function samples and constraint samples based on actual models and thus construct the radial-basis-function-based approximation models. Therefore, an approximate interval optimization problem can be constructed and transformed into the deterministic optimization by the order relation transformation model, which is then solved by the nested optimization method. The optimization results based on uniformly distributed samples are listed in Table 7.16. Obviously, the results are inaccurate, which indicates that the accuracy of the approximation models is relatively low in the entire approximation space. Additionally, the actual penalty function values at optimum for four cases are 7.98, 7.75, 7.83,

**Table 7.15** Optimization results with different initial sample sizes for test function 2 [27]

| Initial sample size | Optimum of **X** | Interval of objective function $[f^L, f^R]$ | Interval of constraint 1 $[g_1^L, g_1^R]$ | Interval of constraint 2 $[g_2^L, g_2^R]$ | Possibility degrees of constraints | Penalty function $f_p$ | Number of iterations | Number of evaluations |
|---|---|---|---|---|---|---|---|---|
| 30 | (4.01, 0.57, −3.00) | [−3.47, 5.71] | [2.35,26.43] | [7.50,44.96] | 0.82,0.87 | 3.98 | 21 | 72 |
| 40 | (4.07, 0.76, −2.78) | [−2.94, 5.65] | [2.64,24.75] | [6.61,40.49] | 0.81,0.83 | 3.99 | 20 | 80 |
| 50 | (4.01, 0.63, −3.00) | [−3.49, 5.51] | [2.20,24.72] | [6.96,43.93] | 0.80,0.85 | 3.92 | 19 | 88 |
| 60 | (4.02, 0.48, −3.00) | [−3.35, 5.95] | [2.61,22.78] | [6.89,43.46] | 0.80,0.85 | 4.03 | 16 | 92 |

**Table 7.16** Optimization results with different sample sizes based on uniformly distributed samples [26]

| Sample size | Optimum of **X** | Bounds of approximate objective function and constraints $\left[\tilde{f}^L, \tilde{f}^R\right][\tilde{g}_i^L, \tilde{g}_i^R]$ $i = 1, 2$ | Bounds of actual objective function and constraints $[f^L, f^R][g_i^L, g_i^R] i = 1, 2$ | Possibility degrees of constraints | Approximate penalty function $\tilde{f}_P$ | Actual penalty function $f_P$ |
|---|---|---|---|---|---|---|
| 72 | (2.69, 0.42, 8.00) | [−6.02, 24.98] | [5.23, 17.37] | 1.00, 1.00 | 9.69 | 7.98 |
| | | [0.86, 30.50] | [8.20, 28.94] | | | |
| | | [ 6.94, 231.80] | [25.87, 262.47] | | | |
| 80 | (2.53, 0.68, 8.00) | [−9.96, 24.81] | [5.02, 16.66] | 1.00, 1.00 | 9.43 | 7.75 |
| | | [1.83, 27.92] | [7.11, 27.33] | | | |
| | | [7.66, 246.91] | [25.90, 262.57] | | | |
| 88 | (2.48, 0.44, 8.00) | [−8.57, 24.76] | [5.10, 16.89] | 1.00, 1.00 | 9.48 | 7.83 |
| | | [1.49, 27.81] | [7.50, 26.97] | | | |
| | | [5.11, 236.17] | [25.75, 262.13] | | | |
| 92 | (2.50, 0.69, 8.00) | [−7.25, 23.95] | [5.00, 16.60] | 1.00, 1.00 | 9.31 | 7.74 |
| | | [2.05, 27.67] | [7.01, 27.12] | | | |
| | | [7.65, 248.98] | [25.89, 262.54] | | | |

and 7.74, while the corresponding values in Table 7.15 are 3.98, 3.99, 3.92, and 4.03, respectively. It indicates that under the same number of samples the proposed local-densifying approximation technique can obtain much better optimization results than the approximation optimization method based on uniformly distributed samples. From the above analysis, we can see that it is actually very difficult to construct an accurate approximation model with dozens of uniformly distributed samples in a 6-dimensional approximation space. A low-accuracy approximation model is thus unable to obtain a satisfying and reliable optimization result. However, the local-densifying approximation technique can guarantee the accuracy of approximation model in some key regions and whereby achieve ideal optimization results even though the computational cost is the same.

### 7.2.4 Application to the Crashworthiness Design of a Thin-Walled Beam of Vehicle Body

The design of the thin-walled beam investigated in Sect. 7.1.7 is used again to demonstrate the effectiveness of the proposed method in engineering application. The uncertain optimization problem is totally the same as the problem in Sect. 7.1.7, and is modeled as follows [27]:

$$\begin{cases} \max\limits_{t,R,d} f_e(t, R, d, \sigma_s, E_t) \\ \text{s.t. } g_f(t, R, d, \sigma_s, E_t) \le [65\text{ kN}, 70\text{ kN}] \\ \quad 0.5\text{ mm} \le t \le 2.5\text{ mm} \\ \quad 1\text{ mm} \le R \le 8\text{ mm} \\ \quad 10\text{ mm} \le d \le 60\text{ mm} \\ \quad \sigma_s \in [294.5\text{ MPa}, 325.5\text{ MPa}] \\ \quad E_t \in [724.85\text{ MPa}, 801.15\text{ MPa}] \end{cases} \tag{7.32}$$

where the uncertainties of the two variables are both 5%.

In the optimization process, the parameters $\xi$, $\phi$, and $\psi$ are set to be 0.0, 1.9, and 2.4, respectively. The weighting factor $\beta$, the penalty factor $\sigma$, and the possibility degree level of the uncertain constraint are set to be 0.5, 100,000, and 0.8, respectively. The allowable errors $\varepsilon_1$ and $\varepsilon_2$ are set to be 0.5 and 0.15, respectively. For the approximate interval optimization in each iteration, the maximum generation numbers of inner and outer layers of IP-GA are both set to be 300.

The initial sample size is set to be 50, and the optimization results are listed in Table 7.17, which shows the algorithm converges after 23 iterations and obtains the optimum design vector $(2.00\text{ mm}, 2.75\text{ mm}, 34.98\text{ mm})^{\mathrm{T}}$. For the optimal design, the corresponding penalty function value is 2.42, the interval of energy absorption caused by the uncertain material properties is [8.30 kJ,9.28 kJ], and the interval of

**Table 7.17** Optimization results of the closed-hat beam subjected to uncertain material properties [27]

| Optimum of ($t$, $R$, $d$) (mm) | Interval of objective function (kJ) | Interval of constraint (kN) | Possibility degree of constraint | Penalty function $f_p$ | Number of iterations | Number of FEM evaluations |
|---|---|---|---|---|---|---|
| (2.00, 2.75, 34.98) | [8.30, 9.28] | [51.10, 69.74] | 0.88 | 2.42 | 23 | 96 |

axial impact force is [51.10 kN,69.74 kN]. When computing the initial samples, the objective function value and the constraint value at the sample can be obtained by one single FEM evaluation, while they are calculated separately in the iterations since the objective function and constraint have different densifying samples. For this example, the entire optimization process costs 96 FEM evaluations.

## 7.3 Summary

In this chapter, the approximation model is introduced to the nonlinear interval optimization, and two types of efficient uncertain optimization methods are proposed. The first one is a nonlinear interval optimization method based on the approximation model management strategy. The whole optimization process is composed of a series of approximate uncertain optimization problems. In the iteration process, the design space and the approximation models are updated by a model management tool. This method applies to optimization problems with variables of small uncertainties. The second one is a nonlinear interval optimization method based on the local-densifying approximation technique. The optimization process of this method is also composed of a series of uncertain optimization problems. However, the approximation space is not updated during the iteration. Instead, a local-densifying technique is used to improve the accuracy of the approximation model in key areas, thus improving the optimization accuracy. This method is not restricted by the uncertainty levels of the variables.

## References

1. Renaud JE, Gabriele GA (1993) Improved coordination in nonhierarchic system optimization. AIAA J 31(12):2367–2373
2. Huang H, Xia RW (1995) Two-level multipoint constraint approximation concept for structural optimization. Struct Optim 9(1):38–45
3. Sui YK, Li SP (2006) The application of improved rsm in shape optimization of two-dimension continuum. Eng Mech 23(10):1–6
4. Roux WJ, Stander N, Haftka RT (1998) Response surface approximations for structural optimization. Int J Numer Meth Eng 42(3):517–534

5. Li G, Wang H, Aryasomayajula SR, Grandhi RV (2000) Two-level optimization of airframe structures using response surface approximation. Struct Multidisciplinary Optim 20(2):116–124
6. Queipo NV, Haftka RT, Shyy W, Goel T, Vaidyanathan R, Kevin Tucker P (2005) Surrogate-based analysis and optimization. Prog Aerosp Sci 41(1):1–28
7. Wang GG, Shan S (2007) Review of metamodeling techniques in support of engineering design optimization. ASME J Mech Des 129(4):370–380
8. Lee HW, Lee GA, Yoon DJ, Choi S, Na KH, Hwang MY (2008) Optimization of design parameters using a response surface method in a cold cross-wedge rolling. J Mater Process Technol 201(1–3):112–117
9. Viana FAC, Simpson TW, Balabanov V, Toropov V (2014) Special section on multidisciplinary design optimization: metamodeling in multidisciplinary design optimization: how far have we really come?. AIAA J 52(4):670–690
10. Rodríguez JF, Renaud JE, Watson LT (1998) Convergence of trust region augmented Lagrangian methods using variable fidelity approximation data. Struct Optim 15(3):141–156
11. Rodríguez JF, Pérez VM, Padmanabhan D, Renaud JE (2001) Sequential approximate optimization using variable fidelity response surface approximations. Struct Multidisciplinary Optim 22(1):24–34
12. Rodríguez JF, Renaud JE, Watson LT (1998) Trust region augmented Lagrangian methods for sequential response surface approximation and optimization. ASME J Mech Des 120:58–66
13. Rodríguez JF, Renaud JE, Wujek BA, Tappeta RV (2000) Trust region model management in multidisciplinary design optimization. J Comput Appl Math 124(1–2):139–154
14. Alexandrov NM, Lewis RM, Gumbert CR, Green LL, Newman PA (2000) Optimization with variable-fidelity models applied to wing design. In: 38th aerospace sciences meeting and exhibit, Reno, NV, U.S.A.
15. Sun CZ (2004) Theoretical and experimental study on the processes of sheet metal forming based on variable blank-holder force. Ph.D. thesis, Shanghai Jiao Tong University, Shanghai
16. Sun RH, Yi HY, Liu QS (2000) Mathematical statistics. Chongqing University Press, Chongqing
17. Jiang C, Han X, Liu GP (2008) A sequential nonlinear interval number programming method for uncertain structures. Comput Methods Appl Mech Eng 197(49–50):4250–4265
18. Morris MD, Mitchell TJ (1995) Exploratory designs for computational experiments. J Stat Plan Inference 43(3):381–402
19. Jiang C, Han X (2007) A new uncertain optimization method based on intervals and an approximation management model. Comput Model Eng Sci 22(2):97–118
20. Liu GR, Quek SS (2003) The finite element method: a practical course. Elsevier Science Ltd., England
21. Kurtaran H, Eskandarian A, Marzougui D, Bedewi NE (2002) Crashworthiness design optimization using successive response surface approximations. Comput Mech 29(4–5):409–421
22. Wang HL (2002) Study on optimail design of auto body structure based on crashworthiness numerical simulation. Ph.D. thesis, Shanghai Jiao Tong University, Shanghai.
23. Belytschko T, Lin JI, Chen-Shyh T (1984) Explicit algorithms for the nonlinear dynamics of shells. Comput Methods Appl Mech Eng 42(2):225–251
24. Mu XF, Yao WX, Yu XQ, Liu KL, Xu F (2005) A survey of surrogate models used in MDO. Chin J Comput Mech 22(5):608–612
25. Jin R, Chen W, Simpson TW (2001) Comparative studies of metamodelling techniques under multiple modelling criteria. Struct Multidisciplinary Optim 23(1):1–13
26. Jiang C (2008) Theories and algorithms of uncertain optimization based on interval. Ph.D. thesis, Hunan University, Changsha.
27. Zhao ZH, Han X, Jiang C, Zhou XX (2010) A nonlinear interval-based optimization method with local-densifying approximation technique. Struct Multidisciplinary Optim 42(4):559–573

# Chapter 8
# Interval Multidisciplinary Design Optimization

**Abstract** This chapter introduces the interval model into the multidisciplinary design optimization (MDO) problem, and whereby constructs an interval MDO model for complex MDO problems with uncertainties. A solution strategy is also given to solve the interval MDO model.

Multidisciplinary design optimization (MDO) has been widely applied to the design of complex engineering systems. "Multidisciplinary" means that multiple engineering disciplines are involved in a design problem, where the result of each discipline analysis is obtained through related theories and simulation tools. Meanwhile, there are complex coupling relationships between disciplines in the optimization problem. Through exploring and utilizing the coupling mechanism of interactions between different disciplines, the MDO methods can address this type of complex problems. In recent years, MDO has become an important research direction in the field of optimization design, and a series of methods [1–5] have been developed in this area. Generally, these methods can be grouped into two categories: single-level optimization method [6–8] and multi-level optimization method [9–11]. The former is mainly used to deal with problems with only a few variables and disciplines, treating the system as a whole. However, the latter carries out the optimizations of different disciplines separately, and then conducts the consistent design at the system level to obtain the optimal solution. The commonly used single-level methods include the individual disciplinary feasible approach (IDF) [6], the all-at-once approach (AAO) [7], the multidisciplinary feasible approach (MDF) [8], etc., and the multi-level optimization methods mainly include the concurrent subspace optimization approach (CSSO) [9], the collaborative optimization approach (CO) [10], the bi-level integrated system synthesis approach (BLISS) [11], etc. The traditional MDO methods mentioned above are developed mainly to solve deterministic problems, where all of the parameters are deterministic. However, there exist many uncertainties in practical engineering problems, including structural sizes, material properties, loads, etc. The combined effects of these uncertain factors can lead to large variations of system performance and even failures of structures. Studies [12] have shown that using conventional MDO methods to deal with uncertain multidisciplinary problems



C. Jiang et al., *Nonlinear Interval Optimization for Uncertain Problems*, Springer Tracts in Mechanical Engineering, https://doi.org/10.1007/978-981-15-8546-3_8

may result in unreliable designs. Therefore, it is necessary to develop MDO models considering uncertain parameters and corresponding algorithms.

In this section, an interval MDO method is proposed by introducing interval model into the MDO problem. It provides an analysis tool for the design optimization of multidisciplinary systems under uncertainties. The main contents of this chapter are organized as follows: an interval MDO model is given in Sect. 8.1; a strategy is proposed to decouple the multidisciplinary analysis of the created interval MDO model based on IDF approach in Sect. 8.2; the decoupled interval optimization model is then converted into a conventional deterministic optimization problem and solved through a two-layer optimization in Sect. 8.3.

## 8.1 An Interval MDO Model

Firstly, a system with three disciplines is taken as an example to illustrate the conventional deterministic MDO problem. As shown in Fig. 8.1, the input of the multidisciplinary system is an $n$-dimensional deterministic design vector $\mathbf{X}$ whose value range is $\Omega^n$, and the outputs are the objective function $f$ and the constraints $g_j$, $j = 1, 2, \ldots, l$. Each solid-line box represents an analysis process of a single discipline. For the convenience of description, each discipline analysis only outputs one constraint and hence here $l = 3$, and general cases involving more constraints

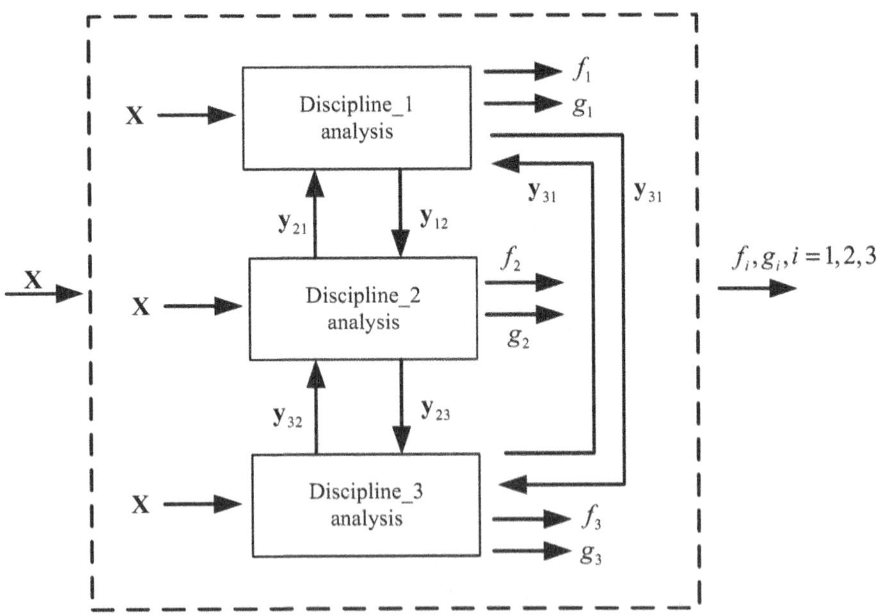


**Fig. 8.1** A deterministic MDO problem [13]

**Table 8.1** The state vectors, objective functions and constraints in the MDO [13]

| State vectors | Objective functions | Constraints |
|---|---|---|
| $\mathbf{y}_{12} = \mathbf{y}_{12}(\mathbf{X}, \mathbf{y}_{21}, \mathbf{y}_{31}), \mathbf{y}_{13} = \mathbf{y}_{13}(\mathbf{X}, \mathbf{y}_{21}, \mathbf{y}_{31})$ | $f_1 = f_1(\mathbf{X}, \mathbf{y}_{21}, \mathbf{y}_{31})$ | $g_1 = g_1(\mathbf{X}, \mathbf{y}_{21}, \mathbf{y}_{31})$ |
| $\mathbf{y}_{21} = \mathbf{y}_{21}(\mathbf{X}, \mathbf{y}_{12}, \mathbf{y}_{32}), \mathbf{y}_{23} = \mathbf{y}_{23}(\mathbf{X}, \mathbf{y}_{12}, \mathbf{y}_{32})$ | $f_2 = f_2(\mathbf{X}, \mathbf{y}_{12}, \mathbf{y}_{32})$ | $g_2 = g_2(\mathbf{X}, \mathbf{y}_{12}, \mathbf{y}_{32})$ |
| $\mathbf{y}_{31} = \mathbf{y}_{31}(\mathbf{X}, \mathbf{y}_{13}, \mathbf{y}_{23}), \mathbf{y}_{32} = \mathbf{y}_{32}(\mathbf{X}, \mathbf{y}_{13}, \mathbf{y}_{23})$ | $f_3 = f_3(\mathbf{X}, \mathbf{y}_{13}, \mathbf{y}_{23})$ | $g_3 = g_3(\mathbf{X}, \mathbf{y}_{13}, \mathbf{y}_{23})$ |

follow the same regulation. $g_j \leq 0$ represents that the design option meets the requirement of the $j$-th discipline. $f_j$ denotes the objective function based on the analysis of the $j$-th discipline, and it can be a part of the system objective function. Therefore we have $f = (\mathbf{X}, f_1, f_2, f_3)$. Unlike the conventional optimization, the state vector $\mathbf{y}_{ji}$ which is the output of the $j$-th constraint and also the input of the $i$-th constraint exists in the MDO problem. In other words, influences from the other disciplines need to be considered during the analysis, and multiple disciplines are coupled under the interaction of the state vectors $\mathbf{y}_{ji}$, hence $\mathbf{y}_{ji}$ is also named the coupling vector. Moreover, $f_j$, $g_j$, and $\mathbf{y}_{ji}$ can be obtained by related theory or analysis tool of the $j$-th discipline, as is listed in Table 8.1.

Therefore, the conventional deterministic MDO problem can be formulated as [6]:

$$\begin{cases} \min\limits_{\mathbf{X}} f(\mathbf{X}, f_j) \\ \text{s.t. } g_j(\mathbf{X}, \mathbf{y}_{ij}) \leq 0, i, j = 1, 2, \ldots, l, \ i \neq j \\ \quad \mathbf{X} \in \Omega^n \end{cases} \tag{8.1}$$

where the objective function can also be written as $f(\mathbf{X}, \mathbf{y}_{ij})$, since $f_j$ is a function with respect to both $\mathbf{X}$ and $\mathbf{y}_{ij}$.

In practical multidisciplinary problems, there usually exist many parametric uncertainties. To meet the reliability requirements of multidisciplinary systems, the effects of these uncertainties on the system performance should be fully taken into consideration in the optimization process. Through describing all uncertain parameters in Eq. (8.1) by intervals, an interval MDO model thus can be created:

$$\begin{cases} \min\limits_{\mathbf{X}} f(\mathbf{X}, \mathbf{U}, \mathbf{y}_{ij}) \\ \text{s.t. } g_j(\mathbf{X}, \mathbf{U}, \mathbf{y}_{ij}) \leq b_j^I = \left[b_j^L, b_j^R\right], i, j = 1, 2, \ldots, l, \ i \neq j, \mathbf{X} \in \Omega^n \\ \quad \mathbf{y}_{ji} = \mathbf{y}_{ji}(\mathbf{X}, \mathbf{U}^c, \mathbf{y}_{ij}), i, j = 1, 2, \ldots, l, \ i \neq j \\ \quad \mathbf{U} \in \mathbf{U}^I = \left[\mathbf{U}^l, \mathbf{U}^R\right], \ U_i \in U_i^I = \left[U_i^L, U_i^R\right], i = 1, 2, \ldots, q \end{cases} \tag{8.2}$$

where $\mathbf{U}$ is the $q$-dimensional uncertain vector described by the interval vector $\mathbf{U}^I$; $b_j^I$ denotes the allowable interval of the $j$-th uncertain constraint and it could be a real number in practical problems. It should be pointed out that the coupling vector

$\mathbf{y}_{ij}$ is theoretically uncertain due to the existence of $\mathbf{U}$. To reduce the complexity of problem, the midpoint $\mathbf{U}^c$ is used to calculate $\mathbf{y}_{ij}$.

As shown in Fig. 8.2, similar to the conventional interval optimization, interval MDO usually involves a two-layer nested optimization: the outer layer for updating the design variables, and the inner layer for calculating the intervals of the uncertain objective function and constraints. The difference is that the interval analysis in the inner layer here involves multiple disciplines, which are coupled via state variables. For example, the output $\mathbf{y}_{12}$ of Discipline_1 is necessary to the analysis in Discipline_2, and vice versa. As a result, the multidisciplinary analysis needs to repeatedly coordinate the analysis results of each discipline to achieve consistency of the state variables. In practical problems, however, time-consuming simulation analysis models are usually necessary for the multidisciplinary analysis. Repeatedly calling those models will lead to extremely low efficiency. Hence, solving interval MDO problems are more challenging than conventional interval optimization problems in terms of computational efficiency.

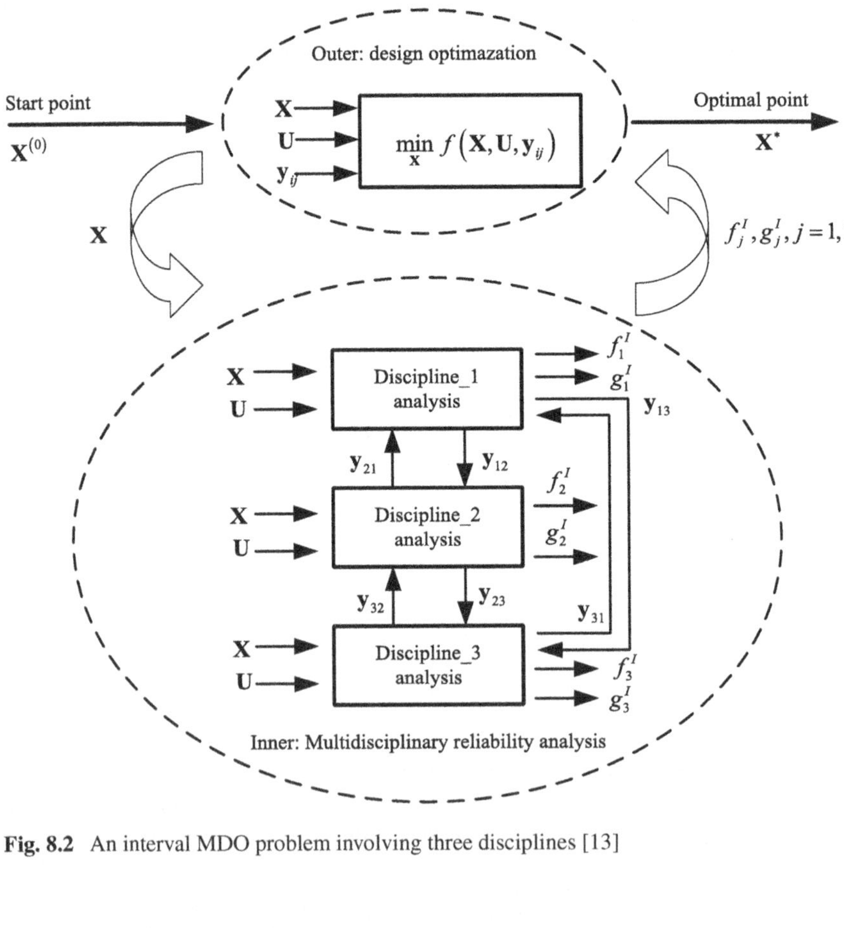


**Fig. 8.2** An interval MDO problem involving three disciplines [13]

## 8.2 Decoupling the Multidisciplinary Analysis

The core issue of the multidisciplinary analysis is how to deal with the coupling relationship between the disciplines, namely, to achieve the consistency design of the state variables in each discipline. Currently, a series of important solving strategies to MDO have been developed, such as the individual disciplinary feasible approach (IDF) [6], the all-at-once approach (AAO) [7], the collaborative optimization approach (CO) [10], the bi-level integrated system synthesis approach (BLISS) [11], etc. Among them, the IDF is one of the most widely applied MDO approaches. In IDF, the multidisciplinary problem is decoupled through adding new design variables and consistency constraint, by which the analysis of each discipline can be carried out independently. The implementation of the IDF approach is simple and convenient. Therefore, the IDF strategy is employed here to deal with the multidisciplinary problem in interval MDO. New design variables are given to corresponding state variables, and the consistency constraint is added to eliminate the difference between the state variables and the added design variables. For example, the design vector $\mathbf{v}_{12}$ is given to $\mathbf{y}_{12}$ which is an output of the analysis in Discipline_1, and a deterministic constraint $\|\mathbf{v}_{ji} - \mathbf{y}_{ji}\| / \|\mathbf{v}_{ji}\| \le \varepsilon_{ji}$, where $\varepsilon_{ji}$ is the allowable error, is added to ensure the consistency between $\mathbf{v}_{12}$ and $\mathbf{y}_{12}$. In this way, Eq. (8.2) can be transformed into a conventional interval optimization problem:

$$\begin{cases} \min\limits_{\mathbf{X},\mathbf{v}_{ij}} f\left(\mathbf{X}, \mathbf{v}_{ij}, \mathbf{U}\right) \\ \text{s.t. } g_j\left(\mathbf{X}, \mathbf{U}, \mathbf{v}_{ij}\right) \le b_j^I = \left[b_j^L, b_j^R\right], i, j = 1, 2, \ldots, l, \ i \ne j, \mathbf{X} \in \Omega^n \\ \mathbf{h}_{ji}^{\mathbf{v}} = \frac{\|\mathbf{v}_{ji} - \mathbf{y}_{ji}\|}{\|\mathbf{v}_{ji}\|} \le \varepsilon_{ji}, i, j = 1, 2, \ldots, l, \ i \ne j \\ \mathbf{y}_{ji} = \mathbf{y}_{ji}\left(\mathbf{X}, \mathbf{U}^c, \mathbf{v}_{ij}\right), i, j = 1, 2, \ldots, l, \ i \ne j \\ \mathbf{U} \in \mathbf{U}^I = \left[\mathbf{U}^I, \mathbf{U}^R\right], \ U_i \in U_i^I = \left[U_i^L, U_i^R\right], i = 1, 2, \ldots, q \end{cases} \tag{8.3}$$

where $\mathbf{v}_{ji}$ is the newly added design vector, $\mathbf{h}_{ji}^v$ refers to the consistency constraint, and $\|\cdot\|$ denotes the norm of a vector.

Through the above analysis, the multidisciplinary problem in Eq. (8.2) can be decoupled. However, the uncertain vector $\mathbf{U}$ still exists in the objective function and constraints in Eq. (8.3). In the next section, Eq. (8.3) will be further transformed into a deterministic optimization problem based on the order relation transformation model proposed in Chap. 3.

## 8.3 Transformation of the Interval Optimization Problem

In Eq. (8.3), for a certain $\mathbf{X}$ and $\mathbf{v}_{ij}$ the objective function value is an interval $f^I\left(\mathbf{X}, \mathbf{v}_{ij}\right) = \left[f^L\left(\mathbf{X}, \mathbf{v}_{ij}\right), f^R\left(\mathbf{X}, \mathbf{v}_{ij}\right)\right]$, whose upper and lower bounds can be expressed as:

$$\begin{aligned}&f^L(\mathbf{X}, \mathbf{v}_{ij}) = \min_{\mathbf{U}} f(\mathbf{X}, \mathbf{v}_{ij}, \mathbf{U}),\ f^R(\mathbf{X}, \mathbf{v}_{ij}) = \max_{\mathbf{U}} f(\mathbf{X}, \mathbf{v}_{ij}, \mathbf{U})\\ &\mathbf{U} \in \Gamma = \left\{\mathbf{U}\middle|U_i^L \le U_i \le U_i^R,\ i = 1, 2, \ldots, q\right\}\end{aligned} \tag{8.4}$$

Using the order relation of interval number "$\le_{cw}$", the objective function in Eq. (8.3) can be transformed into a deterministic multi-objective optimization problem as follows:

$$\min_{\mathbf{X}, \mathbf{v}_{ij}}\left[f^c(\mathbf{X}, \mathbf{v}_{ij}), f^w(\mathbf{X}, \mathbf{v}_{ij})\right] \tag{8.5}$$

With the linear combination method [14], the above formula can be further converted into a single-objective optimization problem:

$$\min_{\mathbf{X}, \mathbf{v}_{ij}} f_d(\mathbf{X}, \mathbf{v}_{ij}) = \left[(1-\beta) f^c(\mathbf{X}, \mathbf{v}_{ij}) + \beta f^w(\mathbf{X}, \mathbf{v}_{ij})\right] \tag{8.6}$$

It should be noted that the three parameters $\xi$, $\phi$, and $\psi$ in Eq. (2.36) here are not introduced into Eq. (8.6) for simplicity. The similar simplification will also be used in dealing with multi-objective optimization problems in subsequent sections.

In addition, based on the possibility degree model of interval number described in Chap. 3, the uncertain constraints in Eq. (8.3) can be transformed into following deterministic constraints:

$$P\left(g_j^I(\mathbf{X}, \mathbf{v}_{ij}) \le b_j^I\right) \ge \lambda_j,\ i, j = 1, 2, \ldots, l,\ i \ne j \tag{8.7}$$

where the interval $g_j^I(\mathbf{X}, \mathbf{v}_{ij})$ can be expressed as:

$$\begin{aligned}&g_j^L(\mathbf{X}, \mathbf{v}_{ij}) = \min_{\mathbf{U}} g_j(\mathbf{X}, \mathbf{U}, \mathbf{v}_{ij}),\ g_j^R(\mathbf{X}, \mathbf{v}_{ij}) = \max_{\mathbf{U}} g_j(\mathbf{X}, \mathbf{U}, \mathbf{v}_{ij})\\ &\mathbf{U} \in \Gamma = \left\{\mathbf{U}\middle|U_i^L \le U_i \le U_i^R,\ i = 1, 2, \ldots, q\right\}\end{aligned} \tag{8.8}$$

Through the above analysis, Eq. (8.3) is finally transformed into the deterministic single-objective optimization problem as follows:

$$\begin{cases}\min\limits_{\mathbf{X}, \mathbf{v}_{ij}} f_d(\mathbf{X}, \mathbf{v}_{ij}) = (1-\beta) f^c(\mathbf{X}, \mathbf{v}_{ij}) + \beta f^w(\mathbf{X}, \mathbf{v}_{ij})\\ \text{s.t.} P\left(g_j^I(\mathbf{X}, \mathbf{v}_{ij}) \le b_j^I\right) \ge \lambda_j\\ \mathbf{h}_{ji}^{\mathbf{v}} = \frac{\|\mathbf{v}_{ji} - \mathbf{y}_{ji}\|}{\|\mathbf{v}_{ji}\|} \le \varepsilon_{ji}\\ \mathbf{y}_{ji} = \mathbf{y}_{ji}(\mathbf{X}, \mathbf{U}^c, \mathbf{v}_{ij})\\ \\ i, j = 1, 2, \ldots, l,\ i \ne j, \mathbf{X} \in \Omega^n\end{cases} \tag{8.9}$$

The above equation still involves a two-layer nested optimization problem. In subsequent example analysis, the outer layer will adopt IP-GA [15] to conduct the design optimization, and the inner layer will use sequential quadratic program (SQP) [16] to calculate the intervals of the objective function and constraints.

## 8.4 Numerical Example and Engineering Application

### 8.4.1 Numerical Example

A numerical example modified from Ref. [17] is investigated. The example has a 3-dimensional design vector $\mathbf{X} = (X_1, X_2, X_3)$ and a 3-dimensional uncertain vector $\mathbf{U} = (U_1, U_2, U_3)$. Also, two disciplines that are coupled by state variables $y_{12}, y_{21}$ are involved. The constraints $g_1$ and $g_2$ and the objective functions $f_1 = 0.5(X_3 + U_3)^2 + X_1^2$ and $f_2 = 0.5(X_3 + U_3)^2 + X_2^2$ are obtained by analyses of Disciplines 2 and 3. The sum of $f_1$ and $f_2$ is regarded as the system objective function. The corresponding interval MDO problem is created as follows:

$$\begin{cases} \min\limits_{\mathbf{X}} f = f_1 + f_2 = (X_3 + U_3)^2 + X_1^2 + X_2^2 \\ \text{s.t. } g_1 = -2X_1 - X_3 + U_1 - U_3 - 2y_{21} \le 0 \\ \quad g_2 = 3X_2 + 5X_3 - U_2 + 5U_3 - 4y_{12} \le 0 \\ \quad y_{12} = X_1 + X_3 + U_3^c + y_{21},\ y_{21} = X_2 + X_3 + U_3^c - y_{12} \\ \quad U_1 \in [4.55, 5.45], U_2 \in [0.55, 1.45], U_3 \in [-0.45, 0.45] \\ \quad 0 \le X_i \le 5.0,\quad i = 1, 2, 3 \end{cases} \tag{8.10}$$

The IDF strategy is employed to decouple the multidisciplinary analysis, and then Eq. (8.10) can be transformed into the following conventional interval optimization problem:

$$\begin{cases} \min\limits_{\mathbf{X}, \mathbf{v}_{ij}} f = (X_3 + U_3)^2 + X_1^2 + X_2^2 \\ \text{s.t. } g_1 = -2X_1 - X_3 + U_1 - U_3 - 2y_{21} \le 0 \\ \quad g_2 = 3X_2 + 5X_3 - U_2 + 5U_3 - 4y_{12} < 0 \\ \quad \frac{\|v_{12} - y_{12}\|}{\|v_{12}\|} \le \varepsilon_{12},\ \frac{\|v_{21} - y_{21}\|}{\|v_{21}\|} \le \varepsilon_{21} \\ \quad y_{12} = X_1 + X_3 + U_3^c + y_{21},\ y_{21} = X_2 + X_3 + U_3^c - y_{12} \\ \quad U_1 \in [4.55, 5.45], U_2 \in [0.55, 1.45], U_3 \in [-0.45, 0.45] \\ \quad 0 \le X_i \le 5.0,\quad i = 1, 2, 3 \end{cases} \tag{8.11}$$

Based on the order relation transformation model, the above equation can be further transformed into a following deterministic optimization problem:

**Table 8.2** Optimization results of the numerical example

| Parameter | Optimization result |
|---|---|
| Design variables $(X_1, X_2, X_3)$ | (2.48, 1.68, 1.70) |
| Added design variables $(v_{12}, v_{21})$ | (3.76, −0.38) |
| State variables $(y_{12}, y_{21})$ | (3.80, −0.38) |
| Interval of objective function $f^I$ | [10.38, 13.44] |
| Interval of constraint 1 $g_1^I$ | [0.00, 1.80] |
| Interval of constraint 2 $g_2^I$ | [−0.19, 5.21] |
| Possible degrees of constraints $(P_1, P_2)$ | (1.000, 0.965) |

$$\begin{cases} \min\limits_{\mathbf{X},\mathbf{v}_{ij}} f_d(\mathbf{X}, \mathbf{v}_{ij}) = (1-\beta) f^c + \beta f^w \\ \text{s.t. } P(g_1 = -2X_1 - X_3 + U_1 - U_3 - 2y_{21} \le 0) \ge \lambda_1 \\ \quad P(g_2 = 3X_2 + 5X_3 - U_2 + 5U_3 - 4y_{12} \le 0) \ge \lambda_2 \\ \quad \frac{\|v_{12}-y_{12}\|}{\|v_{12}\|} \le \varepsilon_{12},\ \frac{\|v_{21}-y_{21}\|}{\|v_{21}\|} \le \varepsilon_{21} \\ \quad y_{12} = X_1 + X_3 + U_3^c + y_{21},\ y_{21} = X_2 + X_3 + U_3^c - y_{12} \\ \quad 0 \le X_i \le 5.0,\ \ i = 1, 2, 3 \end{cases} \tag{8.12}$$

In the optimization process, the parameters are set as follows: $\beta = 0.5$, $\lambda_1 = \lambda_2 = 0.96$, $\varepsilon_{12} = \varepsilon_{21} = 0.05$. The initial design vector is $\left(X_1^{(0)}, X_2^{(0)}, X_3^{(0)}, v_{12}^{(0)}, v_{21}^{(0)}\right) = (1.00, 1.00, 1.00, 1.00, 1.00)$. The optimization results are listed in Table 8.2. It can be found from the table that due to the uncertain vector **U** the objective function value at the optimal solution belongs to the interval [10.38, 13.44], and the intervals of constraint 1 and constraint 2 are [0.00, 1.80] and [−0.19, 5.21], respectively. The possibility degrees of two constraints are 1.000 and 0.965, respectively, both meeting the required levels $\lambda_1 = \lambda_2 = 0.96$. As mentioned above, the proposed method decouples the multidisciplinary analysis by adding new design variables and consistency constraints. The optimal added design vector and corresponding coupling vector are $(v_{12}, v_{21}) = (3.76, -0.38)$ and $(y_{12}, y_{21}) = (3.80, -0.38)$, which are very close and can meet the consistency constraints. It indicates the effectiveness of the IDF-based decoupling strategy in the proposed method.

### *8.4.2 Application to the Aerial Camera Design*

With the help of an unmanned aerial vehicle, the aerial camera system [18] is designed to acquire image information of the ground. As one of the most important electronic devices in the aerial camera system, the aerial camera is required to keep a stable and reliable performance under complex working conditions. In the design of the aerial camera, several electronic modules are highly integrated, including the high-definition image sensor, the signal processor, and the codec. Various extreme working conditions are coupled with each other, such as the vibration of the aerial platform, variation of the ambient temperature, and fluctuation of the power consumption. Consequently, the structural design of the aerial camera generally involves multidisciplinary analysis, and the design results will also play a key role in the overall performance of the camera. As shown in Fig. 8.3, an ultra-high-definition aerial camera is considered as an engineering example. The camera mainly contains the following modules: the lens, the motherboard, the filter and the housing. The camera faces two complex conditions when working: high temperature and vibrations. Through the temperature stress analysis for the high-temperature condition, it is found that significant temperature difference and thermal deformation will emerge inside the camera due to the interaction of the changing ambient temperature and the self-heating components. At the same time, the dynamic analysis for the vibration condition shows that different stress–strain responses of the camera structure will appear under vibrations of different frequencies. When the high temperature and vibration coexist, the thermal deformation inside the camera will exert a significant effect on the dynamic properties. Besides, the stress–strain response caused by the vibration can also affect the temperature–stress response. For example, the deformation of the motherboard will lead to the power dissipation fluctuation of the central processing unit (CPU), affecting the inner temperature response of the camera.

Figure 8.4 illustrates the coupling relationship between the analyses of the above two disciplines. The design variables $\mathbf{X} = (t, l)$ are the thickness $t$ of the housing wall and the length $l$ of the housing side; the design objective is to minimize the camera weight *Mass*. The parameter $P^{s}$ denotes the power dissipation of the image sensor on the motherboard; the parameter $E$ refers to the Young's modulus of the motherboard material. Besides, $D^{\text{CPU}}$ and $P^{\text{CPU}}$ are the state variables. $D^{\text{CPU}}$ means the thermal deformation of CPU from the analysis of the first discipline, and $P^{\text{CPU}}$ is the actual power dissipation of CPU from the analysis of the second discipline. There are 2 constraints: under the ambient temperature of 45 °C, the CPU temperature should be less than the given value $T_0^{\text{CPU}} = 66.0$ °C; under the vibration with the

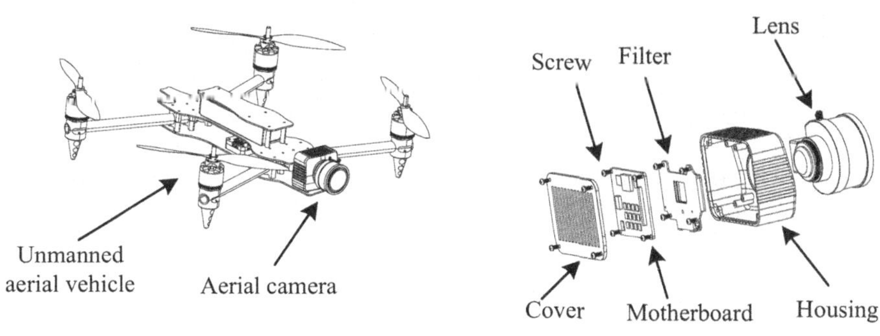


(a) The aerial camera system (b) The exploded-view of the aerial camera

**Fig. 8.3** An ultra-high-definition aerial camera [13]

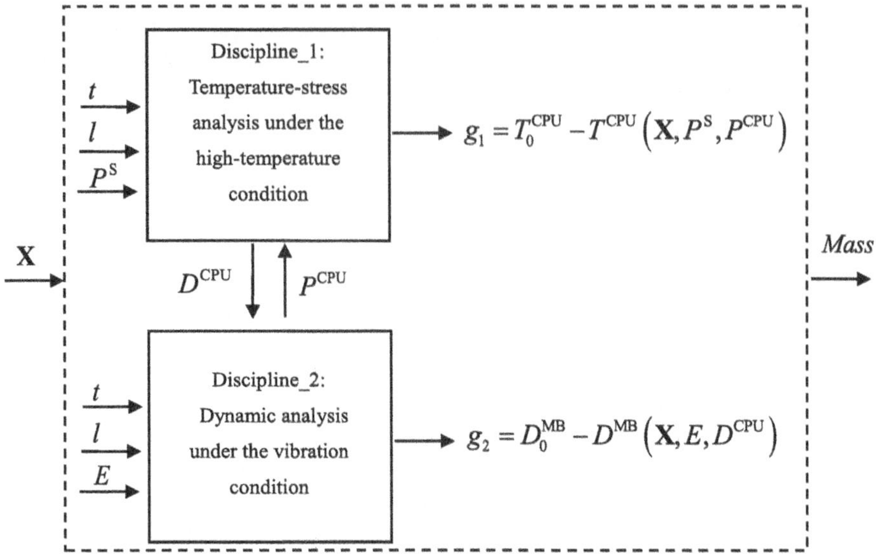


**Fig. 8.4** The multidisciplinary problem in the aerial camera design [19]

frequency range of [30 Hz, 200 Hz], the motherboard deformation should be less than the given value $D_0^{MB} = 1.00$ mm. The parameter vector $\mathbf{U} = (P^S, E)$ is uncertain and described by an interval vector. Therefore, the interval MDO problem can be constructed as follows:

$$\begin{cases} \min\limits_{t,l} Mass = -0.9914 \times t^2 - 3.761 \times t + 0.01086 \times l^2 \\ \qquad\qquad -0.6826 \times l + 0.6346 \times t \times l + 20.55 \\ \text{s.t. } g_1 = T^{CPU} \le T_0^{CPU}, g_2 = D^{MB} \le D_0^{MB} \\ \quad T^{CPU} = T^{CPU}(t, l, P^S, P^{CPU}), D^{MB} = D^{MB}(t, l, E, D^{CPU}) \\ \quad P^{CPU} = P^{CPU}(t, l, E, D^{CPU}), D^{CPU} = D^{CPU}(t, l, P^S, P^{CPU}) \\ \quad T_0^{CPU} = 66.0^\circ\text{C}, D_0^{MB} = 1.00, P^S \in [1.6\text{ W}, 2.4\text{ W}] \\ \quad E \in [1000\text{ MPa}, 12000\text{ MPa}] \\ \quad 1.00\text{ mm} \le t \le 3.00\text{ mm}, 30.00\text{ mm} \le l \le 40.00\text{ mm} \end{cases} \tag{8.13}$$

Through the IDF decoupling strategy and the order relation transformation model for interval optimization, Eq. (8.13) can be changed to the following deterministic optimization problem:

$$\left\{\begin{array}{l} \min\limits_{t,l} Mass = -0.9914 \times t^2 - 3.761 \times t + 0.01086 \times l^2 \\ \qquad -0.6826 \times l + 0.6346 \times t \times l + 20.55 \\ \text{s.t. } P\left(g_1 = T^{\text{CPU}} \le T_0^{\text{CPU}}\right) \ge \lambda_1,\ P\left(g_2 = D^{\text{MB}} \le D_0^{\text{MB}}\right) \ge \lambda_2 \\ \quad \frac{\|v_{12} - D^{\text{CPU}}\|}{\|v_{12}\|} \le \varepsilon_{12},\ \frac{\|v_{21} - P^{\text{CPU}}\|}{\|v_{21}\|} \le \varepsilon_{21} \\ \quad T^{\text{CPU}} = T^{\text{CPU}}\left(t, l, P^{\text{S}}, v_{21}\right),\ D^{\text{MB}} = D^{\text{MB}}(t, l, E, v_{12}) \\ \quad P^{\text{CPU}} = P^{\text{CPU}}(t, l, E, v_{12}),\ D^{\text{CPU}} = D^{\text{CPU}}\left(t, l, P^{\text{S}}, v_{21}\right) \\ \quad T_0^{\text{CPU}} = 66.0^{\circ}\text{C},\ D_0^{\text{MB}} = 1.00\ \text{mm},\ P^{\text{S}} \in [1.6\ \text{W}, 2.4\ \text{W}] \\ \quad E \in [1000\ \text{MPa}, 12000\ \text{MPa}] \\ \quad 1.00\ \text{mm} \le t \le 3.00\ \text{mm},\ 30.00\ \text{mm} \le l \le 40.00\ \text{mm} \end{array}\right. \tag{8.14}$$

where $v_{12}$, $v_{21}$ are the newly added design variables that are corresponding to the state variables $D^{\text{CPU}}$ and $P^{\text{CPU}}$, respectively.

The FEM model for the aerial camera is created as shown in Fig. 8.5. The FEM model contains 7 parts and 204,160 eight-node hexahedron elements. By setting different boundary conditions, the temperature-stress FEM model under high-temperature condition and the dynamic FEM model under vibration condition can be obtained, as shown in Fig. 8.6. To improve the optimization efficiency, the quadratic polynomial response surfaces of the response functions $T^{\text{CPU}}$, $D^{\text{MB}}$, $D^{\text{CPU}}$, $P^{\text{CPU}}$ are created based on 65 samples of these two FEM models. The response surfaces are listed in Table 8.3. To verify the accuracy of the response surfaces, six samples are randomly generated in the design space and the corresponding results obtained through the response surfaces and the FEM models are compared in Table 8.4, which shows the good accuracy of the created response surfaces.

To demonstrate the necessity of interval MDO, two cases are investigated. In Case I, the proposed method is used to solve the design problem, and the corresponding parameters are specified as $\beta = 0.5$, $\lambda_1 = \lambda_2 = 1.0$, $\varepsilon_{12} = \varepsilon_{21} = 0.01$. In Case II, the

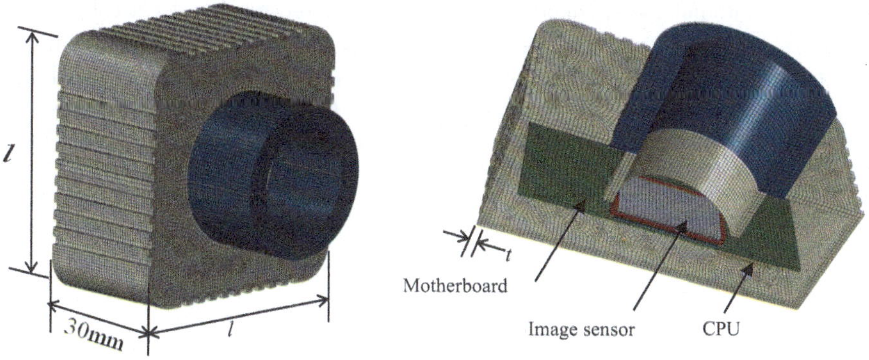


(a) Profile view (b) Sectional view

**Fig. 8.5** The FEM model of the aerial camera [19]

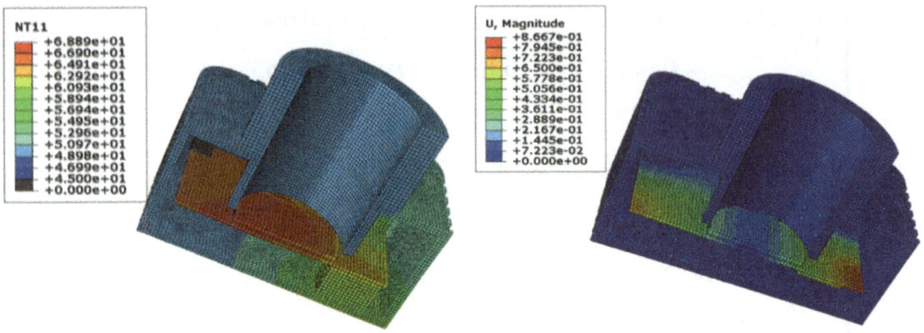


(a) The high-temperature condition (b) The vibration condition

**Fig. 8.6** FEM analyses to the aerial camera under different conditions [19]

uncertainties of the parameters are not taken into consideration, and the parameters are viewed as deterministic (taking midpoint values), hence Eq. (8.13) degrades into a conventional deterministic MDO problem which can be solved with the IDF method. In both cases, the initial design vector is $\left(t^{(0)}, l^{(0)}\right) = (2.5\text{ mm}, 35\text{ mm})$, and the optimization results are listed in Table 8.5. In Case I, the stable solution $(t, l) = (2.43\text{ mm}, 33.05\text{ mm})$ and corresponding camera weight $Mass = 45.86\text{g}$ are obtained by the proposed method. Due to the uncertainties introduced by the power dissipation of the image sensor and Young's modulus of the motherboard material, the CPU temperature and the motherboard deformation are intervals: $T^{\text{CPU}} \in [60.31\,°\text{C}, 65.62\,°\text{C}]$, $D^{\text{MB}} \in [0.77\text{ mm}, 0.94\text{ mm}]$. The possibility degrees of the two constraints are $(P_1, P_2) = (1.00, 1.00)$, which can meet the requirement of $\lambda_1 = \lambda_2 = 1.0$. In Case II, the optimal design vector and the minimal camera weight are $(t, l) = (1.72\text{ mm}, 33.05\text{ mm})$ and $Mass = 36.55\text{g}$, respectively, which are very different from those of Case I. A lighter camera weight is obtained in Case II because uncertainties of the parameters are ignored. If interval disturbance of the two parameters are taken into account under the results of Case II, $T^{\text{CPU}} \in [63.57\,°\text{C}, 68.95\,°\text{C}]$ and $D^{\text{MB}} \in [0.75\text{ mm}, 0.99\text{ mm}]$ can be obtained, with the possibility degrees of the two constraints $(P_1, P_2) = (0.46, 1.00)$. Obviously, the reliability of the first constraint cannot satisfy the requirement. The above analysis indicates that the interval MDO can obtain more reliable optimization results than the conventional deterministic MDO.

## 8.5 Summary

In this chapter, an interval MDO model and corresponding solution method are proposed, providing an analysis tool for the reliability design of complex products and systems. Firstly, the method adopts the IDF approach to decouple the multidisciplinary analysis. Secondly, the interval MDO problem is transformed into a

**Table 8.3** Response surfaces of the four response functions in the aerial camera design [19]

| Function | Response surface |
|---|---|
| $T^{\mathrm{CPU}}$ | $T^{\mathrm{CPU}} = 2.109 \times \left(P^{\mathrm{S}}\right)^2 - 2.563 \times P^{\mathrm{S}} \times v_{21} - 0.4142 \times P^{\mathrm{S}} + 2.559 \times l^2 + 0.0745 \times t \times l$ <br> $-18.58 \times t + 0.003174 \times l^2 - 0.4111 \times l + 4.028 \times (v_{21})^2 + 17.98 \times v_{21} + 81.17$ |
| $D^{\mathrm{MB}}$ | $D^{\mathrm{MB}} = -1.525 \times 10^{-7} \times E^2 + 3.802 \times 10^{-4} \times E \times v_{12} + 3.366 \times 10^{-3} \times E$ <br> $5.150 \times 10^{-3} \times t^2 - 7.875 \times 10^{-4} \times t \times l + 1.233 \times 10^{-2} \times t$ <br> $+1.563 \times 10^{-2} \times l^2 - 1.125 \times l - 2.041 \times 10^{-2} \times (v_{12})^2 - 8.381 \times v_{12} + 3.364$ |
| $D^{\mathrm{CPU}}$ | $D^{\mathrm{CPU}} = 4.387 \times 10^{-2} \times \left(P^{\mathrm{S}}\right)^2 + 4.688 \times 10^{-5} \times P^{\mathrm{S}} \times v_{21} - 0.1446 \times P^{\mathrm{S}} - 2.137 \times 10^{-2} \times t^2$ <br> $-1.055 \times 10^{-4} \times t \times l + 7.399 \times 10^{-2} \times t - 9.340 \times 10^{-5} \times l^2 + 1.560 \times 10^{-2} \times l$ <br> $+1.072 \times 10^{-2} \times (v_{21})^2 + 4.915 \times 10^{-2} \times v_{21} - 0.1738$ |
| $P^{\mathrm{CPU}}$ | $P^{\mathrm{CPU}} = 9.292 \times 10^{-8} \times E^2 - 3.388 \times 10^{-4} \times E \times v_{12} - 2.056 \times 10^{-3} \times E - 6.833 \times 10^{-4} \times t$ <br> $-5.95 \times 10^{-4} \times t \times l + 1.690 \times 10^{-2} \times t - 5.425 \times 10^{-3} \times l^2 + 3.989 \times l$ <br> $+3.401 \times 10^{-3} \times (v_{12})^2 + 7.266 \times v_{12} + 3.971$ |

**Table 8.4** Accuracy test of the response surfaces in the aerial camera design [19]

| $(t, l, v_{12}, v_{21}, P^S, E)$ | Relative error from the FEM results | | | |
|---|---|---|---|---|
| | $T^{CPU}$ (%) | $D^{MB}$ (%) | $D^{CPU}$ (%) | $P^{CPU}$ (%) |
| (2.63, 39.06, 0.28, 0.29, 2.05, 11,302) | 4.35 | 1.17 | 0.27 | 3.66 |
| (1.56, 35.47, 0.77, 0.29, 1.86, 10,021) | 2.00 | 3.54 | 0.58 | 9.70 |
| (2.91, 34.85, 0.68, 0.13, 1.97, 11,902) | 4.58 | 4.93 | 1.72 | 4.05 |
| (2.58, 39.60, 0.59, 0.11, 2.14, 11,023) | 4.41 | 3.23 | 2.08 | 4.33 |
| (2.36, 37.58, 0.65, 0.18, 2.06, 10,458) | 1.46 | 3.26 | 4.05 | 2.08 |
| (2.41, 30.32, 0.37, 0.12, 1.84, 12,398) | 4.31 | 3.44 | 3.26 | 2.34 |

**Table 8.5** MDO results for the aerial camera

| Parameter | Optimization result | |
|---|---|---|
| | Interval MDO | Deterministic MDO |
| Design variables $(t, l)$ (mm) | (2.43, 33.05) | (1.72, 33.05) |
| Thermal deformation of CPU $D^{CPU}$ (mm) | 0.54 | 0.51 |
| Power dissipation of CPU $P^{CPU}$ (W) | 0.20 | 0.20 |
| Objective function $Mass$ (g) | 45.86 | 36.55 |
| CPU temperature $g_1 = T^{CPU}$ (°C) | [60.31, 65.62] | 65.92 |
| Motherboard deformation $g_2 = D^{MB}$ (mm) | [0.77, 0.94] | 0.99 |
| Possibility degrees of constraints $(P_1, P_2)$ | (1.00, 1.00) | – |

conventional deterministic optimization problem based on the order relation transformation model. Finally, a numerical example and an engineering application are used to illustrate the effectiveness of the method. In the construction of the proposed method, the IDF strategy is employed to decouple the multiple disciplines. In the future, other analysis strategies for MDO problems, such as AAO, CO, BLISS, can also be employed to construct the corresponding interval MDO methods.

## References


1. Balling RJ, Sobieszczanski-Sobieski J (1995) An algorithm for solving the system-level problem in multilevel optimization. Struct Optim 9(3):168–177
2. Simpson T, Toropov V, Balabanov V, Viana F (2008) Design and analysis of computer experiments in multidisciplinary design optimization: a review of how far we have come-or not. In: 12th AIAA/ISSMO multidisciplinary analysis and optimization conference, Victoria, British Columbia, Canada, September
3. Tedford NP, Martins JRRA (2010) Benchmarking multidisciplinary design optimization algorithms. Optim Eng 11(1):159–183
4. Martins JRRA, Lambe AB (2013) Multidisciplinary design optimization: a survey of architectures. AIAA J 51(9):2049–2075
5. Hajela P (1999) Nongradient methods in multidisciplinary design optimization-status and potential. J Aircraft 36(1):255–265
6. Cramer EJ, Dennis Jr JE, Frank PD, Lewis RM, Shubin GR (1994) Problem formulation for multidisciplinary optimization. SIAM J Optim 4(4):754–776
7. De La Garza A, Darmofal D (1998) An all-at-once approach for multidisciplinary design optimization. In: 16th AIAA applied aerodynamics conference, Albuquerque, NM, USA, June
8. Allison J, Kokkolaras M, Papalambros P (2005) On the impact of coupling strength on complex system optimization for single-level formulations. In: ASME 2005 international design engineering technical conferences and computers and information in engineering conference, Long Beach, California, September
9. Wujek B, Renaud JE, Batill S (1997) A concurrent engineering approach for multidisciplinary design in a distributed computing environment. Multidis Design Optim State Art 13–16
10. Braun RD, Kroo IM (1995) Development and application of the collaborative optimization architecture in a multidisciplinary design environment. Technical report, NASA Langley Research Center
11. Sobieszczanski-Sobieski J, Altus TD, Phillips M, Sandusky R (2003) Bilevel integrated system synthesis for concurrent and distributed processing. AIAA J 41(10):1996–2003
12. Batill S, Renaud J, Gu X (2000) Modeling and simulation uncertainty in multidisciplinary design optimization. In: 8th Symposium on multidisciplinary analysis and optimization, Long Beach, CA, U.S.A., September
13. Huang ZL, Zhou YS, Jiang C, Zheng J, Han X (2018) Reliability-based multidisciplinary design optimization using incremental shifting vector strategy and its application in electronic product design. Acta Mech Sin 34(2):285–302
14. Hu YD (1990) Practical multi-objective optimization. Shanghai Technological Press, Shanghai
15. Xu YG, Li GR, Wu ZP (2001) A novel hybrid genetic algorithm using local optimizer based on heuristic pattern move. Appl Artifi Intel 15(7):601–631

16. Nocedal J, Wright SJ (1999) Numerical optimization. Springer, New York
17. Du XP, Guo J, Beeram H (2008) Sequential optimization and reliability assessment for multidisciplinary systems design. Struct Multidis Optim 35(2):117–130
18. Sandau R (2009) Digital airborne camera: introduction and technology. Springer, Berlin
19. Huang ZL (2017) Reliability-based design optimization and the applications in electronic product structural design. PhD Hunan

# Chapter 9
# A New Type of Possibility Degree of Interval Number and Its Application in Interval Optimization

**Abstract** This chapter first proposes a new possibility degree model of interval number to realize quantitative comparison for not only overlapping intervals but also separate intervals, and then applies it to the nonlinear interval optimization.

Comparison or ranking between intervals is an essential step during interval optimization [1]. For example, in the order relation transformation model the order relation of interval number and the possibility degree of interval number are used to deal with the uncertain objective function and constraint, respectively, so that the uncertain optimization problem can be transformed into a deterministic optimization problem. As mentioned in Chap. 3, interval ranking models can be generally classified into two categories: the order relation of interval number which is used to qualitatively judge whether an interval is greater or superior to another interval, and the possibility degree of interval number which is used to quantitatively describe the degree that an interval is greater or superior to another interval. In this book, these two kinds of ranking models are also called P-ICR (the preference-based interval comparison relation) and V-ICR (the value-based interval comparison relation), respectively. In previous chapters, the possibility degree of interval number that is used in interval optimization has a value range of [0, 1]. It has been found that by using this model the relative position is well represented for overlapping intervals, but not for separate intervals because the comparison result (either 1 or 0) does not tell how far two intervals are.

In this chapter, we will propose a new possibility degree of interval number that works not only for overlapping intervals but also for separate intervals, which could offer a more effective analysis tool for interval optimization. The outline of this chapter is as follows: firstly, the current researches on the interval ranking relations are briefly reviewed, including introduction and assessment of three main types of possibility degrees of interval number. Then a new type of possibility degree model, named as the reliability-based possibility degree of interval number (RPDI), is proposed, which is then applied to interval optimization problems. Examples including an engineering application will be given at the end.



C. Jiang et al., *Nonlinear Interval Optimization for Uncertain Problems*, Springer Tracts in Mechanical Engineering, https://doi.org/10.1007/978-981-15-8546-3_9

## 9.1 Three Existing Possibility Degree Models of Interval Number and Their Disadvantages

The first study on the P-ICR model dates back to the research by Moore [2, 3], who brought the "<" and "⊂", two conventional operators for real numbers, to interval numbers, introducing two types of partial ordering relations for intervals. His work was further improved by Ishibuchi and Tanaka [4]. They proposed two types of interval ranking models: the first one is based on the preference to the bounds of interval, and the second is based on the preference to midpoint and radius of interval. Later, Chanas and Kuchtathe [5, 6] put these two ranking models together and proposed a uniform formulation. The above methods all belong to the P-ICR, which only tells the information "which one is better" between two intervals, but not "how much better". Therefore, P-ICR cannot quantitatively compare two intervals. On the contrary, V-ICR can quantitatively compare two intervals, which makes it able to play an important role in solving the interval optimization problem, especially in dealing with uncertain constraints. Most of the V-ICRs are based on the probability theory where it is assumed that the given intervals are both random variables with uniform distribution. Hence, the possibility that "how much better" can be viewed as the probability of one interval's random variable larger than the random variable of another one. On this subject, Nakahara's work [7] was one of the first researches that applies the probability method to comparing the interval numbers. Their study laid the groundwork for several follow-up approaches in this field [8–15]. Besides, Sevastjanov [16] proposed an interval comparison model based on the evidence theory in terms of the probability interval of the comparison results. Sevastjanov and Rog [17] developed a two-objective interval comparison approach. Jiang et al. [18] introduced a modified possibility degree of interval number, and then used it in nonlinear interval optimization. The modified method is consistent with the model presented by Kundu [11] while it has a simpler expression. In addition to the probability method, there are also V-ICR models based on other techniques [19–26].

In current V-ICR, the possibility degree of interval number generally comes in three forms [23, 24, 27]:

$$P\left(A^I \le B^I\right) = \min\left\{\max\left\{\frac{B^R - A^L}{2A^w + 2B^w}, 0\right\}, 1\right\} \tag{9.1}$$

$$P\left(A^I \le B^I\right) = \frac{\max\left\{0,\ 2A^w + 2B^w - \max\left\{A^R - B^L, 0\right\}\right\}}{2A^w + 2B^w} \tag{9.2}$$

$$P\left(A^I \le B^I\right) = \frac{\min\left\{2A^w + 2B^w,\ \max\left\{B^R - A^L, 0\right\}\right\}}{2A^w + 2B^w} \tag{9.3}$$

where $A^I$ and $B^I$ are two intervals to be compared; $P\left(A^I \le B^I\right)$ represents the possibility that $A^I$ is less than or equal to $B^I$ (or $B^I$ is greater than or equal to $A^I$). The three forms have been proven to be equivalent to each other [27]. Besides, the possibility degree proposed by Sun and Yao [22] is equivalent to the above three forms, as well as the model presented earlier in Chap. 3. $P(A^I \leq B^I)$ has the following properties:

(1) $0 \leq P(A^I \leq B^I) \leq 1$.
(2) If $A^R \leq B^L$, then $P(A^I \leq B^I) = 1$. It represents that $A^I$ is absolutely less than or equal to $B^I$. On the real axis, $A^I$ lies on the left side of $B^I$.
(3) If $A^L \geq B^R$, then $P(A^I \leq B^I) = 0$. It represents that $A^I$ is absolutely greater than $B^I$. On the real axis, $A^I$ lies on the right side of $B^I$.
(4) (Complementarity) If $P(A^I \leq B^I) = q$, then $P(A^I \leq B^I) = 1 - q$, where $q \in [0, 1]$.
(5) $P(A^I \leq B^I) = \frac{1}{2}$ only if $A^L + A^R = B^L + B^R$.
(6) (Transitivity) For three intervals $A^I$, $B^I$ and $C^I$, if $P(A^I \leq B^I) \geq q$ and $P(B^I \leq C^I) \geq q$, then $P(A^I \leq C^I) \geq q$, where $q \in [0, 1]$.

Figure 9.1 shows how $P(A^I \leq B^I)$ (the dashed line) changes when fixing $A^c$, $A^w$, $B^w$, and moving midpoint $B^c$. The vertical axis represents the value of the possibility degree. As $B^I$ moves right along the horizontal axis, the state of $P(A^I \leq B^I)$ separates into three parts: constant 0 (State 1), a monotonous rise from 0 to 1 (State 2), and constant 1 (State 3). The inflection points occur where $B^R = A^L$ and $B^L = A^R$. $A^I$ partly or completely overlaps $B^I$ in State 2. As $B^I$ moves closer to the right end of $A^I$, $A^I$ is much more likely to be smaller than $B^I$. Hence, $P(A^I \leq B^I)$ increases correspondingly. The above model has been successfully applied to interval optimization due to its simple expression and ease of use. However, it also has some defects so that its application is limited. First of all, the possibility degree of interval number works only if two intervals overlap (State 2). As for separate intervals (State 1 and State 3), it provides very little information since the result always turns out 0 or 1 despite the relative position of two intervals. In practical engineering, such relative

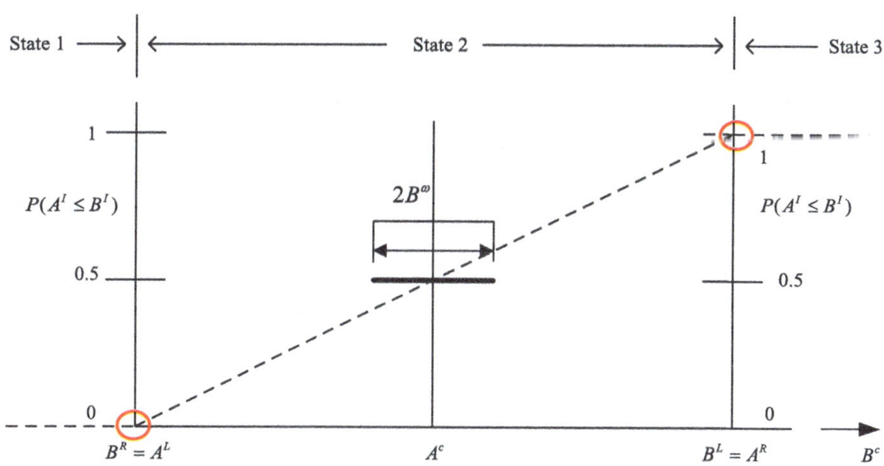


**Fig. 9.1** Variation of the possibility degree of interval number by only changing $B^c$ [28]

position of parameters is usually related to the reliability and safety of a system or structure. To better illustrate this problem, we apply the possibility degree to the strength reliability analysis of a structure. Here, the intervals $A^I$ and $B^I$ represent the structural stress and strength, respectively. The structure is deemed reliable if $A^I$ is smaller than $B^I$. Six cases are considered here [28]:

$$\begin{aligned}
&\text{Case 1}: \ A^I = [150\,\text{MPa}, 170\,\text{MPa}],\ B^I = [200\,\text{MPa}, 220\,\text{MPa}] \\
&\text{Case 2}: \ A^I = [120\,\text{MPa}, 140\,\text{MPa}],\ B^I = [200\,\text{MPa}, 220\,\text{MPa}] \\
&\text{Case 3}: \ A^I = [195\,\text{MPa}, 215\,\text{MPa}],\ B^I = [200\,\text{MPa}, 220\,\text{MPa}] \\
&\text{Case 4}: \ A^I = [185\,\text{MPa}, 205\,\text{MPa}],\ B^I = [200\,\text{MPa}, 220\,\text{MPa}] \\
&\text{Case 5}: \ A^I = [260\,\text{MPa}, 280\,\text{MPa}],\ B^I = [200\,\text{MPa}, 220\,\text{MPa}] \\
&\text{Case 6}: \ A^I = [230\,\text{MPa}, 250\,\text{MPa}],\ B^I = [200\,\text{MPa}, 220\,\text{MPa}]
\end{aligned} \tag{9.4}$$

As seen in Fig. 9.2, $A^I$ lies entirely on the left side of $B^I$ in both Case 1 and Case 2, and the distance between the intervals in Case 2 appears noticeably larger than that in Case 1. The implication is that the stress appears less likely to be larger than the strength in Case 2, and thus Case 2 is more reliable. In the same way, we can infer that Case 4 is more reliable than Case 3, and so do Case 6 and Case 5. Next, we apply the above possibility degree of interval number $P(A^I \leq B^I)$ to analyze this problem. The results are as follows: the possibility degrees of Case 1 and Case 2 are both 1; the possibility degrees of Case 3 and Case 4 are 0.625 and 0.875, respectively; the possibility degrees of Case 5 and Case 6 are both 0. It is easy to conclude that only results of overlapping intervals (Case 3 and Case 4) are reasonable.

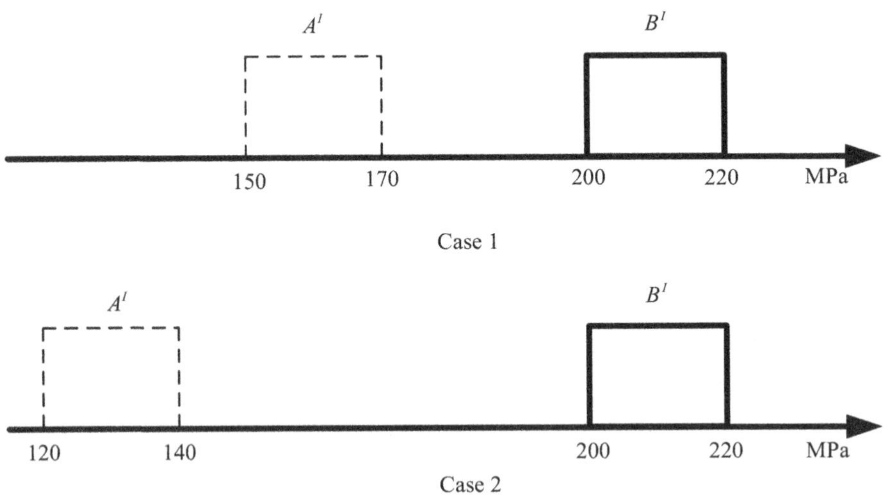


**Fig. 9.2** Relative positions of the stress interval $A^I$ and the strength interval $B^I$ [28]

However, for separate intervals (Case 1 and Case 2, or Case 5 and Case 6) the results cannot tell which case is more reliable since $P(A^I \leq B^I)$ is always 0 or 1. As this example shows, the above possibility degree of interval number cannot quantitatively compare the intervals for all situations, which may hinder its application in practical engineering problems. Therefore, it is necessary to develop a new possibility degree of interval number with more extensive applicability.

Secondly, as shown in Fig. 9.1 there are two inflection points for the possibility degree which may lead to the non-differentiability problem in practical use. For example, consider a simple interval inequality with a variable $x$:

$$[x+1,\ x+2] \leq [8,\ 10] \tag{9.5}$$

We can calculate its possibility degree of interval:

$$P([x+1,\ x+2] \leq [8,\ 10]) = \begin{cases} 1, & x \leq 6 \\ \frac{9-x}{3}, & 6 < x \leq 9 \\ 0, & x > 9 \end{cases} \tag{9.6}$$

It is obvious that Eq. (9.6) is a continuous function with respect to $x$, while it is non-differentiable at $x = 6$ and $x = 9$. If we apply the possibility degree to an interval optimization problem, the resulting optimization problem may not be continuously differentiable, so the gradient-based opotimization methods may become noneffective, bringing difficulties to the problem.

## 9.2 The Reliability-Based Possibility Degree of Interval Number

A new type of possibility degree model termed the reliability-based possibility degree of interval number (RPDI) is proposed in this chapter based on existing researches to overcome the disadvantages of current models. In RPDI, the variation characteristic in Case 2 is extended to Case 1 and Case 3. Thus for intervals $A^I$ and $B^I$ the expression of RPDI can be written as:

$$P_r(A^I \leq B^I) = \frac{B^R - A^L}{2A^w + 2B^w} \tag{9.7}$$

In order to distinguish the new model and the traditional possibility degree models, $P_r$ is used to represent RPDI. If either $A^I$ or $B^I$ degenerates into a real number, denoted as $A$ or $B$, Eq. (9.7) still holds and can be rewritten as:

$$P_r(A \leq B^I) = \frac{B^R - A}{2B^w},\quad P_r(A^I \leq B) = \frac{B - A^L}{2A^w} \tag{9.8}$$

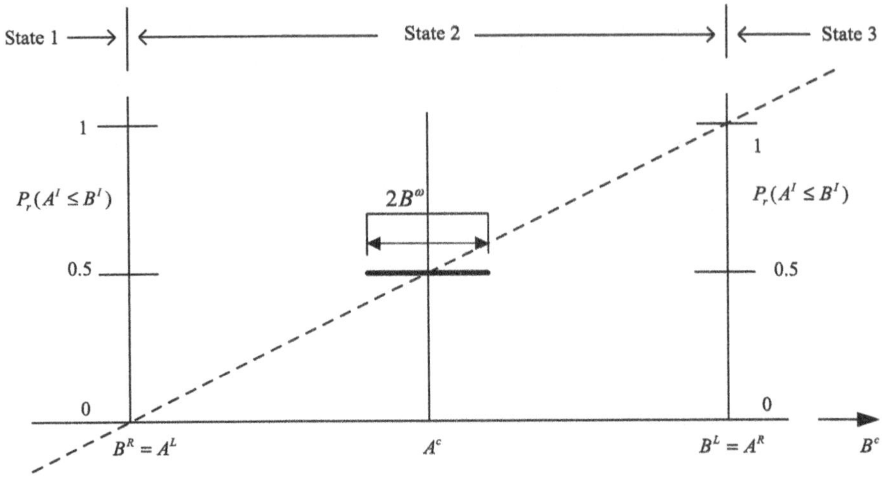


**Fig. 9.3** Variation of the RPDI model by only changing $B^c$ [28]

$P_r(A^I \leq B^I)$ has the following properties:

(1) $-\infty \leq P_r(A^I \leq B^I) \leq +\infty$.
(2) If $A^R \leq B^L$, then $P_r(A^I \leq B^I) \geq 1$. On the real axis, $A^I$ lies on the left side of $B^I$.
(3) If $A^L \geq B^R$, then $P_r(A^I \leq B^I) \leq 0$. On the real axis, $A^I$ lies on the right side of $B^I$.
(4) (Complementarity) If $P_r(A^I \leq B^I) = q$, then $P_r(A^I \leq B^I) = 1 - q$, where $q \in [-\infty, +\infty]$.
(5) $P_r(A^I \leq B^I) = \frac{1}{2}$, only if $A^L + A^R = B^L + B^R$.
(6) (Transitivity) For three intervals $A^I$, $B^I$ and $C^I$, if $P_r(A^I \leq B^I) \geq q$ and $P_r(B^I \leq C^I) \geq q$, then $P_r(A^I \leq C^I) \geq q$, where $q \in [-\infty, +\infty]$.

Figure 9.3 shows how $P_r(A^I \leq B^I)$ changes when fixing $A^c$, $A^w$, $B^w$ and moving the midpoint $B^c$. When $B^I$ moves right along the horizontal axis, $P_r(A^I \leq B^I)$ (the dashed line) grows linearly on the whole real axis, unlike what is shown in Fig. 9.1. It stands to reason since a greater value of $B^c$ will increases the likelihood of $A^I \leq B^I$ if the radius is given. Figure 9.4 shows how $P_r(A^I \leq B^I)$ changes when fixing $A^c$, $A^w$, $B^w$ and moving the radius $B^w$. With $B^w$ rising from 0 to $\infty$, $P_r(A^I \leq B^I)$ (the dashed line) shows two kinds of nonlinear tendencies depending on its value at $B^w = 0$, denoted by $P_{r0}$. If $P_{r0} > 0.5$, $P_r(A^I \leq B^I)$ increases monotonously; if $P_{r0} < 0.5$, it decreases monotonously. In both cases, it approaches 0.5 as $B^w$ tends to $\infty$. Besides, if $P_{r0} = 0.5$, $P_r(A^I \leq B^I)$ stays a constant. As shown in Fig. 9.3, the RPDI works for both overlapping and separate intervals. For the overlapping intervals, both the RPDI and the current possibility degree model are effective. However, for separate intervals, RPDI is still effective (the result is not 0 or 1 anymore but depends on the relative position of the intervals) while the current possibility degree model becomes inaccurate. The problem in Eq. (9.4) is reanalyzed by RPDI and following results can

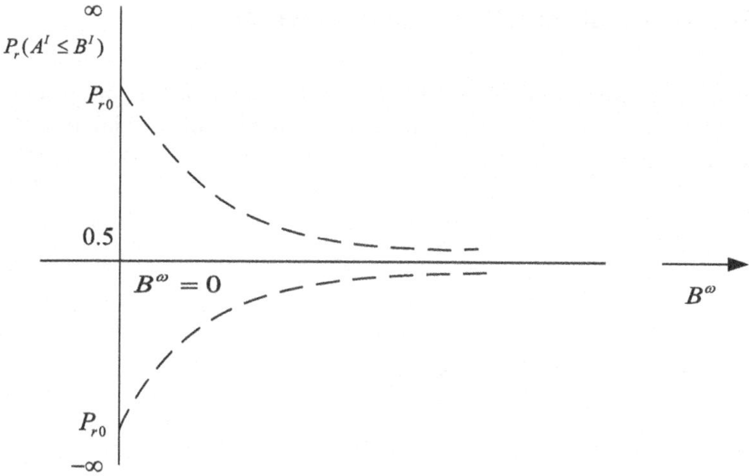


**Fig. 9.4** Variation of the RPDI model by only changing $B^w$ [28]

be obtained: $P_r = 1.75$ for Case 1 and $P_r = 2.5$ for Case 2; $P_r = 0.625$ for Case 3 and $P_r = 0.875$ for Cases 4; $P_r = -1.0$ for Case 5 and $P_r = -0.25$ for Cases 6. The results suggest Case 2, 4, and 6 are more reliable than Case 1, 3, and 5, respectively, which indicates RPDI can successfully reflect the reliability of a system. Besides, a large RPDI value represents a high reliability.

On the other hand, as shown in Fig. 9.3 the RPDI varies continuously and smoothly, so the inflection points do not occur. For example, we use the RPDI to analyze Eq. (9.5) and obtain:

$$P([x+1,\ x+2] \le [8,\ 10]) = \frac{9-x}{3}, \quad -\infty < x < \infty \tag{9.9}$$

where $P_r$ is continuous and differentiable for all real numbers, and this characteristic is very helpful for interval optimization problems.

Above analysis indicates that the RPDI overcomes main disadvantages of current possibility degree of interval number and it has a simpler and more convenient mathematical expression. However, RPDI cannot be viewed as a simple transformation of current possibility degree model even they share similar mathematical expression. Actually, RPDI is an important extension of the current model, since it expands the value range of possibility degree from [0, 1] to $[-\infty, \infty]$ and makes it possible to quantitatively compare any intervals on the real line. Therefore, RPDI can be used as an effective mathematical tool for reliability analysis of systems or structures under interval uncertainties.

## 9.3 Interval Optimization Based on RPDI

This section will use the RPDI model to construct corresponding interval optimization methods. Since in the formulation of subsequent nonlinear interval optimization algorithm the linear interval optimization will be involved in each iteration, we first give two linear interval optimization methods based on the RPDI model in Sect. 9.3.1. Then a nonlinear interval optimization method is presented in the following Sect. 9.3.2.

### *9.3.1 Linear Interval Optimization*

A general linear interval optimization problem can be expressed as follows:

$$\begin{cases} \min\limits_{\mathbf{X}} f(\mathbf{X}, \mathbf{c}) = \sum\limits_{i=1}^{n} c_i^I X_i \\ \text{s.t.} \ \ g_j(\mathbf{X}, \mathbf{a}) = \sum\limits_{i=1}^{n} a_{ij}^I X_i \leq b_j^I, \ \ j = 1, 2, \ldots, l \\ X_i \geq 0, \ \ i = 1, 2, \ldots, n \end{cases} \tag{9.10}$$

where $\mathbf{X}$ is an $n$-dimensional design vector, $f$ denotes the objective function, $g$ denotes the constraint, $l$ is the number of the constraints; $\mathbf{c}$ is an $n$-dimensional coefficient vector in the objective function, and $\mathbf{a}$ is an $n \times l$ coefficient matrix in the constraints; due to uncertainties, the elements of $\mathbf{c}$ and $\mathbf{a}$ are all intervals; $b_j^I$ is the allowable interval of the $j$-th constraint.

The uncertain constraints in Eq. (9.10) can be changed to the following deterministic constraints by using the RPDI model [28]:

$$P_r\left(g_j^I(\mathbf{X}) \leq b_j^I\right) = \frac{b_j^R - g_j^I(\mathbf{X})}{2g_j^w(\mathbf{X}) + 2b_j^w} \geq \lambda_j, \ \ j = 1, 2, \ldots, l \tag{9.11}$$

where $\lambda_i \in [-\infty, +\infty]$ represents the RPDI level of the $j$-th constraint. The value of $\lambda$ determinates the feasible region of $\mathbf{X}$ and a larger value represents a higher reliability requirement on the uncertain constraint. Next, the two different transformation models introduced in Chap. 3 are adopted, respectively, to deal with the above linear interval optimization problem, in which the treatments for the uncertain constraints are the same while the treatments to the uncertain objective function are different.

1. Based on the order relation transformation model

In this approach, the order relation of interval number $\leq_{cw}$ is applied to the objective function, and the RPDI is applied to the constraints, converting Eq. (9.10) into a deterministic multi-objective optimization problem:

$$\begin{cases} \min\limits_{\mathbf{X}} \left[ f^c(\mathbf{X}), f^w(\mathbf{X}) \right] \\ \text{s.t. } P_r\left(g_j^I(\mathbf{X}) \le b_j^I\right) = \dfrac{b_j^R - g_j^L(\mathbf{X})}{2g_j^w(\mathbf{X}) + 2b_j^w} \ge \lambda_j,\ j = 1, 2, \ldots, l \\ X_i \ge 0,\ i = 1, 2, \ldots, n \end{cases} \tag{9.12}$$

Contrary to the nonlinear case, the intervals of the uncertain objective function and constraints at any known $\mathbf{X}$ can be analytically obtained in a linear interval optimization problem:

$$\begin{aligned} & f^L(\mathbf{X}) = \sum_{i=1}^{n} c_i^L X_i,\ f^R(\mathbf{X}) = \sum_{i=1}^{n} c_i^R X_i \\ & g_j^L(\mathbf{X}) = \sum_{i=1}^{n} a_{ij}^L X_i,\ g_j^R(\mathbf{X}) = \sum_{i=1}^{n} a_{ij}^R X_i,\ j = 1, 2, \ldots, l \end{aligned} \tag{9.13}$$

Substituting Eq. (9.13) into Eq. (9.12), we have

$$\begin{cases} \min\limits_{\mathbf{X}} \left[ \dfrac{\sum\limits_{i=1}^{n} c_i^L X_i + \sum\limits_{i=1}^{n} c_i^R X_i}{2}, \dfrac{\sum\limits_{i=1}^{n} c_i^R X_i - \sum\limits_{i=1}^{n} c_i^L X_i}{2} \right] \\ \text{s.t. } \sum\limits_{i=1}^{n} \left[ \lambda_j \left(a_{ij}^R - a_{ij}^L\right) + a_{ij}^L \right] X_i \le b_j^R - \lambda_j \left(b_j^R - b_j^L\right),\ j = 1, 2, \ldots, l \\ X_i \ge 0,\ i = 1, 2, \ldots, n \end{cases} \tag{9.14}$$

Using the linear combination method [29], we further reformulate the problem as a single-objective optimization problem:

$$\begin{cases} \min\limits_{\mathbf{X}} f_d(\mathbf{X}) = \sum\limits_{i=1}^{n} \left[ \left(\frac{1}{2} - \beta\right) c_i^L + \frac{1}{2} c_i^R \right] X_i \\ \text{s.t. } \sum\limits_{i=1}^{n} \left[ \lambda_j \left(a_{ij}^R - a_{ij}^L\right) + a_{ij}^L \right] X_i \le b_j^R - \lambda_j \left(b_j^R - b_j^L\right),\ j = 1, 2, \ldots, l \\ X_i \ge 0,\ i = 1, 2, \ldots, n \end{cases} \tag{9.15}$$

where $0 \le \beta \le 1$ is a weighting factor of the two objective functions. Equation (9.15) is actually a conventional linear programming problem that can be solved by many well-established methods such as the simplex method [30].

2. Based on the possibility degree transformation model

Here a performance interval $V^I = \left[V^L, V^R\right]$ needs to be introduced into the objective function. Besides, the RPDI is used to deal with the constraints, converting Eq. (9.10) into a deterministic optimization problem:

$$\begin{cases} \max\limits_{\mathbf{X}} P_r\left(f^I(\mathbf{X}) \leq V^I\right) = \dfrac{V^R - f^L(\mathbf{X})}{2f^w(\mathbf{X}) + 2V^w} \\ \text{s.t. } P_r\left(g_j^I(\mathbf{X}) \leq b_j^I\right) = \dfrac{b_j^R - g_j^L(\mathbf{X})}{2g_j^w(\mathbf{X}) + 2b_j^w} \geq \lambda_j,\ j = 1, 2, ..., l \\ X_i \geq 0,\ i = 1, 2, ..., n \end{cases} \tag{9.16}$$

Based on Eq. (9.13), Eq. (9.16) can be further rewritten as

$$\begin{cases} \max\limits_{\mathbf{X}} \dfrac{V^R - \sum\limits_{i=1}^{n} c_i^L X_i}{2V^w + \sum\limits_{i=1}^{n} \left(c_i^R - c_i^L\right) X_i} \\ \text{s.t. } \sum\limits_{i=1}^{n} \left[\lambda_j\left(a_{ij}^R - a_{ij}^L\right) + a_{ij}^L\right] X_i \leq b_j^R - \lambda_j\left(b_j^R - b_j^L\right),\ j = 1, 2, ..., l \\ X_i \geq 0,\ i = 1, 2, ..., n \end{cases} \tag{9.17}$$

which is a nonlinear programming problem with linear constraints and can be solved by many existing optimization methods [30].

In Chap. 3, an optimal design vector $\mathbf{X}^*$ that maximizes the possibility degree of objective function to 1 is possible to be found when solving the transformed deterministic optimization problem. However, there are generally multiple solutions that make $P_{\max} = P\left(f^I(\mathbf{X}^*) \leq V^I\right) = 1$. Therefore, the robustness criterion sometimes will be introduced to reconstruct another interval optimization problem for searching a better solution. However, the value range of the RPDI covers $(-\infty, +\infty)$ rather than [0, 1], so the solution to Eq. (9.17) is usually unique. Therefore, it is unnecessary to perform another interval optimization, which illustrates again that the RPDI model has good applicability in interval optimization.

### 9.3.2 Nonlinear Interval Optimization

Nonlinear interval optimization problems are far more complicated than linear interval optimization problems. This section will use RPDI and the order relation transformation model to deal with the nonlinear interval optimization problem in Eq. (4.1). In Chap. 6, the sequential linear programming approach is used to convert the nonlinear interval optimization to a series of linear interval optimization problems. Also, the convergence was ensured with help of the iteration mechanism. Here such an idea is adopted again. The difference is that the linear interval optimization problem in the $s$-th iteration as shown in Eq. (6.1) is converted to a following conventional linear programming problem by the method proposed in Sect. 9.3.1 [31]:

$$
\begin{cases}
\min\limits_{\mathbf{X}} \tilde{f}_d = \sum\limits_{i=1}^{n} \frac{\partial f(\mathbf{X}^{(s)},\mathbf{U}^c)}{\partial X_i}\beta X_i + \beta\left(f\left(\mathbf{X}^{(s)},\mathbf{U}^c\right)\right) - \sum\limits_{i=1}^{n} \frac{\partial f(\mathbf{X}^{(s)},\mathbf{U}^c)}{\partial X_i} X_i^{(s)} \\
\qquad + \sum\limits_{i=1}^{q} \left|\frac{\partial f(\mathbf{X}^{(s)},\mathbf{U}^c)}{\partial U_i}\right|(1-\beta)U_i^w \\
\text{s.t.} \ \sum\limits_{i=1}^{n} \frac{\partial g_j(\mathbf{X}^{(s)},\mathbf{U}^c)}{\partial X_i} X_i \le \sum\limits_{i=1}^{q} \left|\frac{\partial g_j(\mathbf{X}^{(s)},\mathbf{U}^c)}{\partial U_i}\right|\left(1-2\lambda_j\right)U_i^w \\
\qquad + \sum\limits_{i=1}^{n} \frac{\partial g_j\left(\mathbf{X}^{(s)},\mathbf{U}^c\right)}{\partial X_i} X_i^{(s)} + \left(1-\lambda_j\right)b_j^R + \lambda_j b_j^L \\
\max\left[\mathbf{X}_l, \mathbf{X}^{(s)} - \boldsymbol{\delta}^{(s)}\right] \le \mathbf{X} \le \min\left[\mathbf{X}_r, \mathbf{X}^{(s)} + \boldsymbol{\delta}^{(s)}\right]
\end{cases}
\tag{9.18}
$$

Equally, other details of the algorithm are the same with those in Chap. 6. Similarly, we can also adopt RPDI and the possibility degree transformation model to formulate a nonlinear interval optimization method. This book does not elaborate on it due to limited space.

## 9.4 Numerical Example and Engineering Applications

### 9.4.1 Numerical Example

Consider a linear interval optimization problem [28]:

$$\begin{cases} \min_{\mathbf{X}} f(\mathbf{X}, \mathbf{U}) = U_1 X_1 + U_2 X_2 + U_3 X_3 \\ \text{s.t. } g_1(\mathbf{X}, \mathbf{U}) = U_3 X_1 + U_2 X_2 + U_4 X_3 \leq [11.0, 13.0] \\ \quad g_2(\mathbf{X}, \mathbf{U}) = U_5 X_1 + U_6 X_2 + U_7 X_3 \leq [10.0, 12.0] \\ \quad X_1 \geq 1.0,\ X_2 \geq 1.0,\ X_3 \geq 1.0 \\ \quad U_1 \in [-3.0, -2.0],\ U_2 \in [-2.0, -1.0],\ U_3 \in [0.5, 1.5] \\ \quad U_4 \in [1.5, 3.0],\ U_5 \in [0.5, 2.0],\ U_6 \in [1.0, 2.0],\ U_7 \in [-2.0, 0.0] \end{cases} \tag{9.19}$$

We firstly use the order relation transformation model to analyze this problem and the weighting factor $\beta$ is set to be 0.5. Table 9.1 lists the results under six different RPDI levels and the RPDI levels for two constraints are the same in each case. As shown in the table, when the RPDI level grows from 0.0 to 1.8, the multi-objective evaluation function $f_d$ rises from −23.0 to −2.3, which means the design objective is becoming worse. The reason is the smaller feasible region of the transformed deterministic optimization problem caused by a higher RPDI level. An extreme case is when the RPDI level is 2, the feasible region will become an empty set. Obviously, it is impossible to obtain a solution in such a situation. On the other hand, however, a relatively high RPDI level can improve the reliability of design, even though it does influence the objective function performance. Figure 9.5 illustrates the relative positions of the interval of constraint 1 at the optimum design and its allowable interval under different RPDI levels. When $\lambda_1 = 0$, the constraint interval totally lies on the right side of the allowable interval. As $\lambda_1$ increases, it gradually moves left along the coordinate axis. When $\lambda_1 = 1.8$, a far left constraint interval is obtained, which means the reliability of the constraint is highest among these cases.

Secondly, we adopt the possibility degree transformation model to solve the same problem. The performance interval $V^I$ is set to be [3.0, 5.0]. Table 9.2 lists the results under different RPDI levels and the RPDI levels for two constraints are the same in each case. As shown in the table, the RPDI values of objective function and constraints at the optimum design show opposite tendencies with increasing the RPDI levels, which can be also shown in Fig. 9.6. Therefore, if we wish the given performance interval is better satisfied by the objective function, the reliability on constraints should be reduced. In other words, the given RPDI level should be decreased. Besides, in order not to lose universality a case where $\lambda_1 = -0.2$ is also discussed in this example, although a negative RPDI level has no practical meaning since the design obtained by negative RPDI level could be totally unreliable.

### 9.4.2 *Application to a 10-bar Truss*

Consider the 10-bar plane truss in Fig. 9.7 [32]. The objective is to achieve its minimum weight by minimizing the cross-sectional areas of the bars $A_i, i = 1, 2, \ldots, 10$ subject to stress and displacement constraints. The truss is made of aluminum with the density of $2.77 \times 10^3$ kg/m$^3$ and Young's modulus $E$ of

**Table 9.1** Optimization results of the numerical example under different RPDI levels by using the order relation transformation model [28]

| $\lambda_1, \lambda_2$ | Optimum of **X** | Interval of objective function | Interval of constraint 1 | Interval of constraint 2 | $f_d$ | RPDIs of constraints |
|---|---|---|---|---|---|---|
| 0.0, 0.0 | (22.0, 1.0, 1.0) | [−70.0, −46.0] | [13.0, 37.5] | [10.0, 46.0] | −23.0 | 0.00, 0.05 |
| 0.5, 0.5 | (8.5, 1.0, 1.1) | [−29.7, −19.1] | [6.4, 17.6] | [3.0, 19.0] | −9.5 | 0.50, 0.50 |
| 1.0, 1.0 | (4.0, 1.0, 1.2) | [−16.3, −10.2] | [4.3, 11.0] | [0.7, 10.0] | −5.1 | 1.00, 1.00 |
| 1.5, 1.5 | (2.0, 1.0, 1.1) | [−10.1, −6.0] | [3.1, 7.7] | [−0.2, 5.9] | −3.0 | 1.50, 1.50 |
| 1.8, 1.8 | (1.3, 1.0, 1.1) | [−7.8, −4.5] | [2.6, 6.4] | [−0.4, 4.5] | −2.3 | 1.80, 1.80 |
| 2.0, 2.0 | – | – | – | – | – | – |

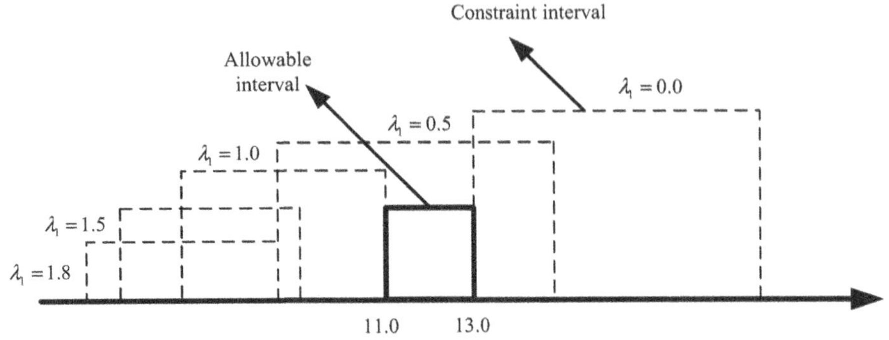


**Fig. 9.5** Relative positions between the interval of constraint 1 at the optimum design and its allowable interval under different RPDI levels [28]

68947 MPa. The length of the horizontal and vertical bars is $L = 9.144$ m. The following constraints are imposed: the maximum vertical displacement of the joint 2 is 12.7 cm; the maximum tensile stress or compressive stress is 517.1 MPa for bar 9, and 172.4 MPa for the other bars. The joint 4 is subjected to a vertical load $P_{4y}$. The joint 2 is subjected to a vertical load $P_{2y}$ and a horizontal load $P_{2x}$. The loads $P_{4y}$, $P_{2y}$, and $P_{2x}$ are uncertain parameters with nominal values of 444.8 kN, 444.8 kN, and 1779.2 kN, respectively. They have the same uncertainty level of 10%.

The axial forces of the bars, $N_i (i = 1, 2, \ldots, 10)$, satisfy the following equations:

$$N_1 = P_{2y} - \frac{\sqrt{2}}{2}N_8,\ N_2 = -\frac{\sqrt{2}}{2}N_{10},\ N_3 = -P_{4y} - 2P_{2y} + P_{2y} - \frac{\sqrt{2}}{2}N_8 \tag{9.20}$$

$$N_4 = -2P_{2y} + P_{2x} - \frac{\sqrt{2}}{2}N_{10},\ N_5 = -2P_{2y} - \frac{\sqrt{2}}{2}N_8 - \frac{\sqrt{2}}{2}N_{10},\ N_6 = \frac{\sqrt{2}}{2}N_{10} \tag{9.21}$$

$$N_7 = \sqrt{2}(P_{4y} + P_{2y}) + N_8,\ N_8 = \frac{a_{22}b_1 - a_{21}b_2}{a_{11}a_{22} - a_{12}a_{21}},\ N_9 = \sqrt{2}P_{2y} + N_{10},\ N_{10} = \frac{a_{11}b_2 - a_{21}b_1}{a_{11}a_{22} - a_{12}a_{21}} \tag{9.22}$$

where

$$\begin{aligned} a_{11} &= \left(\frac{1}{A_1} + \frac{1}{A_3} + \frac{1}{A_5} + \frac{2\sqrt{2}}{A_7} + \frac{2\sqrt{2}}{A_8}\right)\frac{L}{2E},\ a_{11} = a_{21} = \frac{L}{2A_5E} \\ a_{22} &= \left(\frac{1}{A_2} + \frac{1}{A_4} + \frac{1}{A_6} + \frac{2\sqrt{2}}{A_9} + \frac{2\sqrt{2}}{A_{10}}\right)\frac{L}{2E} \end{aligned} \tag{9.23}$$

$$b_1 = \left(\frac{P_{2y}}{A_2} - \frac{P_{4y} + 2P_{2y} - P_{2x}}{A_3} - \frac{P_{2y}}{A_5} - \frac{2\sqrt{2}(P_{4y} + P_{2y})}{A_7}\right)\frac{\sqrt{2}L}{2E} \tag{9.24}$$

**Table 9.2** Optimization results of the numerical example under different RPDI levels by using the possibility degree transformation model [28]

| $\lambda_1, \lambda_2$ | Optimum of **X** | Interval of objective function | Interval of constraint 1 | Interval of constraint 2 | RPDI of the objective function | RPDIs of constraints |
|---|---|---|---|---|---|---|
| −0.2, −0.2 | (39.7, 1.0, 1.0) | [−123.0, −81.3] | [21.8, 64.0] | [18.8, 81.3] | 2.93 | −0.20, −0.11 |
| 0.2, 0.2 | (14.4, 1.0, 1.0) | [−47.3, −30.9] | [9.2, 26.1] | [6.2, 30.9] | 2.84 | 0.20, 0.22 |
| 0.7, 0.7 | (6.1, 1.0, 1.0) | [−22.4, −14.3] | [5.1, 13.7] | [2.1, 14.3] | 2.70 | 0.75, 0.70 |
| 1.2, 1.2 | (3.0, 1.0, 1.0) | [−13.1, −8.1] | [3.5, 9.1] | [0.5, 8.1] | 2.57 | 1.26, 1.20 |
| 1.7, 1.7 | (1.5, 1.0, 1.0) | [−8.4, −5.0] | [2.7, 6.7] | [−0.3, 5.0] | 2.45 | 1.72, 1.70 |

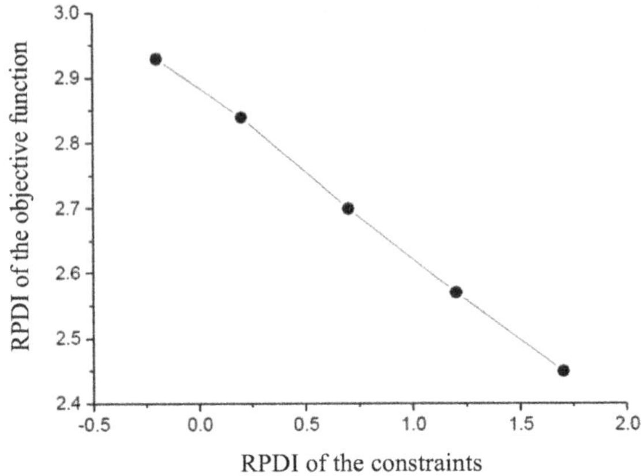


**Fig. 9.6** Relation between the constraint RPDI and the objective function RPDI at the optimum design [28]

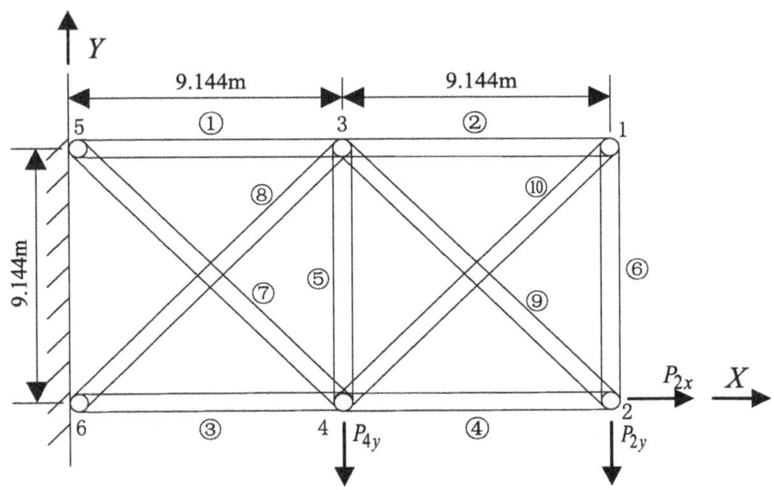


**Fig. 9.7** A 10-bar plane truss [32]

$$b_2 = \left( \frac{2\sqrt{2}(P_{2x} - P_{2y})}{A_4} - \frac{\sqrt{2}P_{2y}}{A_5} - \frac{4P_{2y}}{A_7} - \right) \frac{L}{2E} \tag{9.25}$$

The vertical displacement of the joint 2 can be calculated by:

$$\delta_2 = \left( \sum_{i=1}^{6} \frac{N_i^0 N_i}{A_i} + \sqrt{2} \sum_{i=7}^{10} \frac{N_i^0 N_i}{A_i} \right) \frac{L}{E} \tag{9.26}$$

where $N_i^0$ can be obtained by substituting $P_{4y} = P_{2y} = 0$ and $P_{2y} = 1$ into Eqs. (9.20)–(9.22). Therefore, the interval optimization problem is formulated as [31]:

$$\begin{cases} \min\limits_{\mathbf{A}} Mass(\mathbf{A}) = \sum\limits_{i=1}^{10} (\rho L_i A_i) = \rho L \left( \sum\limits_{i=1}^{6} A_i + \sqrt{2} \sum\limits_{i=7}^{10} A_i \right) \\ \text{s.t. } \sigma_i\left(\mathbf{A}, P_{4y}, P_{2x}, P_{2y}\right) = \dfrac{|N_i|}{A_i} \le \sigma_{i,\text{allow}},\ i = 1, 2, \ldots, 10 \\ \quad \delta_3\left(\mathbf{A}, P_{4y}, P_{2x}, P_{2y}\right) \le 12.7 \\ \quad 0.6425\,\text{cm}^2 \le A_i \le 129.04,\ i = 1, 2, \ldots, 10 \\ \quad P_{4y} \in [400.32\,\text{kN}, 489.28\,\text{kN}],\ P_{2x} \in [1601.28\,\text{kN}, 1957.12\,\text{kN}] \\ \quad P_{2y} \in [400.32\,\text{kN}, 489.28\,\text{kN}] \end{cases} \tag{9.27}$$

where $Mass$ denotes the weight of the truss, and $\sigma_{i,\text{allow}}$ denotes the allowable maximum stress for the $i$-th bar.

The nonlinear interval optimization method formulated in Sect. 9.3.2 is used to solve the above problem. In this example, the multi-objective weighting factor $\beta$ is given as 0.5; the initial cross-sectional areas of the bars are 129.04 cm$^2$; the constraints share the same RPDI level. In order to investigate how RPDI level affects the optimization results, four optimizations are performed with different RPDI levels, 0.8, 0.9, 1.1 and 1.2. The results are listed in Tables 9.3, 9.4, 9.5 and 9.6. It can be found that the given RPDI levels are all satisfied by the optimal design vector for

**Table 9.3** Optimization results of the 10-bar under the RPDI level of 1.2 [31]

| Bar's number | Cross-sectional area (cm$^2$) | Stress interval (MPa) | RPDI of stress constraint |
|---|---|---|---|
| 1 | 114.83 | [42.20, 55.44] | 9.82 |
| 2 | 3.81 | [120.73, 163.83] | 1.20 |
| 3 | 55.29 | [50.82, 151.97] | 1.20 |
| 4 | 97.94 | [119.84, 163.62] | 1.20 |
| 5 | 28.26 | [82.05, 112.53] | 2.96 |
| 6 | 3.81 | [120.73, 163.83] | 1.20 |
| 7 | 72.46 | [135.97, 166.17] | 1.20 |
| 8 | 13.16 | [90.53, 158.65] | 1.20 |
| 9 | 18.07 | [275.87, 337.17] | 3.94 |
| 10 | 5.36 | [120.52, 163.55] | 1.21 |

The displacement interval of $\delta_3$ is [6.15 cm, 11.6 cm], and its RPDI is 1.20; the minimal weight of the truss is 1160.49 kg

**Table 9.4** Optimization results of the 10-bar truss under the RPDI level of 1.1 [31]

| Bar's number | Cross-sectional area ($cm^2$) | Stress interval (MPa) | RPDI of stress constraint |
|---|---|---|---|
| 1 | 113.09 | [42.27, 55.58] | 9.75 |
| 2 | 2.26 | [122.59, 167.93] | 1.10 |
| 3 | 51.68 | [52.82, 161.31] | 1.10 |
| 4 | 94.39 | [121.77, 167.86] | 1.10 |
| 5 | 28.97 | [89.29, 120.82] | 2.64 |
| 6 | 2.26 | [122.73, 167.93] | 1.10 |
| 7 | 71.88 | [138.52, 169.38] | 1.10 |
| 8 | 11.94 | [91.57, 165.24] | 1.10 |
| 9 | 18.78 | [279.52, 341.72] | 3.83 |
| 10 | 3.16 | [122.73, 167.93] | 1.10 |
| The displacement interval of $\delta_3$ is [6.25 cm, 12.1 cm], and its RPDI is 1.11; the minimal weight of the truss is 1119.85 kg | | | |

**Table 9.5** Optimization results of the 10-bar truss under the RPDI level of 0.9 [31]

| Bar's number | Cross-sectional area ($cm^2$) | Stress interval (MPa) | RPDI of stress constraint |
|---|---|---|---|
| 1 | 112.78 | [40.75, 53.57] | 10.27 |
| 2 | 0.65 | [128.18, 177.27] | 0.90 |
| 3 | 44.00 | [56.40, 185.27] | 0.90 |
| 4 | 88.26 | [127.28, 177.41] | 0.90 |
| 5 | 28.20 | [106.39, 140.38] | 1.94 |
| 6 | 0.65 | [128.18, 177.27] | 0.90 |
| 7 | 71.10 | [143.62, 175.55] | 0.90 |
| 8 | 8.97 | [93.43, 181.13] | 0.90 |
| 9 | 19.23 | [288.00, 351.99] | 3.58 |
| 10 | 0.90 | [128.18, 177.27] | 0.90 |
| The displacement interval of $\delta_3$ is [6.25 cm, 12.9 cm], and its RPDI is 0.90; the minimal weight of the truss is 1054.61 kg | | | |

all cases. Also, the lower the RPDI level for the constraints, the lighter the obtained minimum weight and they almost have a linear relationship as shown in Fig. 9.8. When the RPDI level is 1.2, the minimum weight is 1160.49 kg; when the RPDI level falls to 0.8, the minimum weight lowers to 1024.33 kg.

**Table 9.6** Optimization results of the 10-bar truss under the RPDI level of 0.8 [31]

| Bar's number | Cross-sectional area ($cm^2$) | Stress interval (MPa) | RPDI of stress constraint |
|---|---|---|---|
| 1 | 110.39 | [39.16, 50.61] | 11.62 |
| 2 | 0.65 | [131.76, 182.51] | 0.80 |
| 3 | 38.97 | [52.06, 202.30] | 0.80 |
| 4 | 85.68 | [131.14, 182.72] | 0.80 |
| 5 | 29.81 | [113.08, 144.59] | 1.88 |
| 6 | 0.65 | [131.76, 182.51] | 0.80 |
| 7 | 73.04 | [146.24, 178.79] | 0.80 |
| 8 | 5.10 | [88.60, 193.20] | 0.80 |
| 9 | 18.71 | [295.52, 361.23] | 3.37 |
| 10 | 0.90 | [131.76, 182.51] | 0.80 |

The displacement interval of $\delta_3$ is [5.99 cm, 13.4 cm], and its RPDI is 0.90; the minimal weight of the truss is 1024.33 kg

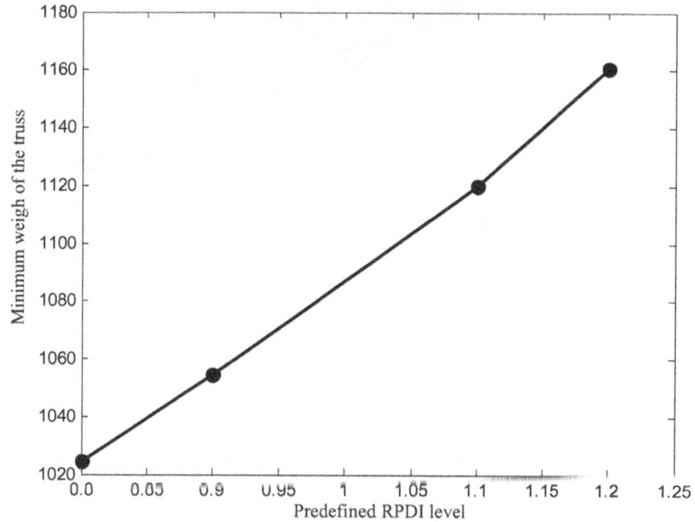


**Fig. 9.8** Relation of the RPDI level of constraints and the minimum weight of the truss [31]

### *9.4.3 Application to the Design of an Automobile Frame*

The frame is the base of an automobile on which parts and unit assemblies are fixed. An automobile frame model is shown in Fig. 9.9 [31]. It consists of two side beams and eight cross beams. The small triangles represent fixed constraints in different directions. $Q_1$, $Q_2$, $Q_3$, and $Q_4$ denote the uniformly distributed forces acting on

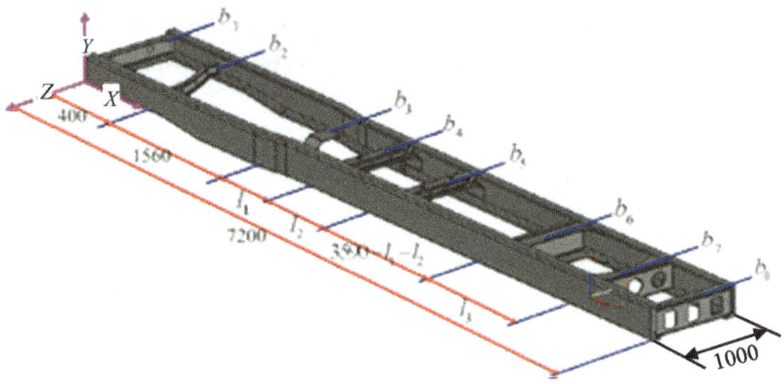


(a) An automobile frame structure (mm)

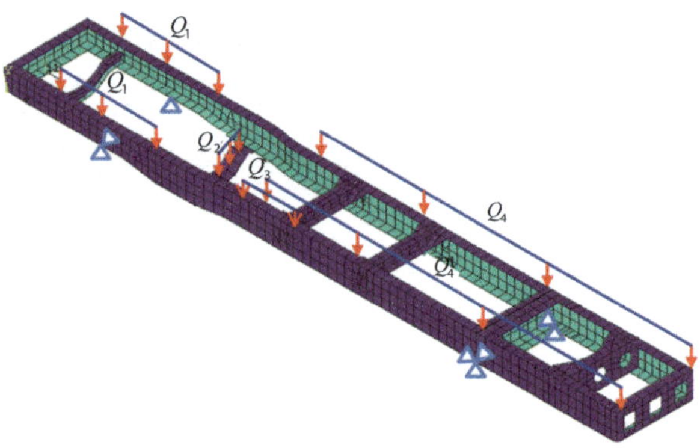


(b) The FEM model of the frame

**Fig. 9.9** The structure and corresponding FEM model of an automobile frame [31]

the frame caused by the cab house, the engine assembly, the tank, and the cargo, respectively. The cross beams are labeled as $b_i,\ i = 1, 2, \ldots, 8$. Four of them, $b_1$, $b_2$, $b_3$ and $b_6$, are fixed. The aim of the problem is to optimize the distances between other cross beams $l_i,\ i = 1, 2, 3$ to achieve the largest stiffness of the frame in *y*-direction while the stress must stay below an allowable value. Because manufacturing and measurement errors are inevitable, the Young's modulus $E$ and the Poisson's ratio $\nu$ of the frame material are interval parameters. The nominal values of $E$ and $\nu$ are $2.0 \times 10^5$MPa and 0.3, respectively. Both of them are given an uncertain level of 10%. The interval optimization problem thus can be formulated as

$$\begin{cases} \min\limits_{\mathbf{l}} d_{\max}(\mathbf{l}, E, \nu) \\ \text{s.t. } \sigma_{\max}(\mathbf{l}, E, \nu) \le 90\,\text{MPa} \\ \quad 500\,\text{mm} \le l_i \le 1200\,\text{mm},\ i = 1, 2, 3 \\ \quad E \in \left[1.8 \times 10^5\,\text{MPa}, 2.2 \times 10^5\,\text{MPa}\right],\ \nu \in [0.27, 0.33] \end{cases} \tag{9.28}$$

**Table 9.7** Optimization results of the automobile frame

| Optimum of $\mathbf{l}$ (mm) | $d_{max}$ (mm) | $\sigma_{max}$ (MPa) | RPDI of the stress constraints |
|---|---|---|---|
| (777.29, 775.83, 825.83) | [1.34, 1.64] | [86.17, 87.17] | 3.83 |

where the objective function $d_{max}$ represents the maximum displacement in the $y$-direction, which is used to represent the vertical stiffness, and the constraint $\sigma_{max}$ represents the maximum equivalent stress on the frame.

The nonlinear interval optimization method formulated in Sect. 9.3.2 is used to solve the above problem. In the optimization process, the multi-objective weighing factor is set to be 0.5; the RPDI level of the stress constraint is set to be 3.8 which is a high-reliability requirement; the initial design vector is set to be $[800\text{ mm}, 800\text{ mm}, 800\text{ mm}]^T$, the FEM is used to calculate $d_{max}$ and $\sigma_{max}$. The optimization results are listed in Table 9.7. As shown in the table, the optimum design vector is $\mathbf{l} = [777.29\text{ mm}, 775.83\text{ mm}, 825.83\text{ mm}]$ and the corresponding maximum displacement in the $y$-direction is only $d^I_{max} = [1.34\text{ mm}, 1.64\text{ mm}]$; the corresponding RPDI value of the stress constraint is 3.83, which is higher than the given RPDI level. Besides, it takes only 8 iterations and 72 evaluations of FEM to reach convergence, which indicates the high computational efficiency of the proposed method.

## 9.5 Summary

This chapter proposes a new type of interval ranking model, namely, the RPDI model that is suitable for both overlapping and separate intervals. Compared with the existing possibility degree models of interval number, the RPDI has a broader range of applications. In this chapter, the RPDI is used to solve both linear and nonlinear interval optimization problems, and its effectiveness is illustrated through several numerical examples. It should be pointed out that the RPDI is more than a simple improvement to current possibility degree models despite their similar expressions. On the contrary, RPDI is a conceptual extension to current models, which has a significant application value to reliability analysis and design of structures or systems.

## References

1. Sengupta A, Pal TK (2009) Fuzzy preference ordering of interval numbers in decision problems. Springer, Berlin Heidelberg
2. Moore RE (1966) Interval analysis. Prentice-Hall, New Jersey
3. Moore RE (1979) Methods and applications of interval analysis. Prenice-Hall, London

4. Ishibuchi H, Tanaka H (1990) Multiobjective programming in optimization of the interval objective function. Eur J Oper Res 48(2):219–225
5. Chanas S, Kuchta D (1996a) Multiobjective programming in optimization of interval objective functions—a generalized approach. Eur J Oper Res 94(3):594–598
6. Chanas S, Kuchta D (1996b) A concept of the optimal solution of the transportation problem with fuzzy cost coefficients. Fuzzy Sets Syst 82(3):299–305
7. Nakahara Y, Sasaki M, Gen M (1992) On the linear programming problems with interval coefficients. Comput Ind Eng 23(1–4):301–304
8. Sevastjanov P, Venberg A (1998) Modelling and simulation of power units work under interval uncertainty. Energy 3:66–70
9. Sevastjanov PV, Rog P (2003) A probabilistic approach to fuzzy and crisp interval ordering. Task Quart 7(1):147–156
10. Wagman D, Schneider M, Shnaider E (1994) On the use of interval mathematics in fuzzy expert systems. Int J Intell Syst 9(2):241–259
11. Kundu S (1997) Min-transitivity of fuzzy leftness relationship and its application to decision making. Fuzzy Sets Syst 86(3):357–367
12. Kundu S (1998) Preference relation on fuzzy utilities based on fuzzy leftness relation on intervals. Fuzzy Sets Syst 97(2):183–191
13. Sevastianov P, Róg P, Venberg A (2001) The constructive numerical method of interval comparison. In: PPAM
14. Sevastianov P, Róg P, Karczewski K (2002) A probabilistic method for ordering group of intervals. Informatyka Teoretyczna I Stosowana 2(2):45–53
15. Yager RR, Detyniecki M, Bouchon-Meunier B (2001) A context-dependent method for ordering fuzzy numbers using probabilities. Inf Sci 138(1):237–255
16. Sevastjanow P (2004) Interval comparison based on dempster-shafer theory of evidence. In: Wyrzykowski R, Dongarra J, Paprzycki M, Waśniewski J (eds) Parallel processing and applied mathematics: 5th international conference, PPAM 2003, Czestochowa, Poland, September 7–10, 2003. Revised Papers. Springer Berlin Heidelberg, Berlin, Heidelberg, pp 668–675
17. Sevastjanov P, Róg P (2006) Two-objective method for crisp and fuzzy interval comparison in optimization. Comput Oper Res 33(1):115–131
18. Jiang C, Han X, Liu GR, Liu GP (2008) A nonlinear interval number programming method for uncertain optimization problems. Eur J Oper Res 188(1):1–13
19. Sengupta A, Pal TK (2000) On comparing interval numbers. Eur J Oper Res 127(1):28–43
20. Sengupta A, Pal TK, Chakraborty D (2001) Interpretation of inequality constraints involving interval coefficients and a solution to interval linear programming. Fuzzy Sets Syst 119(1):129–138
21. Wang YM, Yang JB, Xu DL (2005) A preference aggregation method through the estimation of utility intervals. Comput Oper Res 32(8):2027–2049
22. Sun HL, Yao WX (2008) The basic properties of some typical systems' reliability in interval form. Struct Saf 30(4):364–373
23. Facchinetti G, Ricci RG, Muzzioli S (1998) Note on ranking fuzzy triangular numbers. Int J Intell Syst 13(7):613–622
24. Liu XW, Da QL (1999) A satisfactory solution for interval linear programming. J Syst Eng 14(2):123–128
25. Abbasi MA, Khorram E (2008) Linear programming problem with interval coefficients and an interpretation for its constraints. Iran J Sci Tech (Sci) 32(4):369–390
26. Tseng TY, Klein CM (1989) New algorithm for the ranking procedure in fuzzy decision-making. IEEE Trans Syst Man Cyber 19(5):1289–1296
27. Xu ZS, Da QL (2003) Possibility degree method for ranking interval numbers and its application. J Syst Eng 18(1):67–70
28. Jiang C, Han X, Li D (2012) A new interval comparison relation and application in interval number programming for uncertain problems. Comput Mater Contin 27(3):275–303
29. Hu YD (1990) Practical multi-objective optimization. Shanghai Technological Press, Shanghai
30. Nocedal J, Wright SJ (1999) Numerical optimization. Springer, New York

31. Jiang C, Bai YC, Han X, Ning HM (2010) An efficient reliability-based optimization method for uncertain structures based on non-probability interval model. Comput Mater Contin (CMC) 18(1):21–42
32. Elishakoff I, Haftka RT, Fang J (1994) Structural design under bounded uncertainty—optimization with anti-optimization. Comput Struct 53(6):1401–1405

# Chapter 10
# Interval Optimization Considering the Correlation of Parameters

**Abstract** This chapter first introduces a new type of interval model, i.e., the multidimensional parallelepiped interval model, based on which an interval optimization model for multisource uncertain problems is then established. A solution algorithm is also given for this interval optimization model.

So far, the interval parameters discussed in this book are mutually independent. In practical engineering, however, uncertainties are generally from multiple sources, such as materials, loads, sizes, etc. The uncertain parameters from the same "source" may be correlated, while parameters from different sources are mutually independent. This kind of problem is called the "multisource uncertain problem". The correlation of parameters often has a key or even crucial effect on uncertainty analysis results. Therefore, it is necessary to develop an interval optimization method that can deal with correlated parameters, and it will be very helpful to improve the applicability of interval optimization for complex engineering problems.

This chapter will firstly introduce a new type of interval model proposed by the authors, termed the multidimensional parallelepiped interval model [1, 2]. This model could quantitatively describe the correlation among the interval parameters and thus can solve complex "multisource uncertain problem". Secondly, an interval optimization model and the corresponding solution algorithm are developed based on the multidimensional parallelepiped interval model. Finally, this method is applied to numerical example and practical engineering problems.

## 10.1 Multidimensional Parallelepiped Interval Model

A multidimensional parallelepiped interval model could consider the independence and correlation of uncertain parameters at the same time. The uncertainty domain of the multidimensional parameters is constructed by correlations between parameters and it has the shape of a multidimensional parallelepiped in the parameter space. To create the uncertainty domain $\Omega$, only the intervals of parameters and the correlations between any two parameters are needed. To help better understand the model, we start

 
C. Jiang et al., *Nonlinear Interval Optimization for Uncertain Problems*, Springer Tracts in Mechanical Engineering, https://doi.org/10.1007/978-981-15-8546-3_10

with a two-dimensional problem before proceeding to the general multi-dimensional problem.

### 10.1.1 Two-Dimensional Problem

In the case of a two-dimensional problem, the multidimensional parallelepiped model degenerates to a parallelogram as shown in Fig. 10.1. Its bottom is set to be parallel to the horizontal axis. Theoretically, this parallelogram should be created through enveloping all the samples of $U_1$ and $U_2$. $U_1^I$ and $U_2^I$ here, represent the marginal intervals of two uncertain parameters, namely, the range of two parameters:

$$U_i^I = \left[U_i^L, U_i^R\right], U_i^c = \frac{U_i^L + U_i^R}{2}, U_i^w = \frac{U_i^R - U_i^L}{2}, i = 1, 2 \tag{10.1}$$

The angle $\theta_{12}$ is called the correlation angle because it reflects the correlation or dependence between the two parameters when $U_1^I$ and $U_2^I$ are given. The range of $\theta_{12}$ is [1]:

$$\theta_{12} \in \left[\arctan\frac{U_2^w}{U_1^w}, \pi - \arctan\frac{U_2^w}{U_1^w}\right] \tag{10.2}$$

**Fig. 10.1** The parallelogram interval model [1]

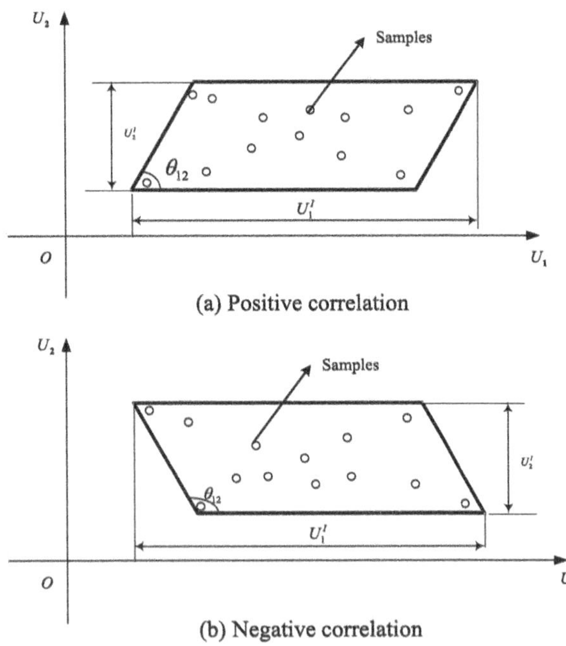

If $\theta_{12} \leq \frac{\pi}{2}$, $U_1$ and $U_2$ are positively correlated, as shown in Fig. 10.1(a); if $\theta_{12} = \frac{\pi}{2}$, $U_1$ and $U_2$ are mutually independent, and the parallelogram model reduces to an interval model; if $\theta_{12} \geq \frac{\pi}{2}$, $U_1$ and $U_2$ are negatively correlated, as shown in Fig. 10.1(b).

## 10.1.2 Multidimensional Problem

For a three-dimensional problem, the uncertainty domain of the parameters has the shape of a parallelepiped, as shown in Fig. 10.2. Similarly, its base surface is set to be parallel to the plane of $U_1OU_2$. $U_1^I$, $U_2^I$, and $U_3^I$ are the marginal intervals of the parameters; $\theta_{12}$, $\theta_{13}$, and $\theta_{23}$ are the correlation angles, which represents the correlation of the corresponding two parameters. For example, if $U_1$ and $U_2$ are correlated and the rest are mutually independent, the uncertainty domain is as shown in Fig. 10.3.

For a general $q$-dimensional problem, the uncertainty domain is a multidimensional parallelepiped. $U_i^I$ is used to denote the marginal interval of the $i$-th uncertain parameter $U_i$. $\theta_{ij}$ is used to denote the correlation angle of $U_i$ and $U_j$. Also, we define a correlation coefficient $\rho_{ij}$ to give a clearer expression of the dependence between $U_i$ and $U_j$:

Fig. 10.2 The parallelepiped interval model [1]

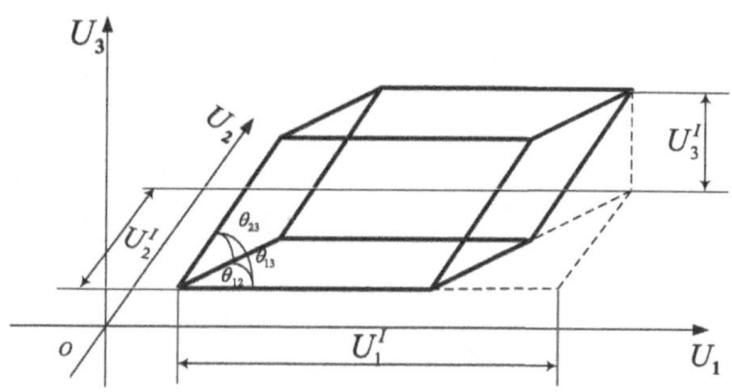


Fig. 10.3 The parallelepiped interval model with correlation only between $U_1$ and $U_2$ [1]

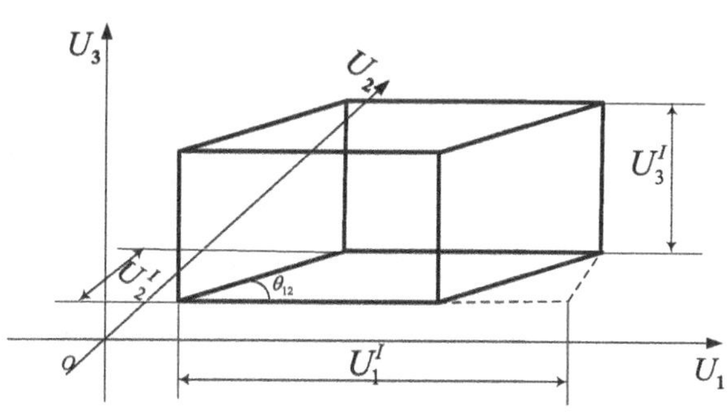

$$\rho_{ij} = \frac{U_j^w}{U_i^w \tan \theta_{ij}} \tag{10.3}$$

where $\theta_{ij} \in \left[\arctan \frac{U_j^w}{U_i^w}, \pi - \arctan \frac{U_j^w}{U_i^w}\right]$. Obviously, the range of $\rho_{ij}$ is $[-1, 1]$. As shown in Fig. 10.4, if $\theta_{ij} = \arctan \frac{U_j^w}{U_i^w}$, then $\rho_{ij} = 1$, denoting these two parameters are completely positively correlated; if $\theta_{ij} = \frac{\pi}{2}$, then $\rho_{ij} = 0$, denoting they are independent; if $\theta_{ij} = \pi - \arctan \frac{U_j^w}{U_i^w}$, then $\rho_{ij} = -1$, denoting they are completely negatively correlated. Therefore, if all the correlation coefficients are

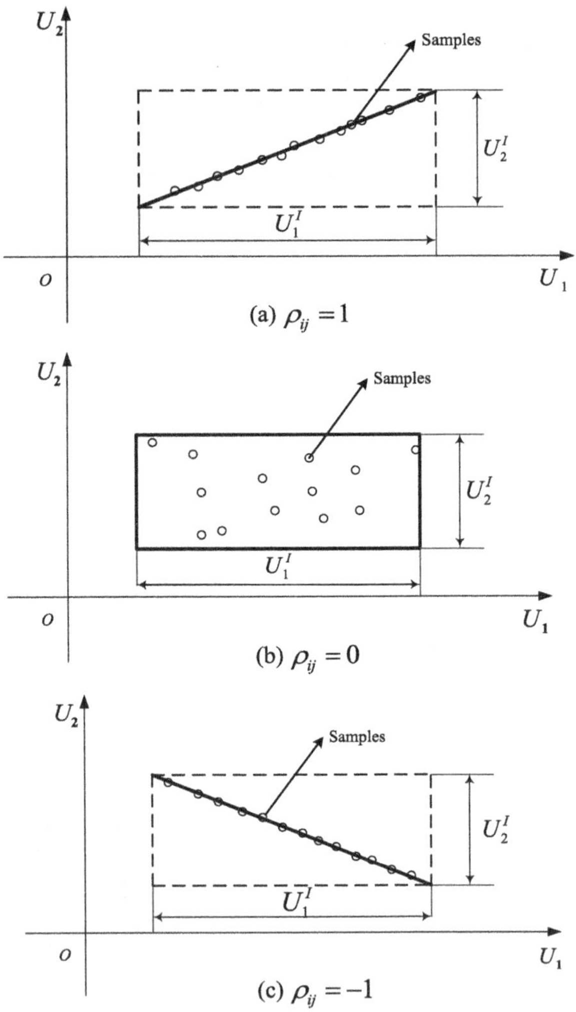


**Fig. 10.4** The parallelogram interval model under some special correlations [1]

zero, the multidimensional parallelepiped interval model will degenerate to the traditional interval model. In other words, the traditional interval model is a special case of the multidimensional parallelepiped interval model.

### 10.1.3 Construction of the Uncertainty Domain

If all the marginal intervals and the correlation angles (or correlation coefficients) are known, a multidimensional parallelepiped can be constructed as the uncertainty domain of the parameters. In many practical problems, the marginal intervals of parameters can be relatively easily obtained from experience. For example, a practical product size in engineering often falls in a specific interval under certain machining precision, and this interval can be determined through the nominal size and the manufacturing tolerance which in general can be both known in advance. In many cases, the correlations among parameters can also be obtained from experience. For example, in a product or structure, the parameters, respectively, from material properties, loads, and geometrical sizes are generally independent of each other because they come from different sources. In the stochastic analysis of sea wave, we know that the wave height and wind speed are highly correlated so a big correlation coefficient should be given. However, the experience could sometimes be so scarce so that the uncertainty domain has to be built based on samples. In the following text, we present a sample-based method to construct the multidimensional parallelepiped for general uncertain problems [1].

Consider a problem with a $q$-dimensional uncertain parameter $U_i, i = 1, 2, \ldots, q$, and $m$ samples for $\mathbf{U}$, denoted by $\mathbf{U}^{(r)}, r = 1, 2, \ldots, m$. Then the procedure to construct the multidimensional parallelepiped interval model can be summarized as:

Step 1: Select any two uncertain parameters $U_i$ and $U_j$, and $i \neq j$.

Step 2: Extract the values of $U_i$ and $U_j$ from samples $\mathbf{U}^{(r)}, r = 1, 2, \ldots, m$, and to obtain a two-dimensional sample set $\left(U_i^{(r)}, U_j^{(r)}\right), r = 1, 2, \ldots, m$.

Step 3: In the two-dimensional parameter space $U_i$–$U_j$, find a minimum-area parallelogram that envelops the sample set $\left(X_i^{(r)}, X_j^{(r)}\right), r = 1, 2, \ldots, m$. Therefore, the marginal intervals $U_i^I$ and $U_j^I$, correlation angle $\theta_{ij}$ and correlation coefficient $\rho_{ij}$ can be obtained.

Step 4: Repeat the above steps for every pair of the uncertain parameters, obtaining marginal intervals and correlation coefficients of all the parameters.

Step 5: The multidimensional parallelepiped interval model can be constructed by the obtained marginal intervals and correlation coefficients.

Directly constructing the multidimensional parallelepiped interval model through samples could be difficult, especially when the dimension of the problem is relatively high. In the above procedure, a complicated $q$-dimensional problem breaks down into $\frac{q(q+1)}{2}$ simple two-dimensional problems and the multidimensional parallelepiped interval model is constructed only through marginal intervals and correlations, which is a much more efficient way for uncertainty modeling. Therefore, it is possible to construct the multidimensional parallelepiped interval model for complex high-dimensional problems. Besides, the marginal intervals are known in many situations, and hence only the correlation angles or correlation coefficients are needed to obtain.

## 10.2 Interval Optimization Based on the Multidimensional Parallelepiped Interval Model

Based on the multidimensional parallelepiped interval model, Eq. (4.1) can be rewritten as a nonlinear interval optimization problem considering parametric correlations [3]:

$$\begin{cases} \min\limits_{\mathbf{X}} f(\mathbf{X}, \mathbf{U}) \\ \text{s.t. } g_i(\mathbf{X}, \mathbf{U}) \leq b_i^I = \left[b_i^L, b_i^R\right],\ i = 1, 2, \ldots, l,\ \mathbf{X} \in \Omega^n \\ \quad \mathbf{U} \in \Omega\left(\mathbf{U}^I, \boldsymbol{\rho}\right) \end{cases} \tag{10.4}$$

where $\mathbf{U}$ is a $q$-dimensional uncertain vector and $\Omega$ is the uncertainty domain of $\mathbf{U}$. $\Omega$ in geometry has the shape of a multidimensional parallelepiped and it is determined by the marginal interval vector $\mathbf{U}^I$ and the correlation coefficient vector $\boldsymbol{\rho}$. It should be pointed out that Eq. (10.4) actually provides a new interval optimization model for parametric correlation problems.

Even the uncertainty domain can be described by multidimensional parallelepiped, it is not easy to explicitly express it, bringing difficulties to subsequent interval optimization. To solve this problem, an affine coordinate is adopted in the next subsection to convert Eq. (10.4) into a conventional interval optimization problem and then it can be further converted into a deterministic optimization to be solved.

### 10.2.1 *Affine Coordinate Transformation*

Apply the affine coordinate transformation to the uncertainty domain $\Omega$ of the parameters $\mathbf{U}$ and the angles between axes become the correlation angles of the corresponding uncertain parameters, instead of $\frac{\pi}{2}$. Besides, the origin in the affine coordinate system coincides with the center of multidimensional parallelepiped interval model, as shown in Fig. 10.5. Based on the affine coordinate theory [4], the following mapping relations between variables in the original coordinates $U_i, i = 1, 2, \ldots, q$ and variables in the affine coordinates $P_i, i = 1, 2, \ldots, q$ can be obtained:

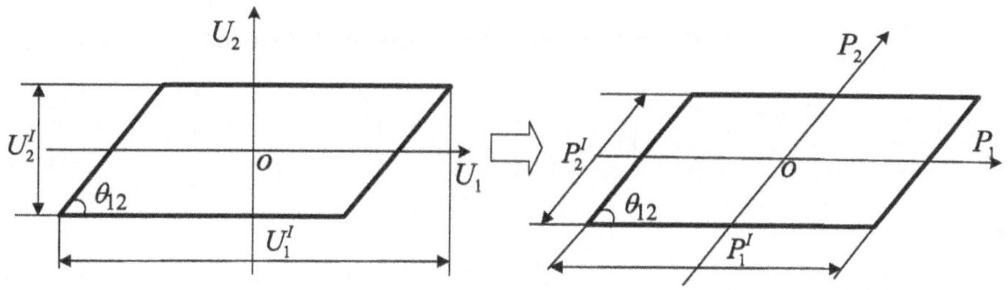


**Fig. 10.5** The affine coordinate transformation of multidimensional parallelepiped interval model [3]

$$\begin{pmatrix} U_1 \\ U_2 \\ \vdots \\ U_q \end{pmatrix} = \mathbf{A} \begin{pmatrix} P_1 \\ P_2 \\ \vdots \\ P_q \end{pmatrix} + \begin{pmatrix} U_1^c \\ U_2^c \\ \vdots \\ U_q^c \end{pmatrix} \tag{10.5}$$

where $\mathbf{U}^c = \left(U_1^c, U_2^c, \ldots, U_q^c\right)$ is the midpoint vector of $\mathbf{U}$ and $\mathbf{A}$ is the transformation matrix:

$$\mathbf{A} = \begin{bmatrix} a_{11} & a_{12} & \ldots & a_{1j} & \ldots & a_{1q} \\ a_{21} & a_{22} & \ldots & a_{2j} & \ldots & a_{2q} \\ \vdots & \vdots & \vdots & \vdots & \vdots & \vdots \\ a_{i1} & a_{i2} & \ldots & a_{ij} & \ldots & a_{iq} \\ \vdots & \vdots & \vdots & \vdots & \vdots & \vdots \\ a_{q1} & a_{q2} & \ldots & a_{qj} & \ldots & a_{qq} \end{bmatrix}^{\mathrm{T}} \tag{10.6}$$

The coefficients of $\mathbf{A}$ can be obtained by:

$$a_{ij} = \begin{cases} 0 & , \ j > i \\ \frac{\cos\theta_k - \sum_{m=1}^{j-1} a_{im}a_{jm}}{a_{jj}} & , \ j < i \\ \sqrt{1 - \sum_{l=1}^{j-1} a_{il}^2} & , \ j = i \end{cases} \tag{10.7}$$

To simplify the expression, $\theta_{ij}$ is represented by $\theta_k$ in Eq. (10.7), where the subscript $k = \frac{(2n-j)(j-1)}{2} + (i-j)$. For clarity, we use a simple three-dimensional case as an example. If $\rho_{12} = \rho_{13} = \rho_{23} = 0.4$, we obtain $\cos\theta_1 = \cos\theta_2 = \cos\theta_3 = 0.3714$ by Eq (10.3) and then calculate all the elements in $\mathbf{A}$ by Eq. (10.7): $a_{12} = a_{12} = a_{23} = 0$, $a_{11} = 1$, $a_{21} = \frac{\cos\theta_1}{a_{11}} = 0.3714$, $a_{22} = \sqrt{1 - a_{11}^2} = 0.9285$, $a_{31} = \frac{\cos\theta_2}{a_{11}} = 0.3714$, $a_{32} = (\cos\theta_3 - a_{31}a_{21})/a_{22} = 0.2514$, and

$a_{33} = \sqrt{1 - a_{31}^2 - a_{32}^2} = 0.8938$. Finally, the transformation matrix **A** is obtained:

$$\mathbf{A} = \begin{bmatrix} 1 & 0 & 0 \\ 0.3714 & 0.9285 & 0 \\ 0.3714 & 0.2514 & 0.8938 \end{bmatrix}^{\mathrm{T}} \tag{10.8}$$

After the affine coordinate transformation, there exists the following mapping relation between the interval radius $U_i^w, i = 1, 2, \ldots, q$ in the original coordinates and the interval radiuses $P_i^w, i = 1, 2, \ldots, q$ in the affine coordinates:

$$\begin{pmatrix} P_1^w \\ P_2^w \\ \vdots \\ P_q^w \end{pmatrix} = [|\mathbf{A}|]^{-1} \begin{pmatrix} U_1^w \\ U_2^w \\ \vdots \\ U_q^w \end{pmatrix} \tag{10.9}$$

where $[|\mathbf{A}|]$ is the matrix after **A**'s elements are replaced by their absolute values. As Fig. 10.5 shows, the above transformation generates the multidimensional parallelepiped interval model to a traditional interval model whose center coincides with the origin of the affine coordinate system.

In the affine coordinates, the interval optimization in Eq. (10.4) is converted to:

$$\begin{cases} \min\limits_{\mathbf{X}} F(\mathbf{X}, \mathbf{P}) \\ \text{s.t. } G_i(\mathbf{X}, \mathbf{P}) \le b_i^I = \left[b_i^L, b_i^R\right],\ i = 1, 2, \ldots, l,\ \mathbf{X} \in \Omega^n \\ \quad \mathbf{P} \in \mathbf{P}^I = \left[\mathbf{P}^L, \mathbf{P}^R\right],\ P_i \in P_i^I = \left[P_i^L, P_i^R\right], i = 1, 2, \ldots, q \end{cases} \tag{10.10}$$

where $F$ and $G$ represent the objective function and the constraint in the affine space, respectively.

The affine coordinate transformation yields a regular interval optimization problem because the interval parameters **P** are mutually independent, and the uncertain domain has the shape of a "multidimensional box" in geometry. Thus the existing nonlinear interval optimization methods can be used to solve this problem.

### *10.2.2 Conversion to a Deterministic Optimization*

The order relation transformation model proposed in Chap. 3 is adopted here to deal with Eq. (10.10), where the order relation $\le_{cw}$ is used to handle the objective function, together with the RPDI [5] presented in Chap. 9 for constraints. Therefore, Eq. (10.10) becomes a deterministic optimization problem [3]:

$$\begin{cases} \min\limits_{\mathbf{X}}(F^c(\mathbf{X}), F^w(\mathbf{X})) \\ \text{s.t. } P_r\left(G_i^I(\mathbf{X}) \le b_i^I\right) = \frac{b_i^R - G_i^L(\mathbf{X})}{2G_i^w(\mathbf{X}) + 2b_i^w} \ge \lambda_i,\ i = 1, 2, \ldots, l,\ \mathbf{X} \in \Omega^n \end{cases} \tag{10.11}$$

Similarly, Eq. (10.11) can degenerate to a single-objective optimization problem eventually:

$$\begin{cases} \min\limits_{\mathbf{X}} F_d(\mathbf{X}) = (1-\beta)F^c(\mathbf{X}) + \beta F^w(\mathbf{X}) \\ \text{s.t. } P_r\left(G_i^I(\mathbf{X}) \le b_i^I\right) = \frac{b_i^R - G_i^L(\mathbf{X})}{2G_i^w(\mathbf{X}) + 2b_i^w} \ge \lambda_i,\ i = 1, 2, \ldots, l,\ \mathbf{X} \in \Omega^n \end{cases} \tag{10.12}$$

Notice that Eq. (10.12) is a two-layer nested optimization problem. Here, the nonlinear interval optimization method based on sequential linear programming proposed in Chap. 6 is adopted to convert the problem into a series of linear interval optimization problems to solve and the convergence is ensured by the iteration mechanism.

## 10.3 Numerical Example and Engineering Applications

### *10.3.1 Numerical Example*

Consider an interval optimization problem as follows [3]:

$$\begin{cases} \min\limits_{\mathbf{X}} f(\mathbf{X}, \mathbf{U}) = U_3X_1 + 7U_2X_2 - U_1X_1X_2 + 100 \\ \text{s.t. } g_1(\mathbf{X}, \mathbf{U}) = U_2U_3X_2 + U_1X_1X_2 \le [790, 810] \\ \quad g_2(\mathbf{X}, \mathbf{U}) = U_1U_2(X_1 - 15)^2 + U_3(X_2 - 20)^2 \le [690, 710] \\ \quad 0 \le X_i \le 20,\ i = 1, 2 \\ \quad \mathbf{U} \in \Omega\left(\mathbf{U}^I, \boldsymbol{\rho}\right) \end{cases} \tag{10.13}$$

where the marginal intervals of the uncertain parameters are $U_1^I = [1.5, 2.5]$, $U_2^I = [5.6, 6.5]$, and $U_3^I = [3.5, 4.5]$, respectively.

To investigate the impacts of parametric correlation on the optimization results, four cases are considered. In Cases 1–3, only one pair of parameters are correlated, while in Case 4, all parameters are correlated to each other. Besides, six different correlation coefficient values are considered in each case. The proposed interval optimization method is used to solve the problem. In this example, the multi-objective weighting factor is $\beta = 0.5$, and two interval constraints are given the same RPDI level, $\lambda_1 = \lambda_2 = 0.9$. The interval optimization results are listed in Tables 10.1, 10.2, 10.3 and 10.4, where $N$ represents the number of evaluations of the objective function, and $P_{r1}$ and $P_{r2}$ are the RPDI values of constraints at optimum design. As the table shows, in this problem, the results are sensitive to the correlation between interval

**Table 10.1** Optimization results of the numerical example for the first correlation case [3]

| $\rho_{12}$ | $X_1$ | $X_2$ | $f^I$ | $g_1^I$ | $g_2^I$ | $P_{r1}$ | $P_{r2}$ | $N$ |
|---|---|---|---|---|---|---|---|---|
| 0.0 | 19.50 | 10.63 | [59.5, 360.6] | [512.8, 826.4] | [468.7, 717.9] | 0.90 | 0.90 | 144 |
| 0.2 | 19.35 | 10.48 | [104.9, 319.4] | [484.2, 831.1] | [456.7, 719.5] | 0.90 | 0.90 | 156 |
| 0.4 | 19.19 | 10.30 | [141.7, 286.8] | [458.3, 830.5] | [448.9, 718.8] | 0.90 | 0.90 | 144 |
| 0.6 | 19.12 | 10.25 | [143.5, 286.9] | [441.9, 835.9] | [443.2, 720.6] | 0.90 | 0.90 | 144 |
| 0.8 | 19.02 | 10.17 | [145.9, 287.0] | [427.1, 833.7] | [440.9, 719.9] | 0.90 | 0.90 | 144 |
| 1.0 | 18.99 | 10.09 | [146.7, 287.1] | [419.1, 837.1] | [438.2, 720.1] | 0.90 | 0.90 | 132 |

**Table 10.2** Optimization results of the numerical example for the second correlation case [3]

| $\rho_{13}$ | $X_1$ | $X_2$ | $f^I$ | $g_1^I$ | $g_2^I$ | $P_{r1}$ | $P_{r2}$ | $N$ |
|---|---|---|---|---|---|---|---|---|
| 0.0 | 19.50 | 10.63 | [59.5, 360.6] | [512.8, 826.4] | [468.7, 717.9] | 0.90 | 0.90 | 144 |
| 0.2 | 19.35 | 10.48 | [83.6, 340.6] | [484.5, 831.0] | [457.3, 719.3] | 0.90 | 0.90 | 156 |
| 0.4 | 19.20 | 10.33 | [88.4, 339.8] | [459.0, 829.8] | [451.8, 718.4] | 0.90 | 0.90 | 156 |
| 0.6 | 19.15 | 10.28 | [89.9, 339.5] | [444.2, 836.2] | [448.4, 719.8] | 0.90 | 0.90 | 156 |
| 0.8 | 19.10 | 10.22 | [91.5, 339.1] | [432.6, 838.4] | [448.2, 720.2] | 0.90 | 0.90 | 156 |
| 1.0 | 19.07 | 10.14 | [92.8, 338.6] | [424.1, 838.4] | [449.7, 719.7] | 0.90 | 0.90 | 144 |

**Table 10.3** Optimization results of the numerical example for the third correlation case [3]

| $\rho_{23}$ | $X_1$ | $X_2$ | $f^I$ | $g_1^I$ | $g_2^I$ | $P_{r1}$ | $P_{r2}$ | $N$ |
|---|---|---|---|---|---|---|---|---|
| 0.0 | 19.50 | 10.63 | [59.5, 360.6] | [512.8, 826.4] | [468.7, 717.9] | 0.90 | 0.90 | 144 |
| 0.2 | 19.45 | 10.59 | [61.1, 360.4] | [506.2, 824.8] | [465.1, 717.5] | 0.90 | 0.90 | 168 |
| 0.4 | 19.43 | 10.57 | [61.7, 360.3] | [503.1, 825.1] | [463.9, 717.6] | 0.90 | 0.90 | 168 |
| 0.6 | 19.43 | 10.57 | [61.6, 360.3] | [502.3, 825.9] | [464.5, 717.7] | 0.90 | 0.90 | 168 |
| 0.8 | 19.45 | 10.57 | [61.2, 360.3] | [503.1, 826.9] | [466.1, 717.9] | 0.90 | 0.90 | 168 |
| 1.0 | 19.47 | 10.58 | [60.6, 360.3] | [504.8, 828.1] | [468.0, 718.1] | 0.90 | 0.90 | 168 |

parameters. For all cases, the RPDI values are equal to or very close to the predetermined level 0.9 for both two constraints. The relationship between optimal objective function value and correlation coefficient for four cases are given in Fig. 10.6. It can be found that the optimal objective function value has a zonal distribution with lower and upper bounds. Besides, different parameter correlations have different impacts on the results. Here, $\rho_{12}$ has the greatest influence on optimization results, followed by $\rho_{12}$, and $\rho_{23}$ has the smallest influence. As the correlation coefficient increases, the midpoint of the objective function at optimum design remains almost unchanged but

**Table 10.4** Optimization results of the numerical example for the fourth correlation case [3]

| $\rho_{12} = \rho_{13} = \rho_{23}$ | $X_1$ | $X_2$ | $f^I$ | $g_1^I$ | $g_2^I$ | $P_{r1}$ | $P_{r2}$ | $N$ |
|---|---|---|---|---|---|---|---|---|
| 0.0 | 19.50 | 10.63 | [59.5, 360.6] | [512.8, 826.4] | [468.7, 717.9] | 0.89 | 0.90 | 144 |
| 0.2 | 19.17 | 10.30 | [134.1, 295.0] | [452.3, 834.8] | [445.1, 720.2] | 0.89 | 0.90 | 144 |
| 0.4 | 18.96 | 10.07 | [166.4, 268.2] | [408.0, 843.8] | [434.5, 721.7] | 0.88 | 0.90 | 132 |
| 0.6 | 18.75 | 9.82 | [170.9, 268.3] | [373.3, 842.6] | [434.7, 720.1] | 0.89 | 0.90 | 120 |
| 0.8 | 18.68 | 9.77 | [172.3, 268.2] | [353.5, 850.8] | [434.2, 721.0] | 0.88 | 0.90 | 120 |
| 1.0 | 18.58 | 9.67 | [174.4, 267.9] | [336.8, 847.1] | [440.0, 720.0] | 0.89 | 0.90 | 120 |

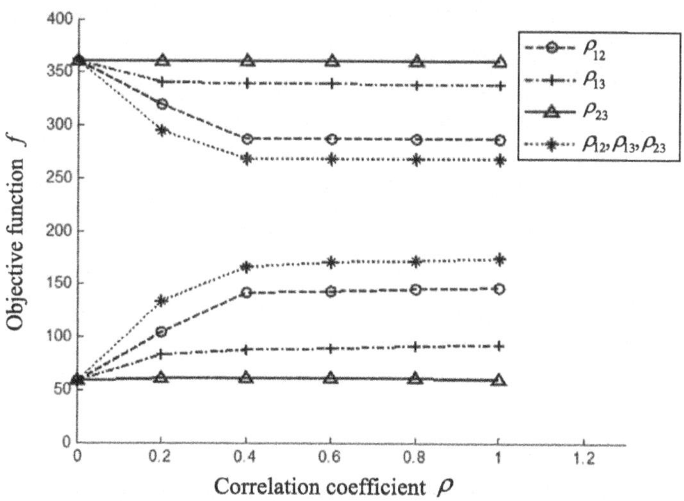


**Fig. 10.6** Optimization results of the objective function of the numerical example under four different correlation cases [3]

the radius becomes smaller. The objective function interval of Case 4 has the most significant variation since all correlations between parameters are considered in this case, which further indicates the influence of correlated parameters on optimization results. In addition, the presented interval optimization method is efficient since only 120–168 evaluations of the objective function are needed during the optimization process for all cases.

### 10.3.2 Application to a 25-bar Truss

Consider a 25-bar truss structure as shown in Fig. 5.3 [6]. The objective is to minimize the weight by optimizing the cross-sectional areas of the bars. The Young's modulus of the truss is $E = 200\,\text{GPa}$; the Poisson's ratio is $\nu = 0.3$; the length of both the horizontal and vertical bars is $L = 15.24\,\text{m}$. Node 1 is subjected to a horizontal load $F_4 = 1300\,\text{kN}$; node 7 is subjected to a vertical load $F_3$; node 9 is subjected to a vertical load $F_2$; node 11 is subjected to a vertical load $F_1$. The cross-sectional areas are $A_1 = 550\,\text{mm}^2$ for bars 1–4, $A_2 = 8100\,\text{mm}^2$ for bars 5–10, $A_3 = 6700\,\text{mm}^2$ for bars 11–15, $A_4$ for bars 16–17, $A_5$ for bars 18–19, $A_6$ for bars 20–21, $A_7$ for bars 22–23, and $A_8$ for bars 24–25. In this problem, the cross-sectional areas $A_i, i = 4, 5, \ldots, 8$ are the design variables; the loads $F_i, i = 1, 2, 3$ are the uncertain parameters, and their marginal intervals are $F_1^I = [1680\,\text{kN}, 1880\,\text{kN}]$, $F_2^I = [2124\,\text{kN}, 2324\,\text{kN}]$, and $F_3^I = [1680\,\text{kN}, 1880\,\text{kN}]$, respectively. The horizontal displacements at nodes 1, 5, 6 and the vertical displacements at nodes 7, 9, 11 cannot cross a given limit. An interval optimization problem thus can be created [3]:

$$
\begin{cases}
\min\limits_{\mathbf{A}} \; Vol(\mathbf{A}, \mathbf{F}) = 2\sqrt{2}\sum\limits_{i=4}^{8} A_i L \\
\text{s.t. } g_1(\mathbf{A}, \mathbf{F}) \le b_i^I, \; i = 1, 2, \ldots, 6 \\
\quad 4500\,\text{mm}^2 < A_4 < 10000\,\text{mm}^2, 4500\,\text{mm}^2 < A_5 < 12000\,\text{mm}^2, \\
\quad 2250\,\text{mm}^2 < A_6 < 10000\,\text{mm}^2, 4500\,\text{mm}^2 < A_7 < 12000\,\text{mm}^2, \\
\quad 4500\,\text{mm}^2 < A_8 < 10000\,\text{mm}^2 \\
\quad b_1^I = [34\,\text{mm}, 38\,\text{mm}], b_2^I = [18\,\text{mm}, 22\,\text{mm}], b_3^I = [41\,\text{mm}, 45\,\text{mm}], \\
\quad b_4^I = [34\,\text{mm}, 38\,\text{mm}], b_5^I = [40\,\text{mm}, 44\,\text{mm}], b_6^I = [21\,\text{mm}, 25\,\text{mm}] \\
\quad \mathbf{F} \in \Omega(\mathbf{F}^I, \boldsymbol{\rho})
\end{cases}
\tag{10.14}
$$

where the objective function $Vol$ denotes the volume of the bar material, the constraints $g_i, i = 1, 2, \ldots, 6$ denote the displacements at the six nodes, and $b_i^I, i = 1, 2, \ldots, 6$ represent the allowable ranges of $g_i, i = 1, 2, \ldots, 6$.

During the optimization, the displacements of nodes are calculated by an FEM model that comprises 25 truss elements. To study how the parametric correlation affects the optimization results, also 4 Cases are investigated: in Cases 1–3, only one pair of parameters is correlated; in Case 4, all parameters are correlated to each other. In each case, 6 different correlation coefficients are considered. In this problem, the multi-objective weighting factor is $\beta = 0.5$, the 6 constraints are given the same RPDI level, 0.9. The optimization results of Case 4 are only provided and listed in Table 10.5, where $N$ represents the number of evaluations of the objective function. Here the results on minimum volume are deterministic due to the absence of interval

Table 10.5 Optimization results of the 25-bar truss for the fourth correlation case [3]

| $\rho_{12} = \rho_{13} = \rho_{23}$ | $Vol/\mathrm{m}^3$ | $P_{r1}$ | $P_{r2}$ | $P_{r3}$ | $P_{r4}$ | $P_{r5}$ | $P_{r6}$ | $N$ |
|---|---|---|---|---|---|---|---|---|
| 0.0 | 1.263 | 1.90 | 0.91 | 0.90 | 0.89 | 0.90 | 0.90 | 220 |
| 0.2 | 1.252 | 1.82 | 1.04 | 0.90 | 0.90 | 0.90 | 1.02 | 240 |
| 0.4 | 1.256 | 1.95 | 1.10 | 0.90 | 0.90 | 0.90 | 1.10 | 280 |
| 0.6 | 1.259 | 2.01 | 1.09 | 0.90 | 0.90 | 0.90 | 1.07 | 320 |
| 0.8 | 1.257 | 2.10 | 1.09 | 0.90 | 0.90 | 0.90 | 1.09 | 280 |
| 1.0 | 1.255 | 2.19 | 1.04 | 0.90 | 0.90 | 0.90 | 1.03 | 300 |

variables in the objective function. The relationship between optimal objective function value and correlation coefficient for four cases are given in Fig. 10.7. As shown in the figure, the correlations between different parameters have a similar impact on the optimization results. As the correlation coefficient increases, the objective function on the whole has a decreasing tendency. Besides, the correlations of parameters have less significant effects on the optimization results compared with the previous example. As shown in Table 10.5, the minimum volume of the structure varies only slightly with the correlation coefficient. In addition, although the problem has 5 design variables and 3 uncertain parameters, it takes only 200–300 evaluations of the objective function to arrive at the optimal designs.

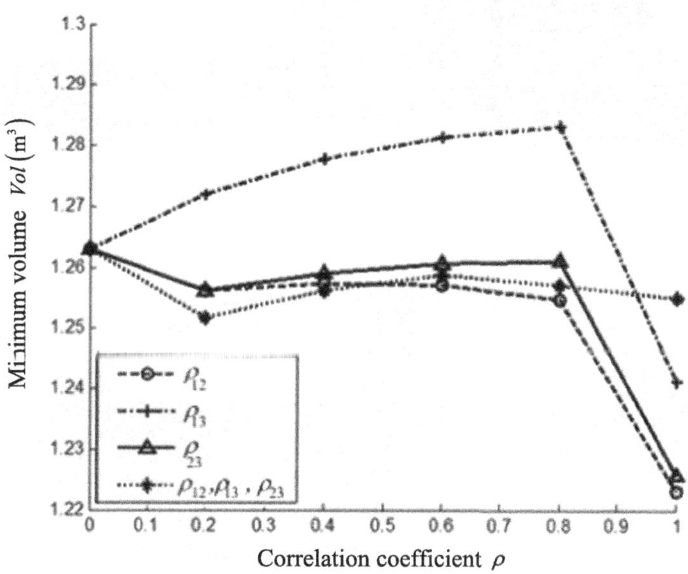


Fig. 10.7 Optimization results of the objective function of the 25-bar truss under four different correlation cases [3]

### *10.3.3 Application to the Crashworthiness Design of Vehicle Side Impact*

The side impact is a key problem to be considered in vehicle safety design [7, 8]. The side impact may easily cause severe damage to occupants due to relatively low strength of side structure of vehicle and small distance between occupants and impact area [9]. The energy-absorbing components in the impact areas, such as the inner and outer panels of side doors, have a significant influence on vehicle safety in side impact [10]. On the other hand, weight reduction is also important in vehicle design. Therefore, minimizing the mass of the vehicle while guaranteeing the crashworthiness of vehicle becomes the main objective in vehicle design [11, 12]. In this example, a vehicle side impact problem is considered as shown in Fig. 10.8. The aim is to minimize the weight *Mass* of inner and outer panels of front and rear doors by optimizing the panels' thicknesses under the safety requirements. On the other hand, the intrusion velocity of measurement point should be constrained to ensure the safety of occupant. Here, the measurement point is located on the B-pillar as shown in Fig. 10.9 and the intrusion velocity should be controlled within a given

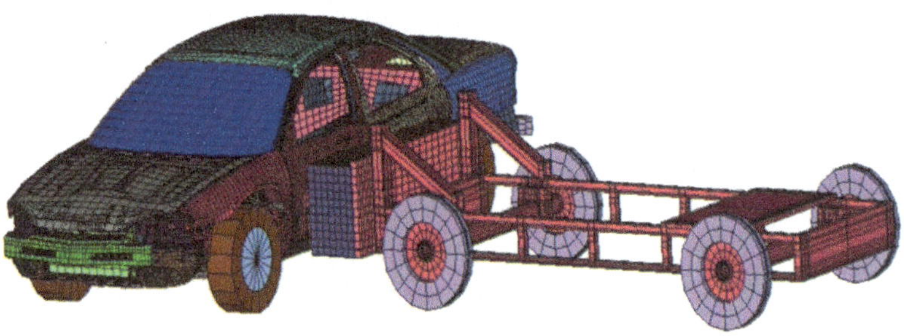

**Fig. 10.8** A side impact problem of vehicle [3]

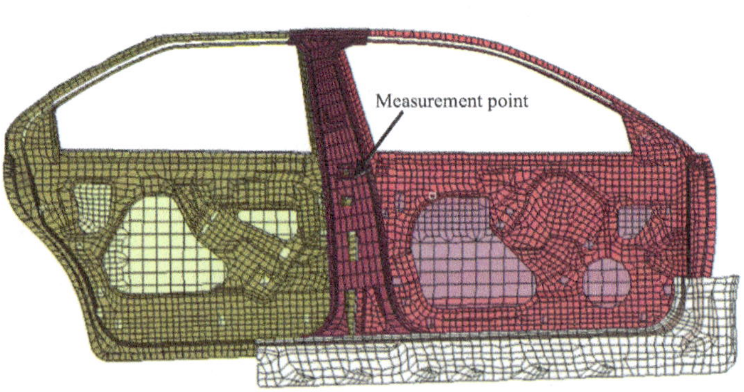


**Fig. 10.9** The measurement point of intrusion velocity on the B-pillar [3]

interval $v^I = [11.4\,\mathrm{m/s}, 12.0\,\mathrm{m/s}]$. Due to errors in manufacturing and measurement, the material properties of the inner and outer panels of front and rear doors are considered as uncertain parameters. For panels of the front door, the marginal interval of Young's modulus is $E_1^I = [190\,\mathrm{GPa}, 210\,\mathrm{GPa}]$ and the marginal interval of density is $Des_1^I = [7.2\times10^3\,\mathrm{kg/m^3}, 8.4\times10^3\,\mathrm{kg/m^3}]$. For the rear door, they are $E_2^I = [190\,\mathrm{GPa}, 210\,\mathrm{GPa}]$ and $Des_2^I = [7.2\times10^3\,\mathrm{kg/m^3}, 8.4\times10^3\,\mathrm{kg/m^3}]$, respectively. Because the front and rear doors are built independently, the material properties of panels from different doors are mutually independent. Thus, only the correlation between the material properties of panels in the same door is considered. From above, an interval optimization problem can be constructed as:

$$\begin{cases} \min\limits_{\mathbf{X}} \ Mass(\mathbf{X}, \mathbf{Des}, \mathbf{E}) \\ \text{s.t. } v_{\max}(\mathbf{X}, \mathbf{Des}, \mathbf{E}) \le v^I \\ \quad 0.6\,\mathrm{mm} \le X_1 \le 1.0\,\mathrm{mm},\ 1.0\,\mathrm{mm} < X_2 < 2.0\,\mathrm{mm}, \\ \quad 0.6\,\mathrm{mm} \le X_3 \le 1.0\,\mathrm{mm},\ 1.0\,\mathrm{mm} < X_4 < 2.0\,\mathrm{mm} \\ \quad (\mathbf{Des}, \mathbf{E}) \in \Omega\left(\mathbf{Des}^I, \mathbf{E}^I, \rho_{12}, \rho_{34}\right) \end{cases} \tag{10.15}$$

where $\rho_{12}$ denotes the correlation coefficient between $E_1^I$ and $Des_1^I$, $\rho_{34}$ denotes the correlation coefficient between $E_2^I$ and $Des_2^I$, the design variable $X_i, i = 1, 2, 3, 4$ are the thicknesses of four panels, and the constraint $v_{\max}$ represents the intrusion velocity.

The FEM model is constructed using 85,671 shell elements and 564 solid elements. The impact velocity is 50 km/h. During the optimization, the multi-objective weighting factor is set to be $\beta = 0.5$, and the RPDI level of the constraint is set to be 1.0. The optimization results are listed in Table 10.6, where $N$ represents the number of FEM evaluations. For the optimal design, the interval of the maximum intrusion velocity is $v_{\max}^I = [11.21\,\mathrm{m/s}, 11.43\,\mathrm{m/s}]$, and the RPDI value is equal to 1.0 as expected. This indicates that the safety requirement on intrusion velocity can be completely satisfied with this optimum design, though uncertainties on the material properties exist. As there are interval parameters in the objective function, the optimal mass is also an interval $[20.13\,\mathrm{kg}, 22.20\,\mathrm{kg}]$. Besides, the efficiency of the proposed interval optimization method is proved again. Only 108 FEM evaluations are used in the optimization even there are 8 design variables and interval variables.

**Table 10.6** Optimization results of the vehicle side impact problem [3]

| $X_1$ | $X_2$ | $X_3$ | $X_4$ | $Mass^I$ | $v_{\max}^I$ | $P_r$ | $N$ |
|---|---|---|---|---|---|---|---|
| 0.93 mm | 2.00 mm | 0.93 mm | 2.00 mm | [20.13 kg, 22.20 kg] | [11.21 m/s, 11.43 m/s] | 1.00 | 108 |

## 10.4 Summary

An interval optimization method considering the correlation of parameters is proposed in this chapter by introducing the multidimensional parallelepiped interval model. The correlated parameters and independent parameters can be described in a uniform frame using the multidimensional parallelepiped interval model, enabling the proposed method to be suitable for interval optimization problems with complex multisource uncertainties. Besides, the interval optimization problem is converted into a deterministic optimization problem using the order relation transformation model and the RPDI model. The analyses on numerical example and engineering applications indicate that the parametric correlations may have a significant impact on the optimization results. Therefore, the errors may be brought if the correlated parameters are simply treated as independent parameters. In addition, the correlations between different pairs of parameters may have a different impact on the results even in the same problem.

## References

1. Jiang C, Zhang QF, Han X, Liu J, Hu DA (2015) Multidimensional parallelepiped model—a new type of non-probabilistic convex model for structural uncertainty analysis. Int J Numer Meth Eng 103(1):31–59
2. Jiang C, Zhang QF, Han X, Qian YH (2014) A non-probabilistic structural reliability analysis method based on a multidimensional parallelepiped convex model. Acta Mech 225(2):383–395
3. Jiang C, Zhang ZG, Zhang QF, Han X, Xie HC, Liu J (2014) A new nonlinear interval programming method for uncertain problems with dependent interval variables. Eur J Oper Res 238(1):245–253
4. Frank Ayres J (1967) Schaum's outline of theory and problems of projective geometry. McGraw-Hill, New York
5. Jiang C, Han X, Li D (2012) A new interval comparison relation and application in interval number programming for uncertain problems. Comput Mat Continua 27(3):275–303
6. Au FTK, Cheng YS, Tham LG, Zeng GW (2003) Robust design of structures using convex models. Comput Struct 81(28–29):2611–2619
7. Zhong ZH, Zhang WG, Cao LB, He W (2003) Automotive crash safety technology. Machinery Industry Press, Beijing
8. Zhu XC (2001) Law and regulation of vehicle collision test and its development. Automob Technol 4:5–10
9. Zhang XR, Su QZ (2008) A research on influencing factors of occupant injury in side impact. Auto motive Eng 3(2):146–150
10. Nelson D, Sparke L (2002) Improved side impact protection: design optimisation for minimum harm. Technical report, SAE Technical Paper 2002-01-0167
11. Uduma K, Wu J, Bilkhu S, Gielow M, Kowsika M (2005) Door interior trim safety enhancement strategies for the SID-IIs dummy. SAE Trans 114:26–33
12. Bosch-Rekveldt M, Griotto G, van Ratingen M, Versmissen T, Mooi H (2005) Head impact location, angle and velocity during side impact: a study based on virtual testing. In: Proceedings of SAE world congress, Detroit, Michigan, April, 2005

# Chapter 11
# Interval Multi-objective Optimization

**Abstract** By employing an interval approach to describe the uncertainty of parameters in multi-objective optimization, this chapter proposes an interval multi-objective optimization model and also an efficient solution algorithm for this optimization model.

Practical engineering problems often involve conflicting design objectives. In other words, there are a lot of multi-objective optimization problems. For examples, a vehicle designer may seek to reduce the vehicle weight while promoting occupant safety; the designer of an electronic device such as radar should consider both mechanical and electronic performances of the device. Through multi-objective optimization, the comprehensive performance of a structure or a system can be optimized, thus the product quality is greatly promoted. Researchers have accomplished a lot in this area [1–8], but most of the achievements focus on the deterministic problems where the parameters have specific values. Another type of problem is the uncertain multi-objective optimization, where uncertain parameters are involved, which is quite important but less studied.

This chapter proposes an interval-based uncertain multi-objective optimization method, expecting to provide a potential analysis tool for practical multi-objective optimization design under complex conditions. The rest of this chapter is organized as follows: Firstly, a mathematical model for interval multi-objective optimization is created. Secondly, the interval multi-objective optimization is transformed into a deterministic optimization problem and then solved by an efficient algorithm. Finally, a numerical example and an engineering application are used to test the proposed method.



C. Jiang et al., *Nonlinear Interval Optimization for Uncertain Problems*, Springer Tracts in Mechanical Engineering, https://doi.org/10.1007/978-981-15-8546-3_11

## 11.1 An Interval Multi-objective Optimization Model

A general deterministic multi-objective optimization problem can be expressed as:

$$\begin{cases} \min_{\mathbf{X}}[f_1(\mathbf{X}), f_2(\mathbf{X}), \ldots, f_m(\mathbf{X})] \\ \text{s.t. } g_i(\mathbf{X}) \le 0, \; i = 1, 2, \ldots, l \;\; \mathbf{X} \in \Omega^n \end{cases} \tag{11.1}$$

where $f(\mathbf{X})$ is the objective function, $g(\mathbf{X})$ is the constraint, and $m$ is the number of objective functions. Generally, the multi-objective optimization problem has multiple, even infinite solutions. Therefore, the target for multi-objective optimization is to obtain an optimal compromise solution in the feasible region.

Currently, there are two main approaches to solve multi-objective optimization problems. The first is to transform the multiple objective functions into a single-objective function according to preference information and then solve it by conventional optimization method to obtain the compromise solution. Such a method is known as the preference-based method. Most traditional multi-objective optimization methods belong to this, such as the weighted scalar-valued criterion [9], the goal programming [10], the efficacy coefficient method [11], etc. This type of method is straightforward, but effective provided that the preference information on the objectives is clear and exact. However, it is not suitable for complex or information-poor engineering problems since the accurate preference information is generally hard to obtain. The second way is to search Pareto-optimal solution set in the feasible region and then select one from the solution set as the optimal compromise solution according to the designer's experience. This approach is known as the production method [12–14]. It directly solves the Pareto-optimal solution set without requirement on preference information. Currently, the production method has become the mainstream method is this area.

When uncertainties are involved in Eq. (11.1), an uncertain multi-objective optimization problem thus will occur. If we use an interval approach to deal with the uncertainties, an interval multi-objective optimization model thus can be created in the following form:

$$\begin{cases} \min_{\mathbf{X}}[f_1(\mathbf{X}, \mathbf{U}), f_2(\mathbf{X}, \mathbf{U}), \ldots, f_m(\mathbf{X}, \mathbf{U})] \\ \text{s.t. } g_i^I(\mathbf{X}, \mathbf{U}) \le b_i^I = \left[b_i^L, b_i^R\right], \; i = 1, 2, \ldots, l \;\; \mathbf{X} \in \Omega^n \\ \quad \mathbf{U} \in \mathbf{U}^I = \left[\mathbf{U}^L, \mathbf{U}^R\right], U_i \in U_i^I = \left[U_i^L, U_i^R\right], i = 1, 2, \ldots, q \end{cases} \tag{11.2}$$

where $\mathbf{U}$ is a $q$-dimensional uncertain vector and its uncertainties are expressed by an interval vector $\mathbf{U}^I$.

## 11.2 Conversion to a Deterministic Multi-objective Optimization

In previous chapters, the optimizations discussed are single-objective problems, where the interval objective function is converted to two deterministic functions with the help of the $\leq_{cw}$-type order relation of interval number. However, this order relation seems not a suitable choice for the multi-objective problem due to the problem's complexity. Otherwise, the converted deterministic optimization will be too complex to solve. Therefore, here the $\leq_c$-type order relation of interval number introduced in Chap. 3 is adopted because it only focuses on the preference to middle point of the objective function. Therefore, the interval objective functions in Eq. (11.2) can be converted to [15, 16]:

$$\min_{\mathbf{X}}\left[f_1^c(\mathbf{X}), f_2^c(\mathbf{X}), \ldots, f_m^c(\mathbf{X})\right] \tag{11.3}$$

where the midpoints of the objective functions can be expressed as:

$$\begin{aligned} & f_i^c(\mathbf{X}) = \frac{f_i^L(\mathbf{X}) + f_i^R(\mathbf{X})}{2} = \frac{\min\limits_{\mathbf{U}} f_i(\mathbf{X}, \mathbf{U}) + \max\limits_{\mathbf{U}} f_i(\mathbf{X}, \mathbf{U})}{2}, i = 1, 2, \ldots, m \\ & \mathbf{U} \in \Gamma = \left\{\mathbf{U} \middle| U_i^L \leq U_i \leq U_i^R, i = 1, 2, \ldots, q\right\} \end{aligned} \tag{11.4}$$

As for the interval constraints in Eq. (11.2), the RPDI model is used to convert them into deterministic constraints like those in Eq. (9.11). Finally, Eq. (11.2) becomes a deterministic multi-objective optimization problem as follows:

$$\begin{cases} \min\limits_{\mathbf{X}}\left[f_1^c(\mathbf{X}), f_2^c(\mathbf{X}), \ldots, f_m^c(\mathbf{X})\right] \\ \text{s.t. } P_r\left(g_i^I(\mathbf{X}) \leq b_i^I\right) \geq \lambda_i, i = 1, 2, \ldots, l, \mathbf{X} \in \Omega^n \\ \quad \mathbf{U} \in \mathbf{U}^I = \left[\mathbf{U}^L, \mathbf{U}^R\right], U_i \in U_i^I = \left[U_i^L, U_i^R\right], i = 1, 2, \ldots, q \end{cases} \tag{11.5}$$

In the above analysis, the order relation $\leq_c$ is used to treat the interval objective functions. Actually, other types of order relations of interval number introduced in Chap. 3 which only focus the single point of interval, such as $\leq_L$, $\leq_R$ and $\leq_w$, can be also adopted, and whereby corresponding deterministic optimization problems like Eq. (11.5) can be created.

## 11.3 Algorithm Flow

The problem in Eq. (11.5) is also a two-layer nested optimization problem, where the outer layer searches for the optimum design vector, and the inner layer calculates the intervals of the uncertain objective functions and constraints. Such intervals here are obtained by the interval structural analysis method introduced in Chap. 5 for improving the computational efficiency. An optimization algorithm based on the multi-objective genetic algorithm is then developed for solving Eq. (11.5). During the optimization process, several design vector individuals are generated by the multi-objective genetic algorithm in the outer layer. In the inner layer, the interval structural analysis method is used to calculate the intervals of objective functions and constraints for each design vector individual for obtaining the midpoints of objective functions and RPDI values of constraints. Therefore, the noninferior solution set can be calculated by solving the transformed deterministic multi-objective optimization problem.

Figure 11.1 shows a flowchart of the proposed interval multi-objective optimization method. The optimization process can be summarized as follows:

Step 1: Initialization: set up the parameters of the outer layer multi-objective genetic algorithm; assign proper RPDI levels to the constraints; set $t = 0$.
Step 2: Generate randomly $N$ design vector individuals through multi-objective genetic algorithm, $\mathbf{X}_i, i = 1, 2, \ldots, N$.
Step 3: For each individual $\mathbf{X}_i$, calculate the upper and lower bounds of uncertain objective functions and constraints by using the interval structural analysis method.
Step 4: Calculate the midpoints of objective functions.
Step 5: Calculate the RPDI values for all constraints.
Step 6: Calculate the noninferior solution set of the current generation.
Step 7: If the termination condition is satisfied, end the program and output the Pareto-optimal solution set; else, set $t = t+1$ and jump to step 2 until convergence.

The micro multi-objective genetic algorithm (μMOGA) [14, 17] is selected as the solver for outer layer optimization since it has relatively good comprehensive performance in accuracy and efficiency and the noninferior solution set it yields is generally uniformly distributed. The μMOGA is developed based on the micro genetic algorithm (μGA) [18]. It has small population size (normally 5–8 individuals), resulting in its high efficiency. The small size, however, may easily cause premature of convergence during evolution. For ensuring the diversity of population and avoiding too early convergence to a local optimum, a restart strategy is adopted. Once too early convergence occurs, it produces an offspring generation that retains parents' optimal individual and population size. Moreover, a detection operator is used to explore the design space of the non-dominated solutions for better convergence efficiency. During the evolution, the μMOGA compares and selects individuals depending on their non-dominated levels and crowding distances. All individuals are ranked in descending order by their non-dominated relations. The higher ranking, the better an individual is. The individuals with the highest ranking are the non-dominated individuals, which are then preserved as the external population. Individuals of the same level are compared by their crowding distance. The individual with larger crowding distance is viewed as a better individual.

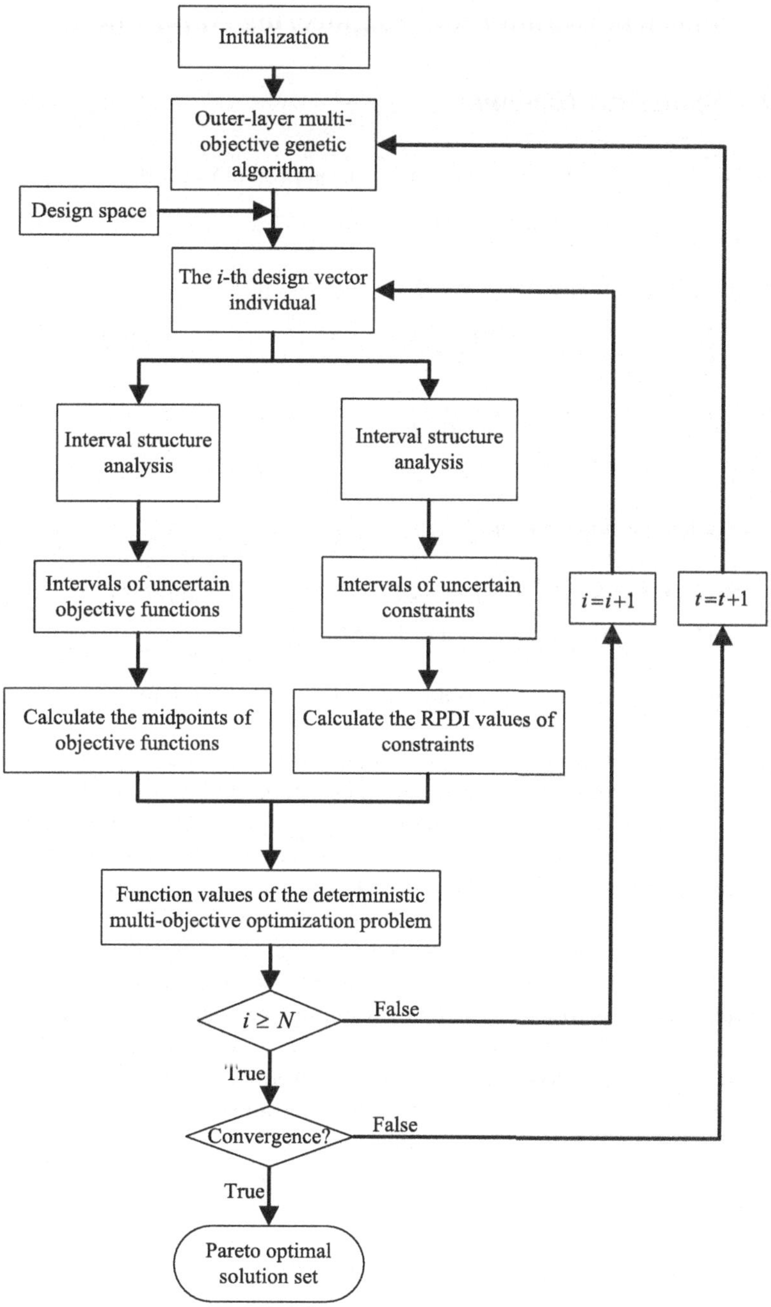


**Fig. 11.1** Algorithm flowchart of the interval multi-objective optimization method [15]

## 11.4 Numerical Example and Engineering Application

### 11.4.1 Numerical Example

Consider an interval multi-objective optimization problem as follows [19]:

$$\begin{cases} \min_{\mathbf{X}}[f_1(\mathbf{X},\mathbf{U}), f_2(\mathbf{X},\mathbf{U})] \\ \text{s.t. } g_1(\mathbf{X},\mathbf{U}) = U_1^2(X_1-2)^3 + U_2X_2 - 2.5 \le [0, 0.3] \\ \quad g_2(\mathbf{X},\mathbf{U}) = U_1^3X_2 + U_2^2X_1 - 3.85 - 8U_2^2(X_2 - X_1 + 0.65)^2 \le [0, 0.3] \\ \quad 0 \le X_1 \le 5,\ 0 \le X_2 \le 3 \\ \quad U_1 \in [0.9, 1.1],\ U_2 \in [0.9, 1.1] \end{cases} \tag{11.6}$$

where the objective functions are:

$$\begin{aligned} f_1(\mathbf{X},\mathbf{U}) &= U_1(X_1 + X_2 - 7.5)^2 + 0.25U_2^2(X_2 - X_1 + 3)^2 \\ f_2(\mathbf{X},\mathbf{U}) &= 0.25U_1^2(X_1 - 1)^2 + 0.5U_2^3(X_2 - 4)^2 \end{aligned} \tag{11.7}$$

In this problem, the interval variables $U_1$ and $U_2$ have the same uncertainty level of 10.0%.

Suppose two constraints have equal RPDI level, $\lambda_1 = \lambda_2 = \lambda$. The proposed method is used to perform optimization under three different RPDI levels, $\lambda = 0.5, 1.0, 1.5$, and the results are plotted in Fig. 11.2a-c, respectively. The dots in the rectangles represent the midpoints of two-objective functions. Under the influence of uncertain variables, each objective function at a specific design vector has a corresponding interval. Therefore, for a two objective problem, each dot corresponds to a rectangle that represents the variation ranges of two objective functions. Figure 11.2d shows a comparison of the noninferior solution sets under three given RPDI levels. For these three cases, the distributions of noninferior solution sets are relatively uniform. As the RPDI level increases, the noninferior solution sets tend to move right because the feasible region of the transformed optimization problem will shrink when RPDI level set to be the interval constraints becomes higher, and thus the objective function performance will decrease.

Next, four uncertainty levels of the two interval variables, 2%, 4%, 6%, and 8% are considered, while under the same RPDI level $\lambda = 1.0$ for constraints. The results are illustrated in Fig. 11.3. As it shows, even the uncertainty levels of variables are different, the noninferior solution sets are similar when the RPDI level of constraints is the same. However, as the uncertainty level grows, the rectangle becomes bigger, suggesting the possible variation of the objective functions caused by the uncertainty of variables tends to become bigger.

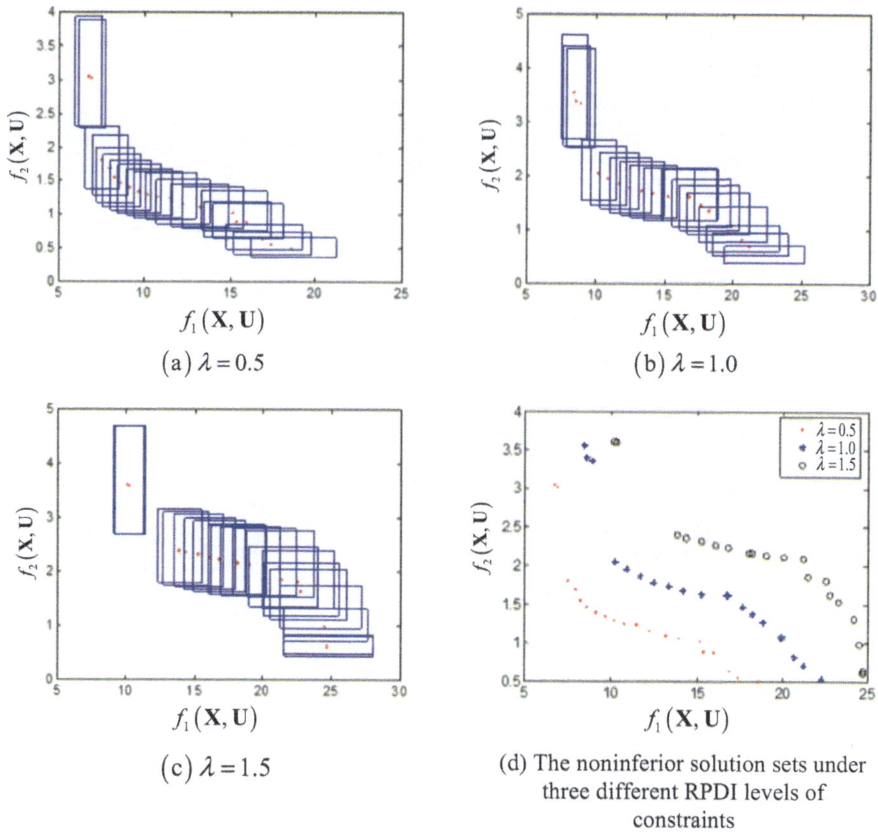


**Fig. 11.2** Optimization results of the numerical example under different RPDI levels [15]

## *11.4.2 Application to the Design of an Automobile Frame*

Consider the automobile frame [20] in Fig. 11.4. This frame is the same as used in Sect. 9.4.3. The frame's longerons are simplified to C-Shaped beams with a thickness of $h_1$. The cross-beams connecting the longerons are also C-shaped, except for $b_3$, which is a plate structure. The thicknesses of $b_1$, $b_2$, and $b_7$ are $h_2$; the thicknesses of $b_3$, $b_4$, and $b_5$ are $h_3$; the thicknesses of $b_6$ and $b_8$ are $h_4$. The aim is to minimize the frame's weight and maximize its stiffness in $y$-direction while the equivalent stress is kept below a given allowable value. The Young's modulus $E$ and the density $\rho$ are uncertain parameters due to manufacturing and measurement errors, and their uncertainty levels are both 5%. Therefore, the interval multi-objective optimization problem can be constructed as:

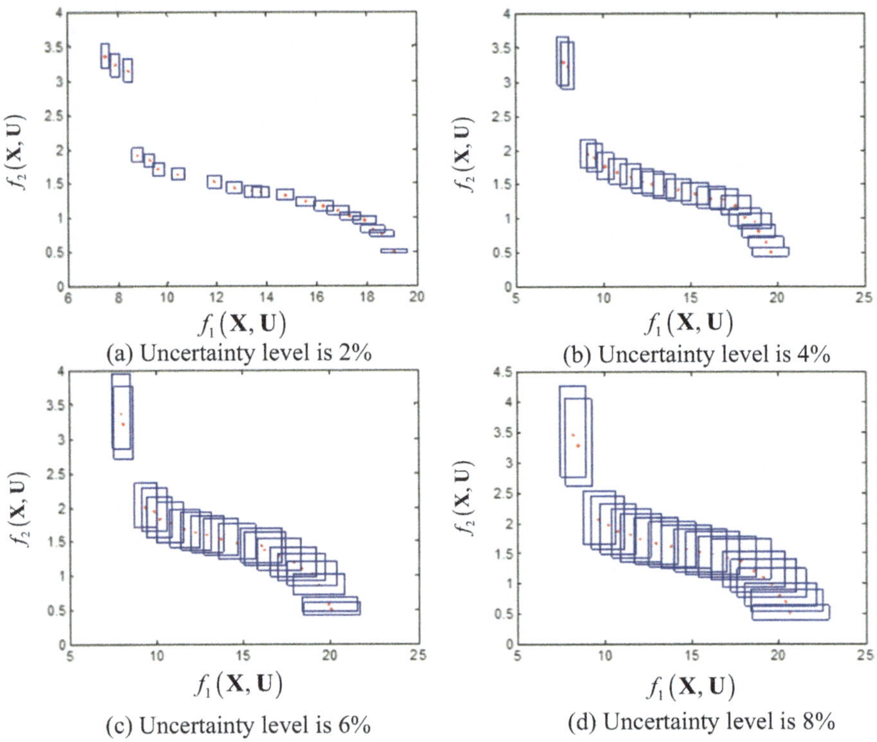


**Fig. 11.3** Optimization results of the numerical example under different uncertainty levels of parameters [15]

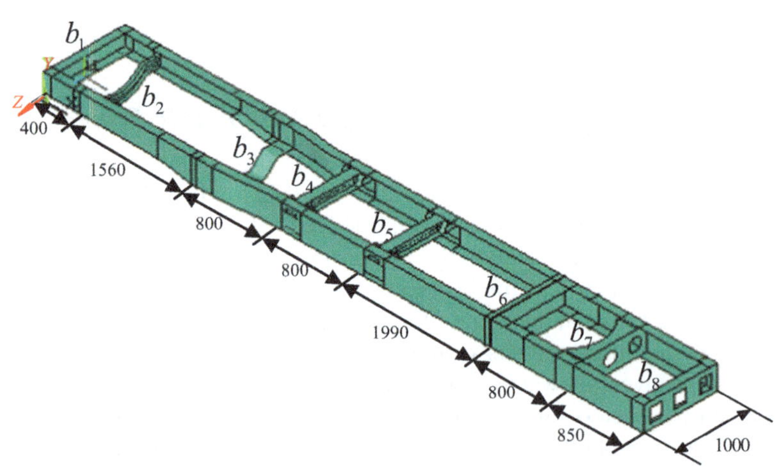


**Fig. 11.4** An automobile frame model (mm) [15]

$$\begin{cases} \min\limits_{\mathbf{h}} \left[d_{\max}(\mathbf{h}, E, \rho), Mass(\mathbf{h}, E, \rho)\right] \\ \text{s.t. } \sigma_{\max}(\mathbf{h}, E, \rho) \le 100\text{MPa} \\ \quad 14\,\text{mm} < h_1 < 18\,\text{mm}, 4\,\text{mm} < h_2 < 8\,\text{mm}, \\ \quad 2\,\text{mm} < h_3 < 6\,\text{mm}, 6\,\text{mm} < h_4 < 10\,\text{mm} \\ \quad E \in \left[1.9 \times 10^5\text{MPa}, 2.1 \times 10^5\text{MPa}\right], \\ \quad \rho \in \left[7.41 \times 10^3\text{kg/m}^3, 8.19 \times 10^3\text{kg/m}^3\right] \end{cases} \tag{11.8}$$

where $d_{\max}$ denotes the maximum displacement of the frame in the $y$-direction, $Mass$ denotes the mass of the frame, and $\sigma_{\max}$ denotes the maximum equivalent stress. $d_{\max}$ and $\sigma_{\max}$ are both calculated by the FEM model.

The stress constraint is given RPDI level of 1.0 during optimization. 20 Pareto solutions are obtained as illustrated in Fig. 11.5. Among these solutions, when $d_{\max}$ peaks at 1.52 mm (midpoint value), $Mass$ is at its minimum of 893.5 kg. Under the influence of uncertain material properties, their intervals are [1.44 mm, 1.59 mm] and $[848.9\,\text{kg}, 938.2\,\text{kg}]$, respectively. On the other way around, when the maximum displacement $d_{\max}$ is at its minimum of 0.22 mm, the mass $Mass$ of the frame peaks at 1201.7 kg. Then their variation intervals brought by uncertain material properties are [0.21 mm, 0.23 mm] and $[1141.6\,\text{kg}, 1261.8\,\text{kg}]$, respectively. Among these 20 Pareto solutions, the decision-maker can pick an optimal compromise solution by given criteria or his preference. Five representative optimal designs are selected and listed in Table 11.1. The last column, titled preference, shows the weighting factor of the two objective functions when the compromise solutions are picked.

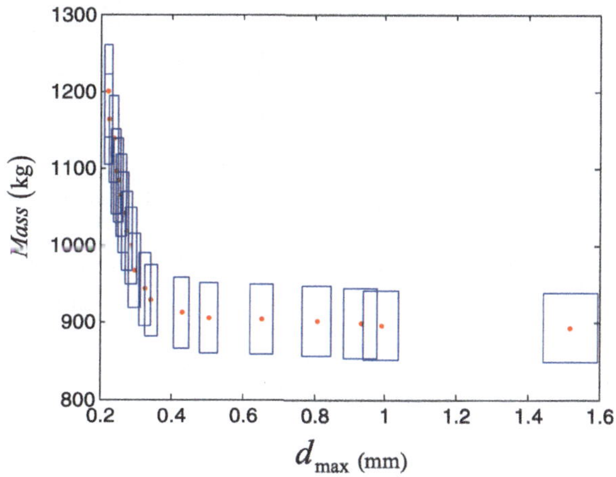


**Fig.11.5** Pareto-optimal solutions of the automobile frame [15]

**Table 11.1** Five solutions of interval multi-objective optimization of the automobile frame [15]

| $h_1$(mm) | $h_2$(mm) | $h_3$(mm) | $h_4$(mm) | $d^I_{max}$ (mm) | $d^c_{max}$(mm) | $Mass^I$(kg) | $Mass^c$(kg) | Preferences |
|---|---|---|---|---|---|---|---|---|
| 14.02 | 4.20 | 2.04 | 6.05 | [1.44, 1.59] | 1.515 | [848.9, 938.2] | 893.55 | (0.0,1.0) |
| 14.01 | 4.16 | 3.18 | 6.02 | [0.48, 0.53] | 0.505 | [861.2, 951.9] | 906.55 | (0.2,0.8) |
| 14.06 | 4.16 | 3.44 | 6.09 | [0.41, 0.45] | 0.430 | [867.4, 958.7] | 913.05 | (0.5,0.5) |
| 14.51 | 4.03 | 5.96 | 6.09 | [0.28, 0.31] | 0.295 | [919.2,1016.0] | 967.60 | (0.8,0.2) |
| 17.97 | 5.00 | 6.00 | 9.37 | [0.21, 0.23] | 0.220 | [1141.6,1261.8] | 1201.70 | (1.0,0.0) |

## 11.5 Summary

This chapter introduces the interval model into multi-objective optimization and constructs an interval multi-objective optimization method. The order relation transformation model with RPDI model is used to transform the uncertain optimization into a deterministic multi-objective optimization problem, which is then solved by a solution algorithm developed based on interval structural analysis method and multi-objective genetic algorithm. In the future, other more efficient optimization algorithms can also be formulated to solve the transformed deterministic multi-objective optimization problem, further improving the computational efficiency.

## References


1. Miettinen K (1999) Nonlinear multiobjective optimization. Kluwer, Boston
2. Zitzler E, Thiele L (1999) Multiobjective evolutionary algorithms: a comparative case study and the strength Pareto approach. IEEE Trans Evol Comput 3(4):257–271
3. Deb K, Pratap A, Agarwal S, Meyarivan T (2002) A fast and elitist multiobjective genetic algorithm: NSGA-II. IEEE Trans Evol Comput 6(2):182–197
4. Yang BS, Yeun YS, Ruy WS (2002) Managing approximation models in multiobjective optimization. Struct Multidiscip Optim 24(2):141–156
5. Zitzler E, Thiele L, Bader J (2010) On set-based multiobjective optimization. IEEE Trans Evol Comput 14(1):58–79
6. Ali M, Siarry P, Pant M (2012) An efficient differential evolution based algorithm for solving multi-objective optimization problems. Eur J Oper Res 217(2):404–416
7. Giagkiozis I, Fleming PJ (2015) Methods for multi-objective optimization: an analysis. Inf Sci 293:338–350
8. Deb K, Abouhawwash M (2016) An optimality theory-based proximity measure for set-based multiobjective optimization. IEEE Trans Evol Comput 20(4):515–528
9. Zadeh L (1963) Optimality and non-scalar-valued performance criteria. IEEE Trans Autom Control 8(1):59–60
10. Charnes A, Cooper WW, Ferguson RO (1955) Optimal estimation of executive compensation by linear programming. Manage Sci 1(2):138–151
11. Keeney RL, Raiffa H (1976) Decisions with multiple objectives: preferences and value tradeoffs. Wiley, London
12. Deb K (2001) Multi-objective optimization using evolutionary algorithms. Wiley, New York
13. Moh J-S, Chiang D-Y (2000) Improved simulated annealing search for structural optimization. AIAA J 38(10):1965–1973
14. Liu GP, Han X, Jiang C (2008) A novel multi-objective optimization method based on an approximation model management technique. Comput Methods Appl Mech Eng 197(33):2719–2731
15. Li XL, Jiang C, Han X (2011) An uncertainty multi-objective optimization based on interval analysis and its application. China Mech Eng 22(9):1100–1106
16. Li XL (2011) Uncertain multi-objective optimization based on non probabilistic convex sets and its application. Master thesis, Hunan University, Changsha
17. Liu GP (2008) Multi-objective optimization methods based on the micro genetic algorithm and applications. PhD thesis, Hunan University, Changsha
18. Krishnakumar K (1990) Micro-genetic algorithms for stationary and non-stationary function optimization. In: 1989 symposium on visual communications, image processing, and intelligent robotics systems. Philadelphia, PA, United States.

19. Li FY, Li GY, Zheng G (2010) Uncertain multi-objective optimization method based on interval. Chin J Solid Mech 31(1):86–93
20. Jiang C, Bai YC, Han X, Ning HM (2010) An efficient reliability-based optimization method for uncertain structures based on non-probability interval model. Comput, Mater Contin (CMC) 18(1):21–42

# Chapter 12
# Interval Optimization Considering Tolerance Design

**Abstract** This chapter proposes an interval optimization method considering tolerance design, which provides not only the optimal design but also the optimal tolerances of the design variables.

In the interval optimization methods studied in the previous chapters, all the uncertain parameters are described by intervals, and specific intervals must be provided for these parameters before modeling the optimization problem. In these methods, the interval variables are separated from the design variables and must be determined before optimization. In the practical engineering, however, there exists another important kind of interval optimization problem, where the intervals of the uncertain parameters cannot be predetermined, and furthermore, the interval variables and the design variables are the same variables. For example, the product size and material property are common design variables in engineering problems, while these design variables generally involve uncertainties due to manufacturing errors. Besides, the intervals of these design variables often cannot be given in advance because of the complexities and diversities of the manufacturing techniques. In such a situation, the manufacturing errors should be taken into account in the design stage to not only obtain the optimal design but also ensure the product performance under larger interval uncertainties. By doing so, the manufacturability can be improved and the cost is reduced. Until now, few studies have been published in this area.

An interval optimization model considering tolerance design and its corresponding solution method is therefore proposed in this chapter. The proposed model can provide not only the optimal design but also the optimal tolerances, which further expands the application of the conventional interval optimization method by relating it to manufacturability. The rest of this chapter is organized as follows: Sect. 12.1 gives the interval optimization model considering the tolerance design; Sect. 12.2 provides a solution method for solving the interval optimization model; finally, a numerical example and engineering applications are analyzed in Sect. 12.3.



C. Jiang et al., *Nonlinear Interval Optimization for Uncertain Problems*, Springer Tracts in Mechanical Engineering, https://doi.org/10.1007/978-981-15-8546-3_12

## 12.1 An Interval Optimization Model Considering Tolerance Design

A conventional deterministic optimization problem is generally expressed as [1, 2]:

$$\begin{cases} \min\limits_{\mathbf{X}} \ f(\mathbf{X}) \\ \text{s.t. } g_i(\mathbf{X}) \le b_i, \ i = 1, 2, \ldots, l \\ \quad \mathbf{X}_l \le \mathbf{X} \le \mathbf{X}_r \end{cases} \tag{12.1}$$

where $\mathbf{X}$ is an $n$-dimensional design vector, with $\mathbf{X}_r$ and $\mathbf{X}_l$, respectively, its upper and lower design bounds, and $b_i$ is the maximum allowable value of the $i$-th constraint.

An optimal design, denoted by $\mathbf{X}_d$, can be obtained by solving the above optimization problem. For practical problems, the structural sizes, material characteristics, loads, etc., are often selected as the design variables. However, there usually exist uncertainties in these variables that are caused by manufacturing errors, measurement errors, etc. In many cases, a minor deviation of $\mathbf{X}_d$ in the manufacturing process can lead to a large fluctuation of the design objective function and constraints, which may cause low performance or even failure of the structure or products. The uncertainties of $\mathbf{X}_d$ obviously can be decreased by improving the manufacturing techniques, while at the same time the manufacturing cost is also increased. In order to solve this problem, the interval method here is introduced to measure the uncertainties of the design variables. In addition, a corresponding interval optimization model is constructed to improve the comprehensive performance of structure or product in terms of design objective, reliability, and manufacturability.

When errors from manufacture and measurement are considered, the real values of $\mathbf{X}$ should belong to an interval vector $\mathbf{X}^I$:

$$X_i^I = \left[X_i^L, X_i^R\right] = \left\{X_i \middle| X_i^L \le X_i \le X_i^R\right\}, i = 1, 2, \ldots, n \tag{12.2}$$

When $X_i^L = X_i^R$, the interval $X_i^I$ will degenerate into a real number $X_i$. Besides, according to Eq. (2.6), $\mathbf{X}^I$ can also be represented in the following form:

$$X_i^I = \left\langle X_i^c, X_i^w \right\rangle = \left\{X_i \middle| X_i^c - X_i^w \le X_i \le X_i^c + X_i^w\right\}, i = 1, 2, \ldots, n \tag{12.3}$$

The similarity of the interval number and the symmetric tolerance is obvious. In practical product design, $X_i^c$ can be viewed as the nominal design of $X_i$, and $X_i^w$ the design tolerance of $X_i$. To match the engineering design behavior, $\mathbf{X}^I$ can also be expressed in the symmetric tolerance form as follows:

$$X_i^I = X_i^c \pm X_i^w, \ i = 1, 2, \ldots, n \tag{12.4}$$

Using the interval approach to describe the uncertainties in the design variables in Eq. (12.1), we then create a new interval optimization model as follows [1, 2]:

$$\begin{cases} \min\limits_{\mathbf{X}^I} \ f(\mathbf{X}^I) \\ \text{s.t.}\ g_i(\mathbf{X}^I) \le b_i,\ i = 1, 2, \ldots, l \\ \quad \mathbf{X}_l \le \mathbf{X}^I \le \mathbf{X}_r \\ \quad X_i^I = [X_i^L, X_i^R],\ \ i = 1, 2, \ldots, n \end{cases} \tag{12.5}$$

Compared with the deterministic optimization in Eq. (12.1), the design variables of the above interval optimization are no longer a real number vector $\mathbf{X}$ but an interval vector $\mathbf{X}^I$. The optimal intervals of the design variables can be obtained by solving Eq. (12.5) to ensure the comprehensive performance of a structure or product under uncertainties.

Since the interval of a design variable can be uniquely determined by its nominal design and tolerance, Eq. (12.5) can also be expressed in an equivalent form:

$$\begin{cases} \min\limits_{\langle \mathbf{X}^c, \mathbf{X}^w \rangle} \ f(\langle \mathbf{X}^c, \mathbf{X}^w \rangle) \\ \text{s.t.}\ g_i(\langle \mathbf{X}^c, \mathbf{X}^w \rangle) \le b_i,\ i = 1, 2, \ldots, l \\ \quad \mathbf{X}_l \le \langle \mathbf{X}^c, \mathbf{X}^w \rangle \le \mathbf{X}_r \end{cases} \tag{12.6}$$

In the above interval optimization model, the number of the design variables is $2n$ rather than $n$ in the deterministic optimization problem. The design variables in Eq. (12.5) consist of $n$ lower bounds and $n$ upper bounds of $\mathbf{X}$, and those in Eq. (12.6) consist of $n$ nominal values and $n$ tolerances of $\mathbf{X}$.

## 12.2 Conversion to a Deterministic Optimization

In this section, a mathematical transformation model is constructed to convert the above interval optimization problem into a conventional deterministic optimization problem. During the conversion process, the performance requirements for design objective, constraint reliability and manufacturability of design variables are comprehensively considered.

In engineering, under the precondition that product performance should be ensured, the smaller the design variables' interval radiuses (i.e. the design tolerances), the higher the required manufacturing precision, and meanwhile the higher the manufacturing cost. Therefore, the tolerance is a crucial design index that reflects the manufacturability of the design. To evaluate the tolerance level of design variables on the whole, a design tolerance index $W$ is defined as [1]:

$$W = \sqrt[n]{\prod_{i=1}^{n} \frac{X_i^w}{\psi_i}} \tag{12.7}$$

where $\psi_i$ is the normalization factor, which can be selected as $\psi_i = \left|X_i^c\right|$. In Eq. (12.7), $W$ is a dimensionless parameter that reflects the global tolerance of all design variables. The larger the $W$, the larger the global tolerance.

In addition, the nominal objective function $f(\mathbf{X}^c)$ can be used to depict the objective function's average performance under parameter uncertainties. By comprehensively considering the global tolerance and the nominal objective function, the uncertain objective function in Eq. (12.6) can be transformed into the following deterministic multi-objective optimization problem:

$$\min_{\mathbf{X}^c,\mathbf{X}^w} \left[f(\mathbf{X}^c), -W\right] \tag{12.8}$$

Through the first objective function, the average performance of the original objective function under uncertainties can be optimized. Through the second objective function, the allowable tolerances of the design variables can be maximized. As the allowable manufacturing error is heightened, the manufacturing cost can be reduced and the manufacturability can be improved.

On the other hand, using the RPDI model [3] that is given in Chap. 9 to deal with the uncertain constraints, Eq. (12.6) can be finally transformed into a deterministic optimization problem as follows:

$$\begin{cases} \min\limits_{\mathbf{X}^c,\mathbf{X}^w} \left(f(\mathbf{X}^c), -\sqrt[n]{\prod\limits_{i=1}^{n} \frac{X_i^w}{\psi_i}}\right) \\ \text{s.t. } P_r\left(g_i^I(\langle\mathbf{X}^c, \mathbf{X}^w\rangle) \le b_i\right) = \frac{b_i - g_i^L}{2g_i^w} \ge \lambda_i,\ i = 1, 2, \ldots, l \\ \mathbf{X}_l \le \langle\mathbf{X}^c, \mathbf{X}^w\rangle \le \mathbf{X}_r \end{cases} \tag{12.9}$$

where the intervals of the constraints can be presented as:

$$g_i^I = \left[g_i^L, g_i^R\right] = \left[\min_{\mathbf{X}\in\langle\mathbf{X}^c,\mathbf{X}^w\rangle} g_i(\mathbf{X}),\ \max_{\mathbf{X}\in\langle\mathbf{X}^c,\mathbf{X}^w\rangle} g_i(\mathbf{X})\right],\ i = 1, 2, \ldots, l \tag{12.10}$$

Solving Eq. (12.9) involves a two-layer nested optimization, where the design variables $\mathbf{X}^c$ and $\mathbf{X}^w$ are optimized in the outer loop and the constraint intervals $\left[g_i^L, g_i^R\right]$, $i = 1, 2, \ldots, l$ are calculated in the inner loop. The improved non-dominated sorting genetic algorithm (NSGA-II) [4] and the sequential quadratic programming method (SQP) [5] are adopted here for the outer and inner optimizations, respectively. Through solving the above optimization problem, the optimal nominal design vector $\mathbf{X}^c$ and the maximum allowable tolerances vector $\mathbf{X}^w$ are both obtained.

Moreover, as mentioned in the previous chapters, the allowable constraint boundaries $b_i$, $i = 1, 2, \ldots, n$ itself may be uncertain in many engineering problems. In such a condition, the interval vector $b_i^I = \langle b_i^c, b_i^w\rangle$, $i = 1, 2, \ldots, l$ can be introduced to describe the uncertainties. For example, the maximum stress of a structure is not allowed to exceed its yield strength, while the precise value of yield strength is generally unavailable due to discrete material characteristics. What we know is often that the yield strength belongs to an interval. Correspondingly, Eq. (12.6) can be expressed as follows:

$$\begin{cases} \min\limits_{\langle \mathbf{X}^c, \mathbf{X}^w \rangle} & f(\langle \mathbf{X}^c, \mathbf{X}^w \rangle) \\ \text{s.t.} & g_i(\langle \mathbf{X}^c, \mathbf{X}^w \rangle) \leq \langle b_i^c, b_i^w \rangle, \ i = 1, 2, \ldots, l \\ & \mathbf{X}_l \leq \langle \mathbf{X}^c, \mathbf{X}^w \rangle \leq \mathbf{X}_r \end{cases} \tag{12.11}$$

The above interval optimization problem can be also transformed into a deterministic optimization problem as shown in Eq. (12.9), except that it adopts Eq. (9.7) to compute the RPDI value of the constraints. For all the following numerical examples in this chapter, only the deterministic constraint boundaries are considered for simplicity.

It should be pointed out that the word "tolerance" here has a broader conception. The above interval optimization model is not merely for tolerance problem in mechanical design. It can also be applied to the analysis and design of loads, material characteristics, etc. Besides, the proposed method can be conveniently extended to asymmetric tolerance problem. At present, there have emerged some extended studies on the proposed interval optimization model [6, 7].

## 12.3 Numerical Example and Engineering Applications

### *12.3.1 Numerical Example*

Consider the following optimization problem with two design variables [1]:

$$\begin{cases} \min\limits_{\mathbf{X}} f(X_1, X_2) = 2X_1 + 21X_2 - X_1X_2 + 100 \\ \text{s.t. } g_1(X_1, X_2) = 3(X_1 - 15)^2 + (X_2 - 20)^2 \leq 220 \\ \quad g_2(X_1, X_2) = X_1X_2 + 12X_2 \leq 430 \\ \quad 10 \leq X_1 \leq 15, \quad 5 \leq X_2 \leq 15 \end{cases} \tag{12.12}$$

Taking the manufacturing errors of $X_1$ and $X_2$ into account, Eq. (12.12) can be transformed into an interval optimization problem, and it can be further transformed into a deterministic multi-objective optimization problem through Eq. (12.9):

$$\begin{cases} \min\limits_{\langle \mathbf{X}^c, \mathbf{X}^w \rangle} \left[ 2X_1^c + 21X_2^c - X_1^c X_2^c + 100, -\sqrt{\frac{X_1^w}{|X_1^c|} \cdot \frac{X_2^w}{|X_2^c|}} \right] \\ \text{s.t. } P_r\left( 3(\langle X_1^c, X_1^w \rangle - 15)^2 + (\langle X_2^c, X_2^w \rangle - 20)^2 \le 220 \right) \ge \lambda_1 \\ \quad P_r(\langle X_1^c, X_1^w \rangle \langle X_1^c, X_1^w \rangle + 12\langle X_2^c, X_2^w \rangle \le 430) \ge \lambda_2 \\ \quad 10 \le X_1^c - X_1^w \le X_1^c + X_1^w \le 25, \;\; 5 \le X_2^c - X_2^w \le X_2^c + X_2^w \le 15 \end{cases} \tag{12.13}$$

where $X_1^c$, $X_2^c$, $X_1^w$, and $X_2^w$ become design variables of the above optimization problem.

In the optimization procedure, the same RPDI levels are given for two constraints. Meanwhile, four different RPDI levels, 0.9, 1.0, 1.1, and 1.2 are adopted to perform the interval optimization separately to analyze the influence of different constraint possibility levels on the optimization results. Parts of Pareto-optimal solutions in these four situations are given in Tables 12.1, 12.2, 12.3 and 12.4 and the distributions of the optimal solution set are shown in Fig. 12.1. As shown in Tables 12.1, 12.2, 12.3 and 12.4, the weight of the nominal objective function decreases in order while the weight of design tolerance index increases. Different from conventional optimization problem, the optimal design obtained here is no longer a specific value but an interval. Through the interval optimization, the optimal nominal design is achieved and the bearable maximum tolerances of all design variables are obtained simultaneously to take the uncertainties involved in the subsequent manufacturing process into account. Besides, the contradictory relationship between two objective functions is obvious.

**Table 12.1** Parts of Pareto-optimal solutions of the numerical example when $\lambda = 0.9$ [1]

| No | $X_1^I$ | $X_2^I$ | $g_1^I$ | $P_{r1}$ | $g_2^I$ | $P_{r2}$ | $f(X^c)$ | $W$ |
|---|---|---|---|---|---|---|---|---|
| 1 | $22.00 \pm 0.15$ | $12.13 \pm 0.41$ | [196, 222] | 0.91 | [397, 429] | 1.04 | 132 | 0.0154 |
| 2 | $21.65 \pm 0.46$ | $12.11 \pm 0.63$ | [168, 224] | 0.93 | [381, 435] | 0.91 | 135 | 0.0332 |
| 3 | $21.22 \pm 0.67$ | $11.86 \pm 1.06$ | [143, 227] | 0.92 | [352, 438] | 0.91 | 140 | 0.0531 |
| 4 | $20.87 \pm 1.00$ | $11.78 \pm 1.10$ | [122, 228] | 0.92 | [340, 436] | 0.93 | 143 | 0.0669 |
| 5 | $20.52 \pm 1.13$ | $11.58 \pm 1.51$ | [106, 231] | 0.91 | [316, 440] | 0.92 | 147 | 0.0848 |
| 6 | $20.23 \pm 1.31$ | $11.42 \pm 1.71$ | [93, 234] | 0.90 | [300, 440] | 0.93 | 149 | 0.0985 |

**Table 12.2** Parts of Pareto-optimal solutions of the numerical example when $\lambda = 1.0$ [1]

| No | $X_1^I$ | $X_2^I$ | $g_1^I$ | $P_{r1}$ | $g_2^I$ | $P_{r2}$ | $f(X^c)$ | $W$ |
|---|---|---|---|---|---|---|---|---|
| 1 | $21.82 \pm 0.26$ | $12.12 \pm 0.43$ | [184, 220] | 1.01 | [392, 428] | 1.06 | 134 | 0.0208 |
| 2 | $21.41 \pm 0.34$ | $11.78 \pm 0.86$ | [164, 219] | 1.02 | [361, 427] | 1.05 | 138 | 0.0342 |
| 3 | $21.08 \pm 0.75$ | $11.87 \pm 0.81$ | [139, 220] | 1.00 | [357, 429] | 1.01 | 141 | 0.0494 |
| 4 | $20.60 \pm 0.96$ | $11.62 \pm 1.14$ | [117, 220] | 1.00 | [331, 428] | 1.02 | 146 | 0.0677 |
| 5 | $20.14 \pm 1.09$ | $11.36 \pm 1.38$ | [102, 217] | 1.03 | [310, 423] | 1.06 | 150 | 0.0811 |
| 6 | $19.70 \pm 1.53$ | $11.28 \pm 1.45$ | [83, 220] | 1.00 | [297, 423] | 1.05 | 154 | 0.0998 |

**Table 12.3** Parts of Pareto-optimal solutions of the numerical example when $\lambda = 1.1$ [1]

| No | $X_1^I$ | $X_2^I$ | $g_1^I$ | $P_{r1}$ | $g_2^I$ | $P_{r2}$ | $f(X^c)$ | $W$ |
|---|---|---|---|---|---|---|---|---|
| 1 | $21.89 \pm 0.25$ | $12.27 \pm 0.23$ | [189, 216] | 1.14 | [405, 427] | 1.16 | 133 | 0.0146 |
| 2 | $21.39 \pm 0.52$ | $12.21 \pm 0.32$ | [159, 209] | 1.22 | [391, 425] | 1.14 | 138 | 0.0255 |
| 3 | $20.75 \pm 0.67$ | $11.60 \pm 1.01$ | [132, 212] | 1.10 | [340, 421] | 1.10 | 144 | 0.0529 |
| 4 | $20.31 \pm 0.88$ | $11.52 \pm 1.11$ | [113, 207] | 1.14 | [327, 419] | 1.12 | 149 | 0.0646 |
| 5 | $19.91 \pm 1.02$ | $11.24 \pm 1.34$ | [101, 207] | 1.12 | [306, 414] | 1.14 | 152 | 0.0780 |
| 6 | $19.23 \pm 1.58$ | $11.08 \pm 1.34$ | [78, 206] | 1.11 | [289, 408] | 1.19 | 158 | 0.0997 |

**Table 12.4** Parts of Pareto-optimal solutions of the numerical example when $\lambda = 1.2$ [1]

| No | $X_1^I$ | $X_2^I$ | $g_1^I$ | $P_{r1}$ | $g_2^I$ | $P_{r2}$ | $f(X^c)$ | $W$ |
|---|---|---|---|---|---|---|---|---|
| 1 | $21.58 \pm 0.17$ | $11.70 \pm 0.48$ | [184, 214] | 1.21 | [375, 411] | 1.52 | 136 | 0.0182 |
| 2 | $21.10 \pm 0.47$ | $11.83 \pm 0.47$ | [154, 204] | 1.31 | [371, 413] | 1.41 | 141 | 0.0297 |
| 3 | $20.73 \pm 0.70$ | $11.65 \pm 0.68$ | [135, 205] | 1.20 | [352, 412] | 1.29 | 145 | 0.0444 |
| 4 | $20.14 \pm 0.78$ | $11.20 \pm 1.06$ | [117, 203] | 1.20 | [318, 404] | 1.31 | 150 | 0.0606 |
| 5 | $19.67 \pm 0.96$ | $11.11 \pm 1.27$ | [99, 199] | 1.22 | [302, 404] | 1.26 | 154 | 0.0748 |
| 6 | $19.19 \pm 1.19$ | $10.89 \pm 1.40$ | [86, 197] | 1.20 | [285, 398] | 1.28 | 158 | 0.0894 |

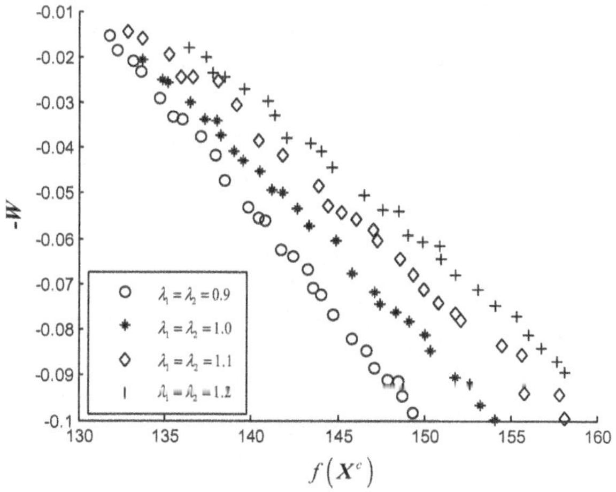


**Fig. 12.1** Pareto-optimal solutions of the numerical example under different RPDI levels [1]

When the weight of design tolerance index increases, the optimal design tolerance index has an increasing trend, which means the manufacturability of the design is improved and the manufacturing cost can be reduced. However, the optimal nominal objective function deteriorates at the same time. As shown in Table 12.2 with the case of $\lambda_1 = \lambda_2 = 1.0$, the allowable tolerances of $X_1$ and $X_2$ in the first group are $\pm 0.26$ and $\pm 0.43$, respectively, while the allowable tolerances in the sixth group become $\pm 1.53$ and $\pm 1.45$, respectively, approximately six and three times the first group, which indicates a significant improvement of manufacturability. Meanwhile, the optimal nominal objective function increases from 134 in the first group to 154 in the sixth group, leading to a worse average objective performance. Therefore, it is important to balance the manufacturability and the nominal design objective in practical application. A reasonable weight should be chosen for the design.

In addition, different RPDI levels have obvious impact on the interval optimization results. As shown in Fig. 12.1, as $\lambda$ increases, the Pareto frontier gradually goes away from the origin of objective function space, which means the optimal design tolerance index decreases while the optimal nominal objective function value increase. The reason is that $\lambda$ describes the reliability of the interval constraints. A larger $\lambda$ corresponds to a smaller feasible region of the deterministic constraints in Eq. (12.13) and will cause the deteriorations of tolerances and nominal objective function. Therefore, the computational results indicate that not only the design objective and the manufacturability, but also the constraint reliability under parameter uncertainties are considered in the proposed interval optimization model.

## 12.3.2 Application to a Cantilever Beam

Consider a cantilever beam as shown in Fig. 12.2 [8]. The length of the structure is $L = 1.0\,\text{m}$. The cross section is a rectangle whose width and height are $b$ and $h$, respectively. The beam is subjected to a horizontal load $P_x = 50\,\text{kN}$ and a vertical load $P_y = 25\,\text{kN}$ at the free end. The maximum stress at the fixed end of the cantilever beam can be analytically expressed as:

$$\sigma = \frac{6P_xL}{b^2h} + \frac{6P_yL}{bh^2} \tag{12.14}$$

Taking $b$ and $h$ as design variables, to minimize the volume $Vol$ of the cantilever beam under the premise that the maximum stress cannot exceed an allowable stress $\sigma_s = 250\,\text{MPa}$, an optimization problem can be established as follows:

$$\begin{cases} \min\limits_{b,h} \; Vol(b,h) \\ \text{s.t. } \sigma(b,h) = \frac{6P_xL}{b^2h} + \frac{6P_yL}{bh^2} \le \sigma_s \\ 5\,\text{cm} \le b \le 20\,\text{cm} \\ 5\,\text{cm} \le h \le 20\,\text{cm} \end{cases} \tag{12.15}$$

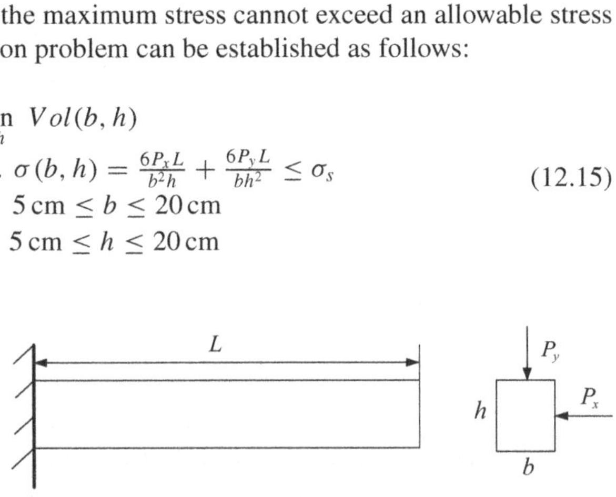


**Fig. 12.2** A cantilever beam [8]

As the design variables $b$ and $h$ involve uncertainties in the practical manufacturing process, a corresponding interval optimization model should be formulated. Similarly, four different RPDI levels, 0.9, 1.0, 1.1, and 1.2, are investigated and corresponding distributions of Pareto-optimal solution sets are shown in Fig. 12.3. The optimization results show some similarities to the first example: under a given RPDI level the optimal nominal objective function value and the optimal design tolerance index have an opposite trend. The larger the design tolerance index, the smaller the nominal objective function. Besides, when the RPDI level decreases, which means the reliability requirement given to the stress constraint becomes higher, the optimal nominal objective function and the optimal design tolerance index will deteriorate simultaneously. Table 12.5 gives parts of Pareto-optimal solutions when $\lambda = 1.0$. As shown in the results, the intervals of the maximum stress for these solutions is completely under the allowable stress $\sigma_s = 250$ MPa, which means the structure has a high reliability.

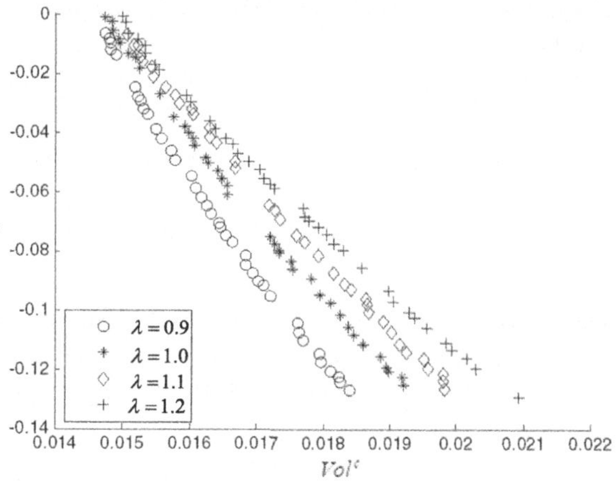


**Fig. 12.3** Pareto-optimal solutions of the cantilever beam under different RPDI levels [1]

**Table 12.5** Parts of Pareto-optimal solutions of the cantilever beam when $\lambda = 1.0$ [1]

| No | $b^I$ (cm) | $h^I$ (cm) | $\sigma^I$ (MPa) | $P_r$ | $Vol(b^c, h^c)$ (m$^3$) | $W$ |
|---|---|---|---|---|---|---|
| 1 | $13.60 \pm 0.91$ | $12.67 \pm 1.08$ | [158.5, 248.9] | 1.01 | 0.0172 | 0.075 |
| 2 | $13.60 \pm 0.93$ | $12.70 \pm 1.11$ | [156.9, 249.6] | 1.00 | 0.0173 | 0.078 |
| 3 | $13.73 \pm 1.01$ | $12.62 \pm 1.09$ | [154.8, 249.5] | 1.01 | 0.0173 | 0.080 |
| 4 | $14.06 \pm 1.02$ | $12.33 \pm 1.11$ | [153.2, 248.6] | 1.01 | 0.0173 | 0.081 |
| 5 | $13.81 \pm 1.12$ | $12.69 \pm 1.09$ | [150.7, 248.4] | 1.02 | 0.0175 | 0.083 |
| 6 | $13.95 \pm 1.06$ | $12.58 \pm 1.23$ | [148.9, 249.7] | 1.00 | 0.0175 | 0.086 |

### 12.3.3 Application to the Crashworthiness Design of Vehicle Side Impact

In the vehicle side impact, the main factors affecting occupant safety are intrusion and intrusion velocity of the sidewall structure [9]. The sides of a vehicle are where rigidity and strength are the weakest. During a side impact, the side door, the B-pillar, and the sidewall structure play the major load-bearing roles. An FEM model of a type of automobile with 720,383 shell elements, including 148,040 elements in the moving deformable barrier, is shown in Fig. 12.4. A crash analysis is performed according to the side crash regulation in US-NCAP. The initial collision velocity of the moving deformable barrier is set as 62 km/h. During the crash, the B-pillar's maximum intrusion $Intr$ is a crucial evaluation parameter of safety so that it should be restricted in an allowable range. In this problem, the inner plate thickness $t_1$ and external plate thickness $t_2$ of the B-pillar are chosen as design variables to make their mass minimized and simultaneously the B-pillar's maximum intrusion is chosen as a constraint to ensure the safety. Therefore, an optimization problem can be established as follows:

$$\begin{cases} \min\limits_{t_1, t_2} \ Mass(t_1, t_2) \\ \text{s.t.} \ Intr(t_1, t_2) \leq Intr_a \\ \quad 1.0\,\text{mm} \leq t_1 \leq 2.0\,\text{mm} \\ \quad 1.0\,\text{mm} \leq t_2 \leq 2.0\,\text{mm} \end{cases} \tag{12.16}$$

where $Intr_a = 350\,\text{mm}$ is the allowable value of B-pillar's maximum intrusion.

Consider the manufacturing errors of $t_1$ and $t_2$, a corresponding interval optimization problem can be constructed. In order to promote optimization efficiency, the

**Fig. 12.4** A side impact problem of vehicle and the B-pillar structure [1]

Latin hypercube design (LHD) is adopted to select 10 samples in the design space and whereby construct the quadratic polynomial response surface of the maximum intrusion $Intr$ based on FEM analyses [1]:

$$Intr(t_1, t_2) = 436.54 - 41.20t_1 - 11.56t_2 - 35.92t_1t_2 + 23.08t_1^2 + 7.99t_2^2 \tag{12.17}$$

The interval optimization is then solved based on the above approximation model to save the computational cost. The RPDI level is set as $\lambda = 1.0$ and the distribution of Pareto-optimal solution set is shown in Fig. 12.5. As shown in the figure, the optimal total mass and the tolerance have an opposite trend. In practical problems, the designer can choose an ideal design according to engineering requirements. For example, if the minimal mass $Mass(t_1^c, t_2^c)$ of B-pillar is expected to be 7.0 kg, the corresponding interval optimization results can be obtained as in Table 12.6. In such a situation, the optimal thicknesses of $t_1$ and $t_2$ are 1.68 mm and 1.72 mm, respectively, and the maximum allowable manufacturing tolerances are $\pm0.05$ mm and $\pm0.09$ mm, respectively. Under this design, the interval of the B-pillar's maximum intrusion is $Intr^I = [315.0\,\text{mm}, 350.0\,\text{mm}]$, which completely satisfies the design requirement $Intr_a = 350\,\text{mm}$.

**Fig. 12.5** Pareto-optimal solutions of the vehicle side impact problem [1]

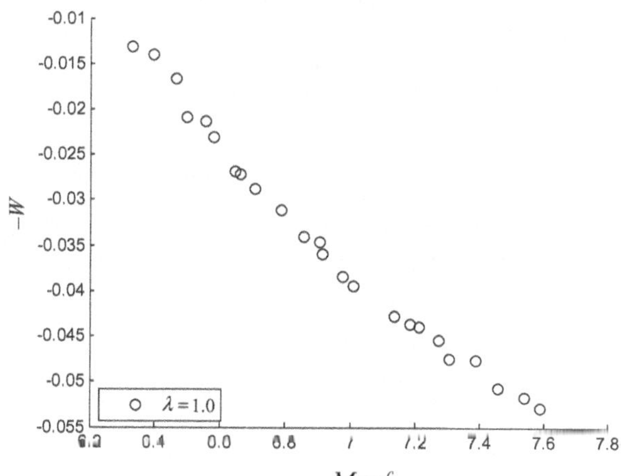


**Table 12.6** An optimization solution of the vehicle side impact problem [1]

| $t_1^I$ (mm) | $t_2^I$ (mm) | $Intr^I$ (mm) | $P_r$ | $Mass(t_1^c, t_2^c)$ (kg) | $W$ |
|---|---|---|---|---|---|
| $1.68 \pm 0.05$ | $1.72 \pm 0.09$ | [315.0, 350.0] | 1.00 | 7.0 | 0.038 |

## 12.4 Summary

This chapter proposes a new interval optimization method considering the tolerance design that comprehensively considers the optimality of objective function, the manufacturability of design and the reliability of constraints, extending the application of conventional interval optimization method. The optimality of objective is ensured by optimizing the nominal objective function. A larger allowable manufacturing error to improve manufacturability and reduce cost is obtained by maximizing the design tolerance index, and the reliability under parameter uncertainties is guaranteed by handling the uncertain constraints using the RPDI model. In a practical situation, the proposed method is able to offer an optimal design comprehensively considering the product performance, manufacturability, and reliability requirements. In future, a series of efficient solution algorithms are needed to develop for the proposed interval optimization model, further promoting its engineering applicability.

## References

1. Jiang C, Xie HC, Zhang ZG, Han X (2015) A new interval optimization method considering tolerance design. Eng Optim 47(12):1637–1650
2. Zhang ZG (2014) An interval optimization method consider the dependency and tolerance of uncertain variables. Master degree, Hunan University, Changsha
3. Jiang C, Han X, Li D (2012) A new interval comparison relation and application in interval number programming for uncertain problems. Comput Mater Continua 27(3):275–303
4. Deb K, Agrawal S, Pratap A, Meyarivan T (2000) A fast elitist non-dominated sorting genetic algorithm for multi-objective optimization: NSGA-II. In: International conference on parallel problem solving from nature, Berlin, Heidelberg
5. Nocedal J, Wright SJ (1999) Numerical optimization. Springer, New York
6. Xie HC (2014) Optimization design for key structure performance indicators of vehicle based on interval uncertainty. PhD thesis, Hunan University, Changsha
7. Xie HC, Jiang C, Zhang ZG, Yu S (2014) Vehicle ride comfort optimization based on interval analysis. Autom Eng 36(9):1127–1131
8. Du XP (2008) Saddlepoint approximation for sequential optimization and reliability analysis. ASME J Mech Des 130(1):011011–011022
9. Zhang XR, Su QZ (2008) A research on influencing factors of occupant injury in side impact. Autom Eng 30(2):146–150

# Chapter 13
# Interval Differential Evolution Algorithm

**Abstract** By introducing the interval model into the existing differential evolution, this chapter proposes a novel interval differential evolution algorithm, which can directly solve the original interval optimization problem rather than transforming it to a deterministic optimization problem first.

In the interval optimization model discussed in the previous chapters, the uncertain optimization problem should be transformed into a deterministic problem to solve by using mathematical tools such as order relation of interval number and possibility degree of interval number. Such transformation is a common treatment used in conventional stochastic programming and fuzzy programming. In recent years, how to develop a direct solution method without transformation of the original problem to furthermore simplify the optimization procedure has attracted a lot of attention in the field of interval optimization. At present, there have been some studies on this issue [1].

Based on current differential evolution (DE) algorithm, an interval differential evolution (IDE) algorithm is presented in this chapter to effectively solve the interval optimization problem [2]. Being different from the methods in previous chapters, the IDE algorithm can directly solve the original problem without transforming it into a deterministic problem, which provides some new ideas for interval optimization. The rest of this chapter is organized as follows: firstly, the fundamentals of classic DE algorithm is introduced; secondly, the IDE algorithm is presented; finally, the effectiveness of the proposed method is verified by numerical examples and an engineering application.



C. Jiang et al., *Nonlinear Interval Optimization for Uncertain Problems*, Springer Tracts in Mechanical Engineering, https://doi.org/10.1007/978-981-15-8546-3_13

## 13.1 Fundamentals of the Differential Evolution Algorithm

A general deterministic optimization problem can be expressed as

$$\begin{cases} \min\limits_{\mathbf{X}} f(\mathbf{X}) \\ \text{s.t. } g_i(\mathbf{X}) \le b_i,\ i = 1, 2, \ldots, l \\ \quad \mathbf{X}_l \le \mathbf{X} \le \mathbf{X}_r \end{cases} \tag{13.1}$$

where $\mathbf{X}$ is an $n$-dimensional design vector; $\mathbf{X}_l$ and $\mathbf{X}_r$ are lower and upper bounds of $\mathbf{X}$, respectively; $b_i$ is the maximum allowable value of the $i$-th constraint.

DE algorithm, belonging to a type of stochastic searching algorithm, is able to efficiently solve the above problem. It is first presented by Storn and Price [3] to solve the Chebyshev polynomial fitting problem. So far, DE is widely applied in engineering optimization, path planning, informatics, operational research, etc., and has shown excellent performance [4–10]. In general, the DE algorithm has the following advantages: (1) DE algorithm is based on real number coding, which is relatively simple and easy to program compared to other evolutionary algorithms. (2) Relatively less control parameters are used in the DE algorithm. For example, a classic DE algorithm only has three control parameters: population size *NP*, scaling factor $F$, and crossover probability *CR*, which makes the algorithm performance easy to control. (3) The space complexity of the DE algorithm is relatively low, which facilitates the solution of some large-scale problems. The basic idea of the DE algorithm can be summarized as follows: (1) Generate *NP* initial individuals. (2) In each iterative step, perform mutation operation and crossover operation to generate new individuals and use the greedy criterion to select better *NP* individuals from consecutive two generations to form the next generation. (3) Through iteration, the optimization result can be obtained when the convergence is reached. The flowchart of the DE algorithm is shown in Fig. 13.1.

The major contents of the DE algorithm include the initial population generation strategy, the mutation strategy, the crossover strategy, and the selection strategy. The main difference to other population-based evolution algorithm lies in the crossover strategy and the selection strategy. Next, a brief introduction to these strategies is given.

### 13.1.1 Initial Population Generation Strategy

At generation $t = 0$, the algorithm will randomly generate *NP* $n$-dimensional real number vector individuals $\mathbf{X}_{k,0} = \left(X_{k,1,0}, X_{k,2,0}, \ldots, X_{k,n,0}\right),\ k = 1, 2, \ldots, NP$, in the design space:

$$X_{k,j,0} = X_{j,l} + \text{rand}\,(0,1) \times \left(X_{j,r} - X_{j,l}\right),\ j = 1, 2, \ldots, n \tag{13.2}$$

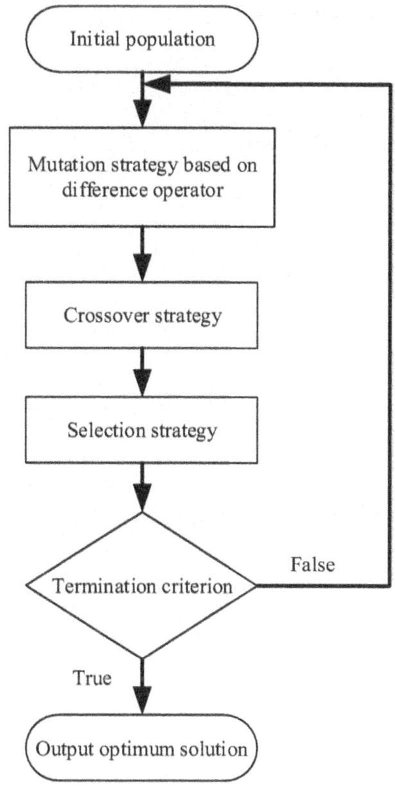


**Fig. 13.1** Basic flowchart of the DE algorithm [2]

where rand (0,1) represents a real number randomly generated between 0 and 1.

### 13.1.2 Mutation Strategy

In the $t$-th generation, for each target vector individual $\mathbf{X}_{k,t} = (X_{k,1,t}, X_{k,2,t}, \ldots, X_{k,n,t})$, generate a corresponding mutant vector $\mathbf{V}_{k,t} = (V_{k,1,t}, V_{k,2,t}, \ldots, V_{k,n,t})$. According to the mutation strategy, the algorithm randomly selects two individuals $\mathbf{X}_{r_2,t}$ and $\mathbf{X}_{r_3,t}$ in the current population and then defines their difference vector:

$$\Delta\mathbf{X} = \mathbf{X}_{r_2,t} - \mathbf{X}_{r_3,t} \tag{13.3}$$

as the searching direction. Finally, the $\Delta\mathbf{X}$ is multiplied by a scaling factor $F$ and then added to another target vector individual to generate a mutant individual. The most common used difference mutation strategies include [3]:

(1) DE/rand/1

$$\mathbf{V}_{k,t} = \mathbf{X}_{r_1,t} + F \times \left(\mathbf{X}_{r_2,t} - \mathbf{X}_{r_3,t}\right) \tag{13.4}$$

(2) DE/best/1

$$\mathbf{V}_{k,t} = \mathbf{X}_{\text{best},t} + F \times \left(\mathbf{X}_{r_1,t} - \mathbf{X}_{r_2,t}\right) \tag{13.5}$$

(3) DE/rand/2

$$\mathbf{V}_{k,t} = \mathbf{X}_{r_1,t} + F \times \left(\mathbf{X}_{r_2,t} - \mathbf{X}_{r_3,t}\right) + F \times \left(\mathbf{X}_{r_4,t} - \mathbf{X}_{r_5,t}\right) \tag{13.6}$$

(4) DE/current-to-rand/1

$$\mathbf{V}_{k,t} = \mathbf{X}_{k,t} + F \times \left(\mathbf{X}_{r_1,t} - \mathbf{X}_{k,t}\right) + F \times \left(\mathbf{X}_{r_2,t} - \mathbf{X}_{r_3,t}\right) \tag{13.7}$$

(5) DE/current-to-best/1

$$\mathbf{V}_{k,t} = \mathbf{X}_{k,t} + F \times \left(\mathbf{X}_{\text{best}} - \mathbf{X}_{k,t}\right) + F \times \left(\mathbf{X}_{r_1,t} - \mathbf{X}_{r_2,t}\right) \tag{13.8}$$

where $r_1$, $r_2$, $r_3$, $r_4$, and $r_5$ represent the individuals that are different from the $k$-th one of the current population and $\mathbf{X}_{\text{best},t}$ is the best individual of the current population.

### 13.1.3 Crossover Strategy

In the DE algorithm, a generally used binomial crossover strategy [3] is:

$$Y_{k,j,t} = \begin{cases} V_{k,j,t}, & \text{if } \text{rand}_j(0, 1) \le CR \text{ or } j = j_{\text{rand}} \\ X_{k,j,t}, & \text{otherwise} \end{cases} \tag{13.9}$$

where *CR* refers to the crossover probability, $j_{\text{rand}}$ is an arbitrary integer in $[1, n]$, $\mathbf{Y}_{k,t}$ represents the $k$-th trial vector individual in the $t$-th generation. Besides, there are also other crossover strategies, such as the exponential crossover [3, 11], the orthogonal crossover [12], etc.

### 13.1.4 Selection Strategy

The purpose of the selection strategy is to select *NP* better individuals in target vector individual $\mathbf{X}_{k,t}$ and trial vector individual $\mathbf{Y}_{k,t}$ and pass them to the $(t+1)$-th generation. For an unconstrained problem, the strategy can be expressed as follows:

$$\mathbf{X}_{k,t+1} = \begin{cases} \mathbf{Y}_{k,t}, & \text{if } f(\mathbf{Y}_{k,t}) \le f(\mathbf{X}_{k,t}) \\ \mathbf{X}_{k,t}, & \text{otherwise} \end{cases} \tag{13.10}$$

In the optimization process, the crossover, mutation, and selection strategies are iteratively performed until the convergence is reached. Different to the unconstrained optimization, the DE algorithm should take objective function and constraints into account simultaneously in its selection strategy when handling the constrained optimization problem. In other words, a constraint-handling technique must be adopted. Commonly used constraint-handling techniques include the penalty function [13], the feasibility rule [14], etc.

## 13.2 Formulation of the Interval Differential Evolution Algorithm

This section presents an IDE algorithm based on the existing DE to directly solve the nonlinear interval optimization problems. Firstly, a model named *satisfaction value of interval possibility degree* is proposed based on the RPDI discussed in Chapter 9 to deal with the interval constraints. Secondly, an interval preferential rule is established for individual selection based on the satisfaction value of interval possibility degree. Finally, the computational procedure of the proposed IDE algorithm is given.

### *13.2.1 Satisfaction Value of Interval Possibility Degree and Treatment of Uncertain Constraints*

The objective function value and the constraint function value in the conventional deterministic optimization problem are specific real numbers. Therefore, whether the objective function value is satisfactory and whether the constraints are violated can be judged directly through their values. However, in interval optimization, the objective function value and the constraint function value usually belong to intervals due to the involvement of interval parameters. In order to construct satisfaction value of interval possibility degree to handle the interval constraints, here a left truncated possibility degree of interval number is first proposed based on RPDI model to quantitatively describe the degree that an interval is greater or superior to another interval. For intervals $A^I$ and $B^I$, the left truncated possibility degree of interval number, $P_r^{'}(A^I \le B^I)$, is defined as [2]:

$$P_r^{'}(A^I \le B^I) = \max\left(\frac{B^R - A^L}{2A^w + 2B^w}, 0\right) \tag{13.11}$$

When $A^I$ or $B^I$ degrades into a real number, $P_r^{'}(A^I \le B^I)$ can be rewritten as [2]:

$$P_r^{'}(A^I \le b) = \max\left(\frac{b - A^L}{2A^w}, 0\right), \quad P_r^{'}(a \le B^I) = \max\left(\frac{B^R - a}{2B^w}, 0\right) \tag{13.12}$$

$P_r^{'}(A^I \le B^I)$ has the following properties:

(1) $P_r^{'}(A^I \le B^I) \in [0, +\infty)$;
(2) If $A^R \le B^L$, then $P_r^{'}(A^I \le B^I) \ge 1$ and $A^I$ is completely on the left side of $B^I$ on the real number axis;
(3) If $A^L \le B^R$, then $P_r^{'}(A^I \le B^I) = 0$ and $A^I$ is completely on the right side of $B^I$ on the real number axis.

Figure 13.2 describes the variation tendency of $P_r^{'}(A^I \le B^I)$. When fixing $A^c$, $A^w$, $B^w$ and moving the midpoint $B^c$, $P_r^{'}(A^I \le B^I)$ shows two states: at Stage 1, $P_r^{'}(A^I \le B^I) = 0$, and when entering Stage 2, it is monotonically increasing from 0 to $\infty$. The inflection point occurs at $B^R = A^L$.

Based on the left truncated possibility degree of interval number, the conception of satisfaction value of interval possibility degree is presented. For intervals $A^I$ and $B^I$, the satisfaction value of interval possibility degree $R_\lambda$ is defined as [2]:

$$R_\lambda(A^I \le B^I) = \begin{cases} P_r^{'}(A^I \le B^I)/\lambda, & \text{if } P_r^{'}(A^I \le B^I) \le \lambda \\ 1, & \text{otherwise} \end{cases} \tag{13.13}$$

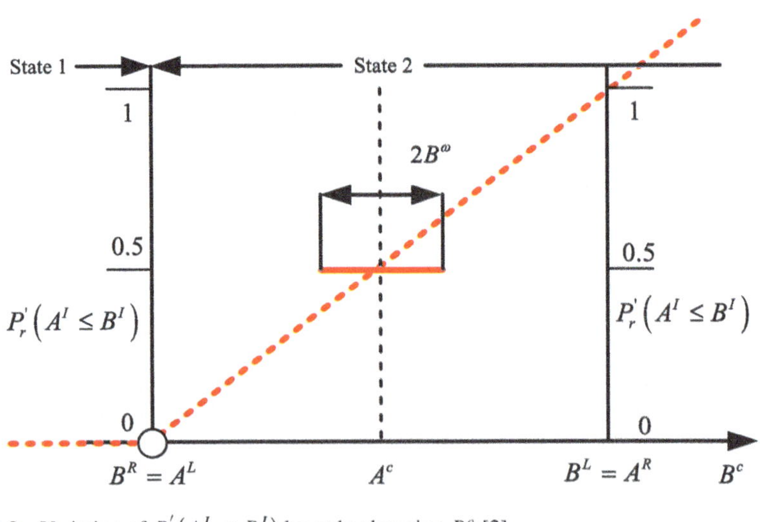


**Fig. 13.2** Variation of $P_r^{'}(A^I \le B^I)$ by only changing $B^c$ [2]

where $\lambda > 0$ is a predetermined possibility degree level. The satisfaction value of interval possibility degree is used to quantitatively describe the degree that "$A^I \le B^I$" can satisfy the given possibility degree level $\lambda$. Therefore, it is also called the satisfaction value of interval possibility degree under the level $\lambda$. $R_\lambda$ has the following properties:

(1) $R_\lambda(A^I \le B^I) \in [0, 1]$.
(2) $P_r'(A^I \le B^I) \le \lambda$ means the given $\lambda$ level for possibility degree is not satisfied. For this situation, $R_\lambda(A^I \le B^I)$ is set as $P_r'(A^I \le B^I)/\lambda$.
(3) If $P_r'(A^I \le B^I) > \lambda$ means the given $\lambda$ level for possibility degree is satisfied. For this situation, $R_\lambda(A^I \le B^I)$ is set as 1.

It should be pointed out that the satisfaction values of interval possibility degree may vary under different $\lambda$ levels for two intervals $A^I$ and $B^I$.

The proposed satisfaction value of interval possibility degree in Eq. (13.13) can be used to handle the interval constraints. For any interval constraint $g_i(\mathbf{X}, \mathbf{U}) \le b_i^I$ in Eq. (4.1), the value of $R_\lambda$ can be calculated by [2]:

$$R_{\lambda_i}(g_i^I(\mathbf{X}) \le b_i^I) = \begin{cases} P_r'(g_i^I(\mathbf{X}) \le b_i^I)/\lambda_i, & \text{if } P_r'(g_i^I(\mathbf{X}) \le b_i^I) \le \lambda_i \\ 1, & \text{otherwise} \end{cases} \tag{13.14}$$

where $\lambda_i > 0$ is the predetermined possibility degree or RPDI level of the $i$-th constraint. As mentioned in the previous chapters, $\lambda_i$ represents the reliability or safety level that the $i$-th interval constraint should reach under uncertainties and a lager $\lambda_i$ means a higher reliability requirement.

In order to further illustrate the above method, it is applied to a simple problem of strength reliability analysis. Let $g^I$ and $b^I$ be the stress and the strength of a structure, respectively, and consider five cases as follows:

$$\begin{aligned}
&\text{Case 1}: \ g^I = [150\text{MPa}, 170\text{MPa}],\ b^I = [200\text{MPa}, 240\text{MPa}] \\
&\text{Case 2}: \ g^I = [170\text{MPa}, 190\text{MPa}],\ b^I = [200\text{MPa}, 240\text{MPa}] \\
&\text{Case 3}: \ g^I = [190\text{MPa}, 220\text{MPa}],\ b^I = [200\text{MPa}, 240\text{MPa}] \\
&\text{Case 4}: \ g^I = [230\text{MPa}, 270\text{MPa}],\ b^I = [200\text{MPa}, 240\text{MPa}] \\
&\text{Case 5}: \ g^I = [250\text{MPa}, 270\text{MPa}],\ b^I = [200\text{MPa}, 240\text{MPa}]
\end{aligned} \tag{13.15}$$

For this problem, if the possibility degree level $\lambda$ is set to be 1.2, the $R_\lambda$ values of the above five cases are 1.0, 0.97, 0.60, 0.10, and 0, respectively, which means only Case 1 can reach the given $\lambda$ level. Whereas if the possibility degree $\lambda$ is reduced to 1.0, the $R_\lambda$ value of the five cases become 1.0, 1.0 0.71, 0.13, and 0, respectively. In this situation, both Case 1 and Case 2 can satisfy the requirements of interval possibility level.

When multiple interval constraints in Eq. (4.1) are considered, an overall satisfaction value of interval possibility degree, denoted as $R_t$, can be obtained after calculating the $R_\lambda$ value for each constraint [2]:

$$R_t = \sum_{i=1}^{l} R_{\lambda_i} \tag{13.16}$$

By comparing $R_t$ and the number of the interval constraints $l$, we can judge whether a vector individual in DE population meets the constraints:

(1) When $R_t < l$, there is at least one interval constraint not satisfying the given possibility degree level and this vector individual is called an infeasible solution. In such a situation, the larger the $R_t$, the larger degree that the vector individual can satisfy all the interval constraints. Such individuals should be given priority to pass into the next generation.
(2) When $R_t = l$, all the given possibility degree levels are satisfied by the interval constraints. In such a situation, the vector individual is called a feasible solution.

### 13.2.2 Selection Strategy Based on an Interval Preferential Rule

In the frame of the DE algorithm, how to select better individuals from target vector individuals and trial vector individuals is crucial to the overall performance of the algorithm. This chapter develops an interval preferential rule based on the satisfaction value of interval possibility degree model to select better individuals. The flowchart of this rule is shown in Fig. 13.3. Let $\mathbf{X}_1$ and $\mathbf{X}_2$ be arbitrary two individuals in the evolution population with the overall satisfaction values of interval possibility degree $R_t(\mathbf{X}_1)$ and $R_t(\mathbf{X}_2)$, the objective function midpoint values $f^c(\mathbf{X}_1)$ and $f^c(\mathbf{X}_2)$, and the objective function radiuses $f^w(\mathbf{X}_1)$ and $f^w(\mathbf{X}_2)$, respectively. The individual selection procedure can be summarized as follows [2]:

(1) When $R_t(\mathbf{X}_1) < l$ and $R_t(\mathbf{X}_2) = l$, select the feasible solution individual $\mathbf{X}_2$ into the next population, and vice versa. This rule means the feasible solution individual will be given priority to be selected into the next generation.
(2) When $R_t(\mathbf{X}_1) = R_t(\mathbf{X}_2) = l$. If $f^c(\mathbf{X}_1) < f^c(\mathbf{X}_2)$, select $\mathbf{X}_1$. In the opposite situation, select $\mathbf{X}_2$. This rule means if two individuals are both feasible solutions, the one with smaller objective function midpoint value should be selected to guarantee the nominal performance of the objective function.
(3) When $R_t(\mathbf{X}_1) = R_t(\mathbf{X}_2) = l$ and $f^c(\mathbf{X}_1) = f^c(\mathbf{X}_2)$, if $f^w(\mathbf{X}_1) < f^w(\mathbf{X}_2)$, select $\mathbf{X}_1$. Otherwise, select $\mathbf{X}_2$. This rule means when the two individuals are both feasible and have the same objective function midpoint values, the one with smaller objective function radius should be selected to ensure the robustness of the objective function.
(4) When $R_t(\mathbf{X}_1) = R_t(\mathbf{X}_2) = l$, $f^c(\mathbf{X}_1) = f^c(\mathbf{X}_2)$ and $f^w(\mathbf{X}_1) = f^w(\mathbf{X}_2)$, randomly select one into the next generation.
(5) When $R_t(\mathbf{X}_1) < l$ and $R_t(\mathbf{X}_2) < l$, if $R_t(\mathbf{X}_1) > R_t(\mathbf{X}_2)$, select $\mathbf{X}_1$ and otherwise select $\mathbf{X}_2$. This rule means when the two individuals are both infeasible, the individual with larger overall satisfaction value of interval possibility degree

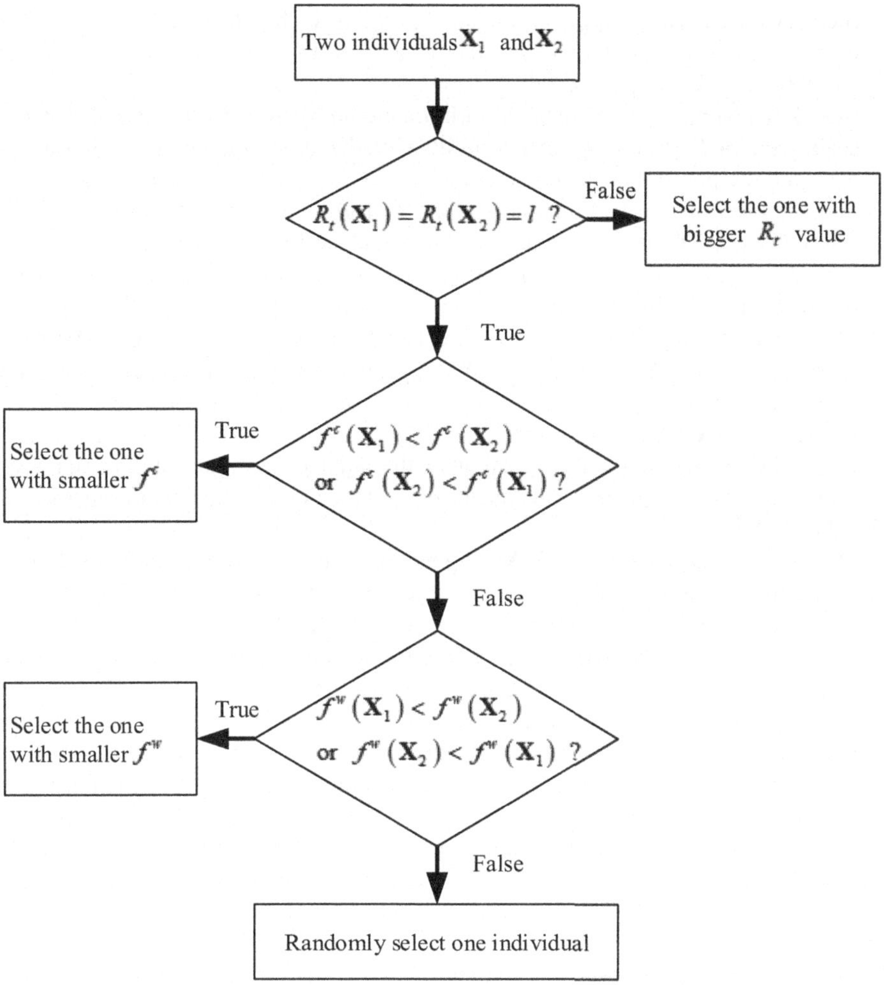


**Fig. 13.3** The selection strategy based on an interval preferential rule [2]

should be selected since a larger $R_t$ means the given interval possibility degree levels are better satisfied.

## *13.2.3 Algorithm Flow*

Based on the above analyses, an IDE algorithm is then developed in this chapter to solve the interval optimization problem. Its flowchart can be summarized as follows:

Step 1: Construct an interval optimization problem as shown in Eq. (4.1) according to the actual situation, then define the design variables $\mathbf{X} \in \Omega^n$ and the interval variables $\mathbf{U} \in \mathbf{U}^I = [\mathbf{U}^L, \mathbf{U}^R]$.
Step 2: Set parameters of the DE algorithm, including the population size $NP$, the scaling factor $F$, the crossover probability $CR$, the maximum number of iterations $t_{\max}$, and the possibility degree levels $\lambda_i, i = 1, 2, ..., l$ for all the constraints.
Step 3: Generate $NP$ initial individuals to form the initial population $P_0 = \{\mathbf{X}_{1,0}, \mathbf{X}_{2,0}, ..., \mathbf{X}_{NP,0}\}$ in the design space, and set $t = 0$.
Step 4: In the $t$-th generation, adopt the optimization method to calculate the intervals of objective function and constraints, $[f^L(\mathbf{X}_{k,t}), f^R(\mathbf{X}_{k,t})]$ and $[g_i^L(\mathbf{X}_{k,t}), g_i^R(\mathbf{X}_{k,t})]$, $i = 1, 2, K, l$, for each individual $\mathbf{X}_{k,t}$ in the $t$th population $P_t = \{\mathbf{X}_{1,t}, \mathbf{X}_{2,t}, ..., \mathbf{X}_{k,t}, ..., \mathbf{X}_{NP,t}\}$. Then calculate the objective function midpoint and radius $f^c(\mathbf{X}_{k,t}), f^w(\mathbf{X}_{k,t})$ and constraints' midpoints and radiuses $g_i^c(\mathbf{X}_{k,t}), g_i^w(\mathbf{X}_{k,t}), i = 1, 2, ..., l$.
Step 5: For each individual $\mathbf{X}_{k,t}$, calculate the satisfaction values of interval possibility degree for all the interval constraints and then obtain the overall satisfaction value $R(\mathbf{X}_{k,t})$.
Step 6: For each individual $\mathbf{X}_{k,t}$, generate the trial vector individual $\mathbf{Y}_{k,t}$ using the mutation and crossover strategies. The population consisting of trial vector individuals is denoted as $Y_t = \{\mathbf{Y}_{1,t}, \mathbf{Y}_{2,t}, ..., \mathbf{Y}_{k,t}, ..., \mathbf{Y}_{NP,t}\}$. For each trial vector individual, adopt the optimization method to calculate the intervals of objective function and constraints $[f^L(\mathbf{Y}_{k,t}), f^R(\mathbf{Y}_{k,t})]$ and $[g_i^L(\mathbf{Y}_{k,t}), g_i^R(\mathbf{Y}_{k,t})]$, $i = 1, 2, ..., l$, whereby the objective function midpoint and radius $f^c(\mathbf{Y}_{k,t}), f^w(\mathbf{Y}_{k,t})$, and the constraints' midpoints and radiuses $g_i^c(\mathbf{Y}_{k,t}), g_i^w(\mathbf{Y}_{k,t}), i = 1, 2, ..., l$. Then calculate the overall satisfaction value $R_t(\mathbf{X}_{k,t})$ for each trial individual.
Step 7: Combine the current population and the trial population, select $NP$ individuals according to the interval preferential rule given in Sect. 13.2.2, denoted as $P_{t+1} = \{\mathbf{X}_{1,t+1}, \mathbf{X}_{2,t+1}, ..., \mathbf{X}_{NP,t+1}\}$, and pass them to the $(t+1)$-th generation. Set $t := t + 1$.
Step 8: Check if the convergence condition $t \le t_{\max}$ is satisfied. If the convergence is reached, stop the iteration and export the optimal individual. Otherwise, go to Step 4 to continue the iteration.

The flowchart of the proposed IDE algorithm is given in Fig. 13.4. Different from the methods in the previous chapters, the IDE algorithm directly solves the original interval optimization problem rather than transforming it into a deterministic problem, which simplifies the solving procedure. On the other hand, the two-layer nested optimization still exists in the IDE algorithm, with the outer layer for searching the optimal design using DE algorithm, and the inner layer for calculating the intervals of the objective function and constraints for different individuals. In practical use, the conventional optimization methods such as the sequential quadratic programming (SQP) [15] can be adopted for inner layer optimization. Also, some approximation analysis methods such as the interval structural analysis introduced in Chap. 5 can be used for inner layer optimization to promote optimization efficiency.

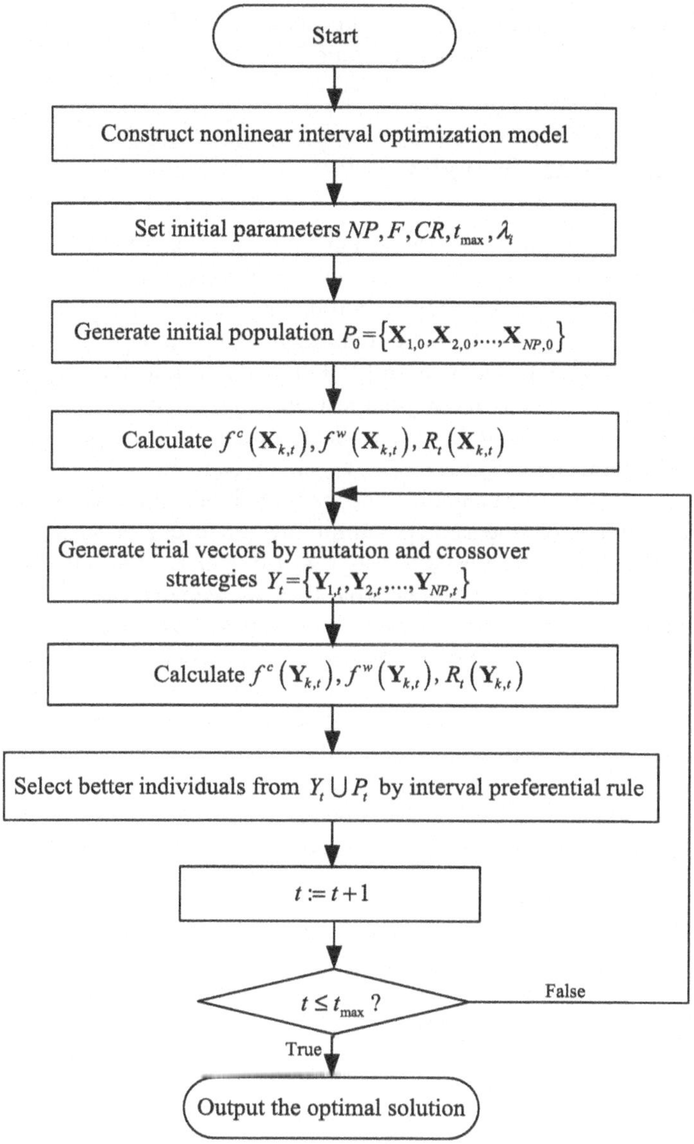


**Fig. 13.4** Algorithm flowchart of the IDE algorithm [2]

## 13.3 Numerical Examples and Engineering Application

In the numerical examples and engineering application analyzed next, SQP is used in inner layer optimization, and the rand/1 mutation strategy and binominal crossover strategy are adopted to generate trial vector individuals. The parameter control strategy in the jDE algorithm [16] is used for parameters $F$ and $CR$. The population size is $NP = 50$ and the maximum generations are $t_{\max} = 500$.

### 13.3.1 Numerical Examples

In this section, 10 test functions, denoted as $f_1 - f_{10}$, are analyzed. These test functions are modified from deterministic test functions published in 2006 IEEE Congress on Evolutionary Computation [17] by introducing interval variables. As shown in Table 13.1, these functions show different degrees of complexities. Their specific expressions can be seen in Appendix A [2]. For distinguishing the original functions and the modified functions, the original test functions are denoted as $f_1^{'} - f_{10}^{'}$. Just setting the interval variable in $f_1 - f_{10}$ as the midpoints of these intervals will yield $f_1^{'} - f_{10}^{'}$. The optimization results of $f_1 - f_{10}$ are given in Table 13.2.

Firstly, $f_{10}$ is used as an example to be analyzed by the proposed IDE algorithm. During the optimization, the same possibility degree level $\lambda$ is set for two interval constraints. Besides, four cases, $\lambda = 0.6$, 0.8, 1.0, and 1.2 respectively are tested. The optimization results are shown in Table 13.3. As seen in the table, the difference between the results considering interval uncertainties and the results using original deterministic constraints is obvious. Besides, as $\lambda$ increases, the optimal interval of objective function also changes, and the objective function midpoint gradually increases. The reason is that a larger $\lambda$ level means the reliability requirements on the interval constraints are higher, which will shrink the feasible region for the design variables and therefore the corresponding nominal objective function value will also deteriorate. In addition, the $R_y$ values of two constraints are both 1.0, which means the possibility degree level requirements are completely satisfied with the constraints.

**Table 13.1** Characteristics of the 10 test functions [2]

| Functions | Number of design variables | Number of interval variables | Number of constraints |
|---|---|---|---|
| $f_1$ | 13 | 3 | 9 |
| $f_2$ | 10 | 2 | 1 |
| $f_3$ | 5 | 3 | 6 |
| $f_4$ | 2 | 3 | 2 |
| $f_5$ | 10 | 4 | 8 |
| $f_6$ | 2 | 2 | 2 |
| $f_7$ | 7 | 3 | 4 |
| $f_8$ | 2 | 2 | 1 |
| $f_9$ | 5 | 3 | 3 |
| $f_{10}$ | 3 | 3 | 2 |

**Table 13.2** Optimization results of the original deterministic test function $f_1^{'} - f_{10}^{'}$ [2]

| Functions | Optimal objective function value | Optimum of **X** |
|---|---|---|
| $f_1'$ | −15 | (1, 1, 1, 1, 1, 1, 1, 1, 1, 3, 3, 3, 1) |
| $f_2'$ | −1.00 | (0.32, 0.32, 0.32, 0.32, 0.32, 0.32, 0.32, 0.32, 0.32, 0.32) |
| $f_3'$ | −30,665.54 | (78, 33, 30, 45, 78) |
| $f_4'$ | −6961.81 | (14.10, 0.84) |
| $f_5'$ | 24.306 | (2.17, 2.36, 8.77, 5.095, 0.99, 1.43, 1.32, 9.82, 8.28, 8.37) |
| $f_6'$ | −0.095825 | (1.23, 4.24) |
| $f_7'$ | 680.63 | (2.33, 1.95, -0.48, 4.37, -0.62, 1.04, 1.59) |
| $f_8'$ | 0.7499 | (−0.707, 0.50) |
| $f_9'$ | 0.053941 | (−1.72, 1.60, 1.83, −0.766, −0.76) |
| $f_{10}'$ | −2.44 | (3.09, 0.82, -3.00) |

**Table 13.3** Optimization results of the test function $f_{10}$ under different possibility degree levels [2]

| λ | Intervals of objective function | Midpoint of objective function | Radius of objective function | Interval of constraint 1 | Interval of constraint 2 | Possibility degrees of constraints | Optimum of **X** |
|---|---|---|---|---|---|---|---|
| 0.6 | [−4.86, 1.62] | −1.62 | 3.24 | [−0.80, 18.20] | [5.60, 28.29] | 1.0, 1.0 | (3.35, 0.68, −3.00) |
| 0.8 | [2.51, 8.35] | 5.43 | 2.92 | [4.00, 18.50] | [6.88, 42.49] | 1.0, 1.0 | (2.44, 0.88, 3.99) |
| 1.0 | [4.50, 14.90] | 9.70 | 5.20 | [7.00, 26.27] | [15.00, 115.73] | 1.0, 1.0 | (2.63, 0.60, 6.96) |
| 1.2 | [16.77, 51.91] | 34.34 | 17.57 | [19.25, 80.01] | [22.19, 160.57] | 1.0, 1.0 | (6.00, 3.18, 8.00) |

Secondly, other 9 test functions are also analyzed under four cases of λ using the IDE algorithm. In the optimization process, the λ possibility degree levels are the same for all the constraints a test function. The optimization results given in Table 13.4 shows remarkable similarity to the results of $f_{10}$. Moreover, the convergence curves of the objective function midpoint and the objective function radiuses are illustrated in Fig. 13.5. It can be seen that these two parameters both become stable after about 200 iterations, which demonstrates the relatively high convergence speed of the IED algorithm.

**Table 13.4** The optimization results of uncertain optimization functions [2]

| Function | $\lambda$ | Interval of objective function | Midpoint of objective function | Radius of objective function | Optimum of **X** |
|---|---|---|---|---|---|
| $f_1$ | 0.6 | [−15.00, −11.00] | −13 | 2 | (1.0, 1.0, 1.0, 1.0, 1.0, 1.0, 1.0, 1.0, 1.0 1.75, 3.50, 1.75, 1.0) |
| | 0.8 | [−13.81,−9.81] | −11.81 | 2 | (1.0, 1.0, 1.0, 1.0, 1.0, 1.0, 1.0, 1.0, 1.0 1.48, 2.85, 1.48, 1.0) |
| | 1 | [−13.09, −9.09] | −11.09 | 2 | (1.0, 1.0, 1.0, 1.0, 1.0, 1.0, 1.0, 1.0, 1.0 1.32, 2.45, 1.32, 1.0) |
| | 1.2 | [−12.39, −8.39] | −10.39 | 2 | (1.0, 1.0, 1.0, 1.0, 1.0, 1.0, 1.0, 1.0, 1.0 1.16, 2.08, 1.16, 1.0) |
| $f_2$ | 0.6 | [−1.15, −0.77] | −0.96 | 0.19 | (0.32, 0.31, 0.32, 0.31, 0.31, 0.32, 0.31, 0.31, 0.32, 0.31) |
| | 0.8 | [−0.48, −0.32] | −0.4 | 0.08 | (0.29, 0.29, 0.29, 0.29, 0.29, 0.29, 0.29, 0.29, 0.29, 0.29) |
| | 1 | [−0.28, −0.18] | −0.23 | 0.05 | (0.27, 0.27, 0.27, 0.27, 0.27, 0.27, 0.27, 0.27, 0.27, 0.27) |
| | 1.2 | [−0.16, −0.10] | −0.13 | 0.03 | (0.25, 0.26, 0.26, 0.26, 0.25, 0.25, 0.27, 0.27, 0.26, 0.26) |
| $f_3$ | 0.6 | [−31009.82, −29484.18] | −30247 | 762.82 | (78.00, 33.00, 29.68, 45.00, 44.75) |
| | 0.8 | [−29681.74, −27772.66] | −28727.2 | 954.54 | (78.00, 33.00, 35.19, 45.00, 39.65) |
| | 1 | [−28430.71, −26188.29] | −27309.5 | 1121.21 | (78.00, 33.00, 39.37, 45.00, 34.83) |
| | 1.2 | [−26111.42, −23313.18] | −24723.3 | 1410.12 | (78.00, 33.00, 45.00, 29.23, 27.00) |
| $f_4$ | 0.6 | [−7239.67, −5895.45] | −6567.56 | 672.11 | (14.25, 1.20) |
| | 0.8 | [−3597.74, −2897.24] | −3247.49 | 350.25 | (15.03, 5.00) |
| | 1 | [−3600.00, −2900.00] | −3250 | 350 | (15.00, 5.00) |

(continued)

**Table 13.4** (continued)

| Function | λ | Interval of objective function | Midpoint of objective function | Radius of objective function | Optimum of **X** |
|---|---|---|---|---|---|
| | 1.2 | [−3598.82, −2899.18] | −3249 | 349.82 | (14.97, 5.00) |
| $f_5$ | 0.6 | [13.68, 18.26] | 15.97 | 2.29 | (1.37, 2.78, 8.63, 5.07, 0.98, 1.45, 0.78, 9.71, 9.69, 7.09) |
| | 0.8 | [14.89, 20.13] | 17.51 | 2.62 | (2.53, 2.84, 8.44, 5.09, 0.93, 1.36, 0.84, 9.62, 9.42, 7.28) |
| | 1 | [17.96, 23.8] | 20.88 | 2.92 | (1.68, 2.96, 8.32, 5.11, 0.55, 1.41, 0.86, 9.49, 9.24, 7.25) |
| | 1.2 | [36.74, 42.82] | 39.78 | 3.04 | (1.72, 3.16, 7.11, 5.65, 0.24, 2.96, 1.08, 9.40, 9.05, 7.28) |
| $f_6$ | 0.6 | [−0.69, −47] | −0.58 | 0.11 | (0.71, 3.74) |
| | 0.8 | [−0.12, −0.08] | −0.1 | 0.02 | (1.23, 4.24) |
| | 1 | [−0.0128, −0.0072] | −0.01 | 0.0028 | (1.35, 4.05) |
| | 1.2 | [−1.16E-6, −1.16E-6] | −1.16E-06 | 0 | (1.40, 4.00) |
| $f_7$ | 0.6 | [624.82, 724.82] | 683.82 | 59 | (2.38, 1.93, 2.98E-5, 4.29, 5.26E-6, 1.05, 1.57) |
| | 0.8 | [629.34, 747.80] | 688.57 | 59.23 | (2.22, 1.91, 1.61E-5, 4.30, 1.61E-6, 1.05, 1.58) |
| | 1 | [632.23, 751.01] | 691.62 | 59.39 | (2.13, 1.89, 6.06E-6, 4.30, 4.97E-6, 1.04, 1.59) |
| | 1.2 | [635.04, 754.14] | 694.59 | 59.55 | (2.04, 1.88, 6.00E-7, 4.30, 4.49E-7, 1.04, 1.59) |
| $f_8$ | 0.6 | [0.45, 0.55] | 0.5 | 0.05 | (0.50, 0.50) |
| | 0.8 | [0.63, 0.77] | 0.7 | 0.07 | (−0.62, 0.436) |

(continued)

**Table 13.4** (continued)

| Function | λ | Interval of objective function | Midpoint of objective function | Radius of objective function | Optimum of **X** |
|---|---|---|---|---|---|
| | 1 | [0.765, 0.935] | 0.85 | 0.085 | (−0.689, 0.3889) |
| | 1.2 | [0.9, 1.1] | 1 | 0.1 | (−0.75, 0.34) |
| $f_9$ | 0.6 | [9.42E-4, 9.42E-4] | 9.42E-04 | 0 | (−1.58, −1.52, 1.26,−1.52, 1.52) |
| | 0.8 | [3.186 E-3, 3.894 E-3] | 3.54E-03 | 3.54E-04 | (−1.52, −1.48, 1.19, −1.44, 1.47) |
| | 1 | [6.71E-3, 8.20 E-3] | 7.46E-03 | 7.46E-04 | (−1.46, −1.45, 1.14, −1.39, 1.45) |
| | 1.2 | [1.30E-2, 1.60E-2,] | 1.45E-02 | 1.45E-03 | (−1.41, −1.43, 1.10, −1.34, 1.43) |
| $f_{10}$ | 0.6 | [−4.86, 1.62] | −1.62 | 3.24 | (3.35, 0.68, −3.00) |
| | 0.8 | [2.51, 8.35] | 5.43 | 2.92 | (2.44, 0.88, 3.99) |
| | 1 | [4.50, 14.9] | 9.7 | 5.2 | (2.63, 0.60, 6.96) |
| | 1.2 | [16.77, 51.91] | 34.34 | 17.57 | (6.00, 3.18, 8.00) |

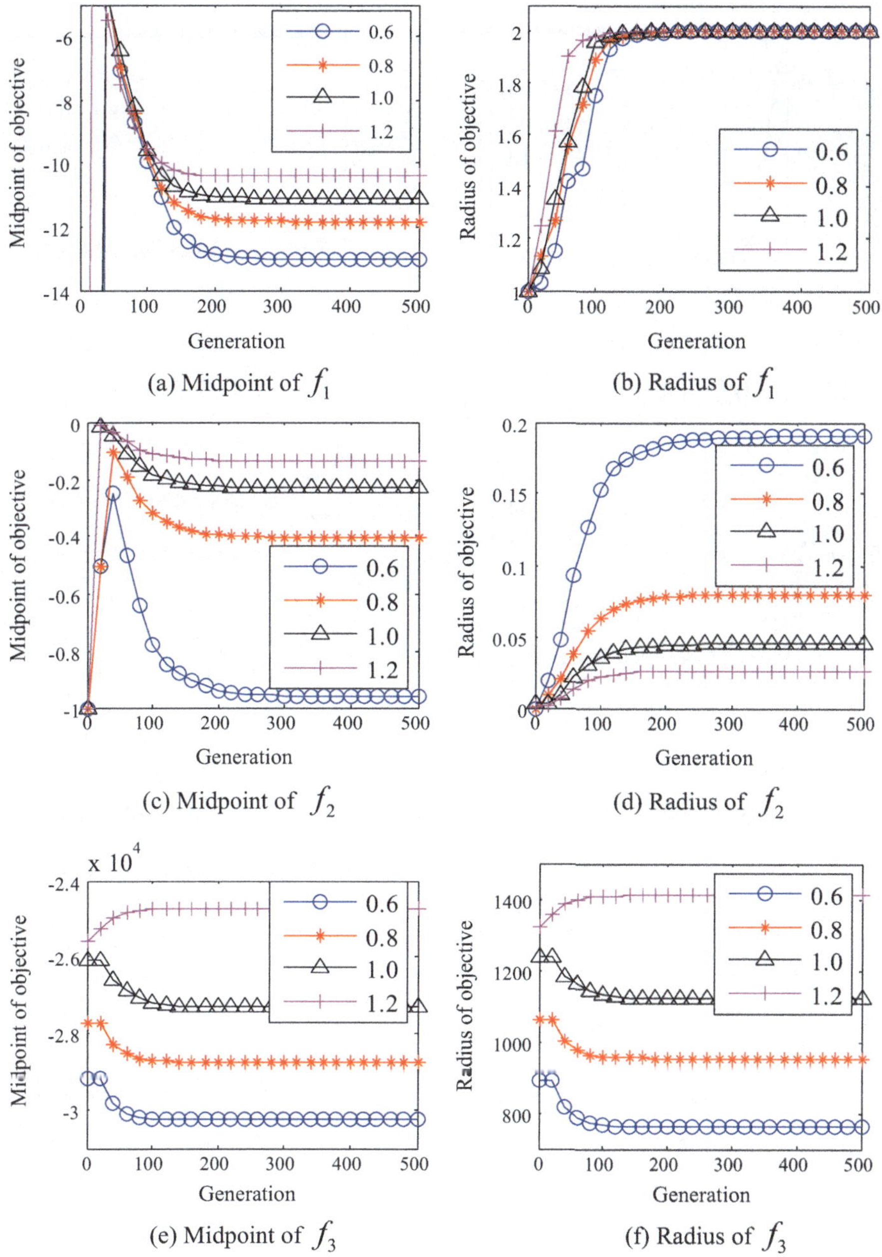


**Fig. 13.5** Convergence curves of the test functions $f_1 - f_{10}$ [2]

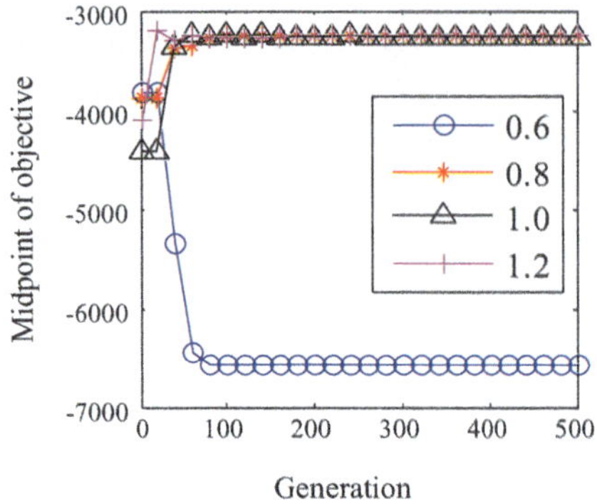


(g) Midpoint of $f_4$

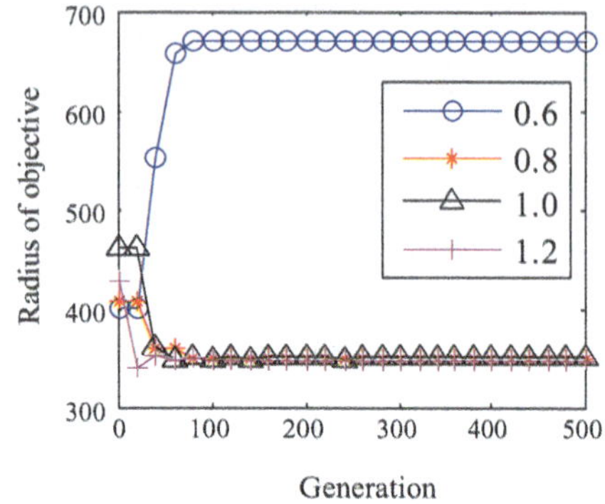


(h) Radius of $f_4$

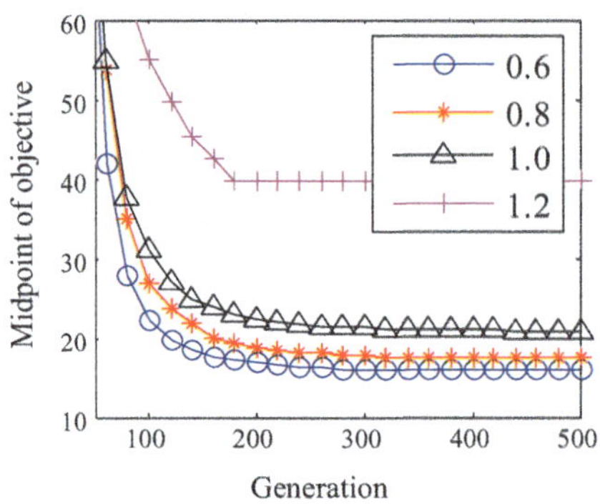


(i) Midpoint of $f_5$

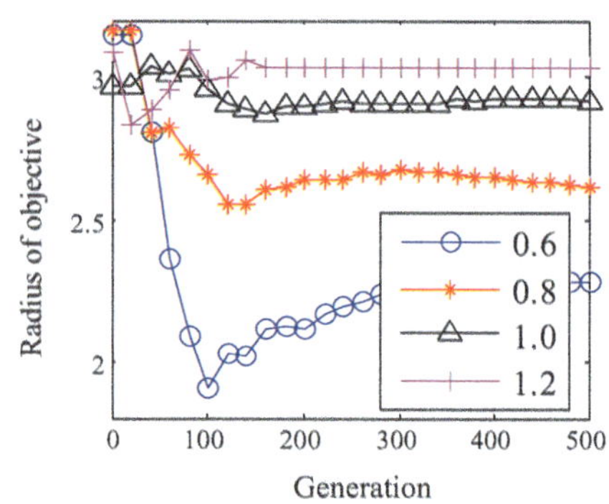


(g) Radius of $f_5$

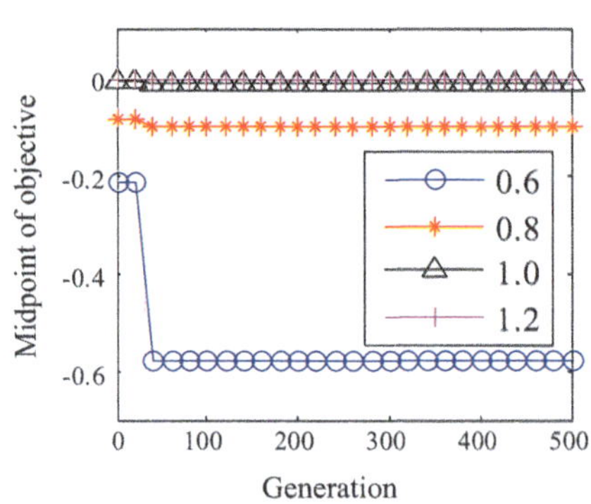


(k) Midpoint of $f_6$

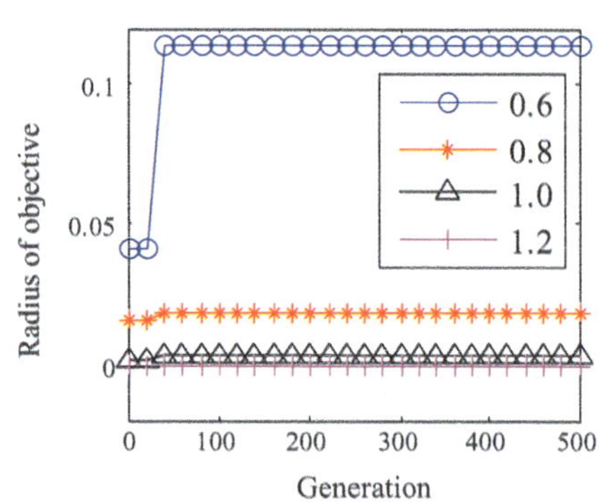


(l) Radius of $f_6$

**Fig. 13.5** (continued)

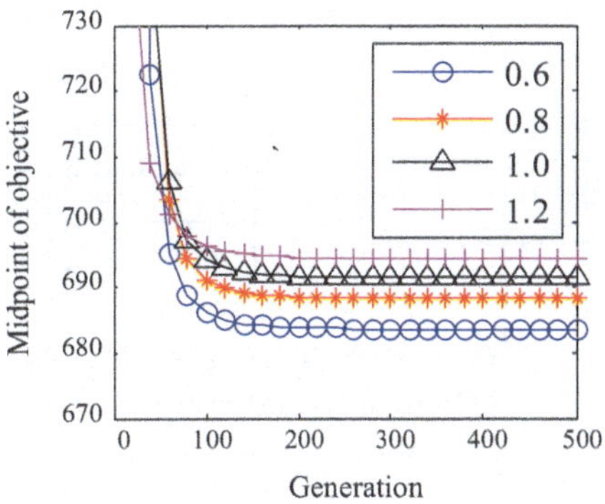


(m) Midpoint of $f_7$

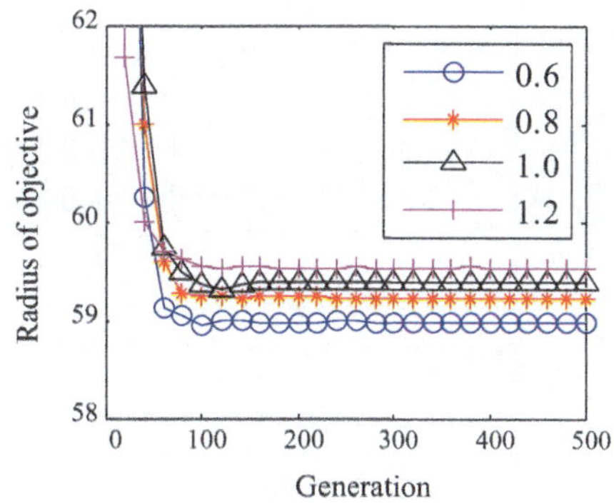


(n) Radius of $f_7$

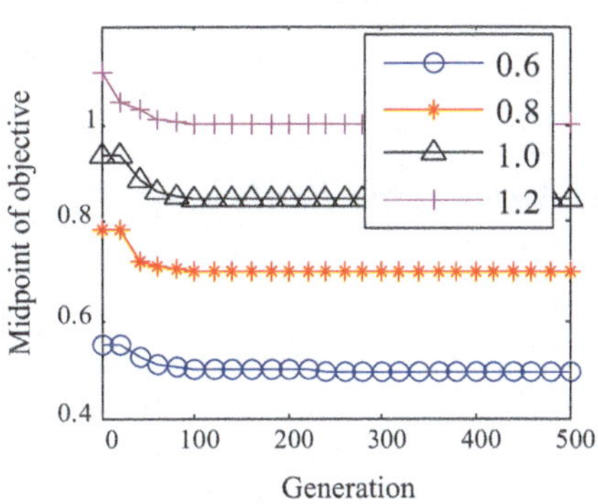


(o) Midpoint of $f_8$

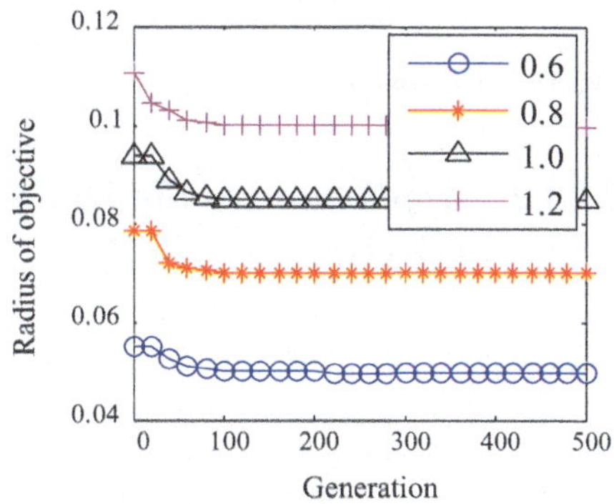


(p) Radius of $f_8$

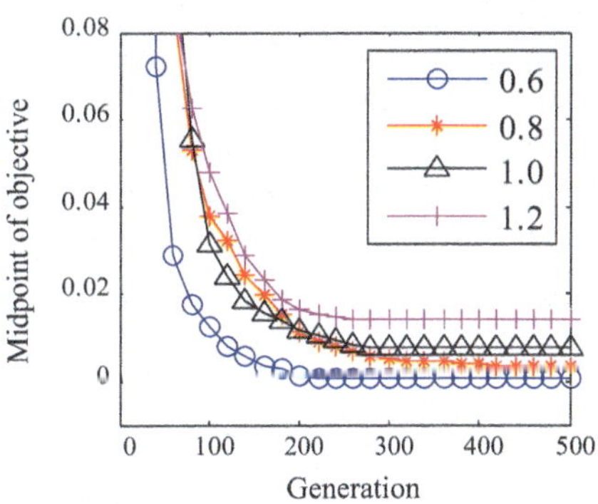


(q) Midpoint of $f_9$

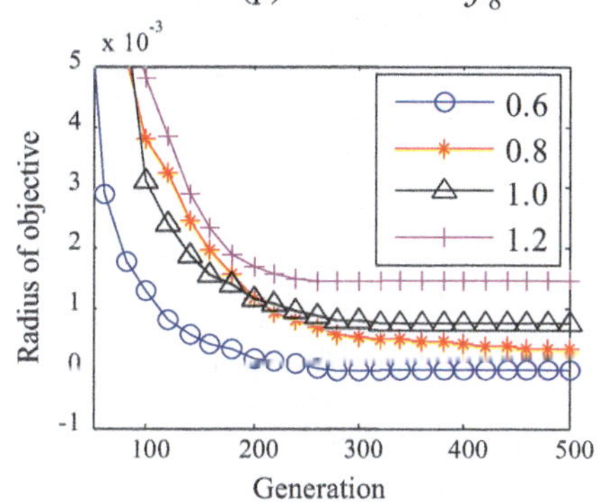


(r) Radius of $f_9$

**Fig. 13.5** (continued)

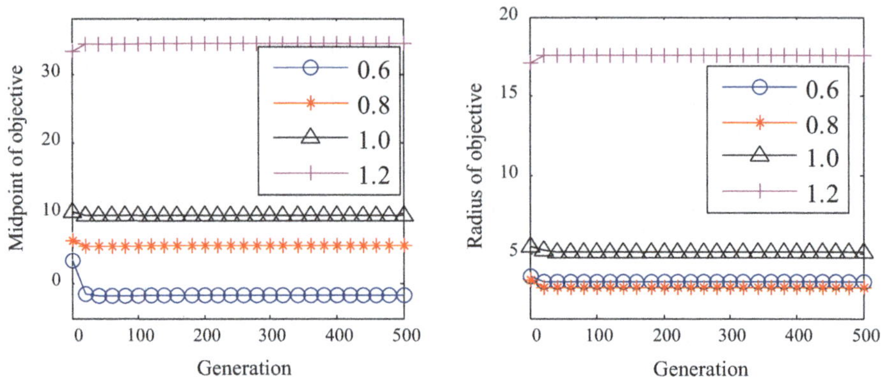


(s) Midpoint of $f_{10}$ (t) Radius of $f_{10}$

**Fig. 13.5** (continued)

## *13.3.2 Application to the Design of Augmented Reality Glasses*

Augmented reality (AR) glasses are high-tech equipment developed in the smart wear field in recent years. It has shown a lot of potentials in several areas such as education, health care, security, and aviation due to its integration of many functions, including computation, communication, location, and photograph. Like other wearable smart devices, many design requirements, especially the comfortability and security, should be taken into account in the design stage of AR glasses.

As illustrated in Fig. 13.6, consider the AR glasses consisting of 5 modules: frame, micro-camera, micro-projector, battery, and controller. Among these modules, the lightweight design and the thermal design have the most crucial influence on comfortability and security of AR glasses. The AR glasses are often used in some uncertain situations, such as varying environmental temperature and power consumption,

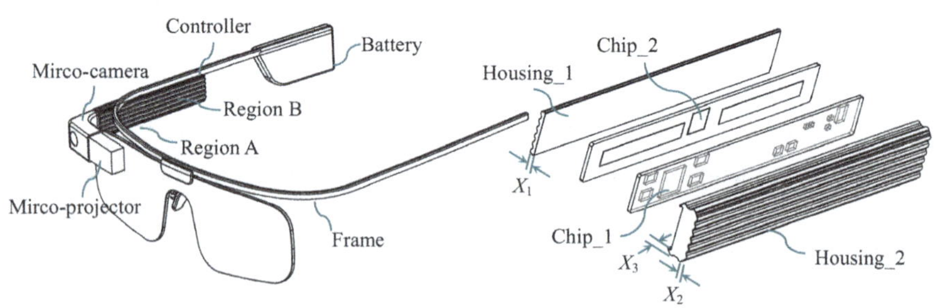


(a) AR glasses (b) Exploded view of controller

**Fig. 13.6** The structure of the AR glasses [18]

leading to a relatively high uncertainty in temperature response of the controller. Considering these uncertain factors, the environment temperature $U_1$, the ambient air velocity $U_2$, the power consumptions $U_3$ and $U_4$ of Chip_1 and Chip_2, $U_4$ are treated as interval parameters. On the other hand, three structural sizes of the controller housing $X_1$, $X_2$, and $X_3$ are selected as design variables. The optimization objective is to minimize the weight of the controller housing that consists of housing_1 and housing_2 with material densities of $\rho_1$ and $\rho_2$, respectively. Three constraints are considered: (1) the surface temperature of region A $T_A$ cannot exceed a given interval $b_1^I = [32\ ^\circ\mathrm{C}, 35\ ^\circ\mathrm{C}]$ for comfortability; (2) the surface temperature of region B $T_B$ cannot exceed a given interval $b_2^I = [35\ ^\circ\mathrm{C}, 42\ ^\circ\mathrm{C}]$ for security; (3) The core temperature of chip_1 $T_C$ cannot exceed a given interval $b_3^I = [55\ ^\circ\mathrm{C}, 65\ ^\circ\mathrm{C}]$ for reliability. Therefore, an interval optimization problem of this engineering problem can be constructed as follows:

$$\begin{cases} \min\limits_{\mathbf{X}} f(\mathbf{X}) = 630\rho_1 X_1 + 630\rho_2 X_2 + 420\rho_1 X_3 + 20\rho_1 X_1 X_3 + 20\rho_1 X_2 X_3 \\ \text{s.t. } g_1(\mathbf{X}, \mathbf{U}) = T_A(\mathbf{X}, \mathbf{U}) \le b_1^I = \left[32\ ^\circ\mathrm{C}, 35\ ^\circ\mathrm{C}\right] \\ \quad g_2(\mathbf{X}, \mathbf{U}) = T_B(\mathbf{X}, \mathbf{U}) \le b_2^I = \left[35\ ^\circ\mathrm{C}, 42\ ^\circ\mathrm{C}\right] \\ \quad g_3(\mathbf{X}, \mathbf{U}) = T_C(\mathbf{X}, \mathbf{U}) \le b_3^I = \left[55\ ^\circ\mathrm{C}, 65\ ^\circ\mathrm{C}\right] \\ \quad \mathbf{X} = (X_1, X_2, X_3), \mathbf{U} = (U_1, U_2, U_3, U_4) \\ \quad \rho_1 = 0.0014 g/\mathrm{mm}^3, \ \ \rho_1 = 0.0027 g/\mathrm{mm}^3 \\ \\ \quad 0.80\mathrm{mm} \le X_i \le 2.40\mathrm{mm}, \ \ i = 1, 2, 3 \end{cases} \tag{13.17}$$

An FEM model of the controller is established and illustrated in Fig. 13.7. The FEM model includes 4 components that are discretized into 22,928 8-node thermal coupling hexahedral elements. The responses of three constraints can be obtained simultaneously through a single FEM analysis. For the parameterization and high efficiency, the quadratic polynomial response surfaces of three constraints are established using 100 FEM simulation samples [2]:

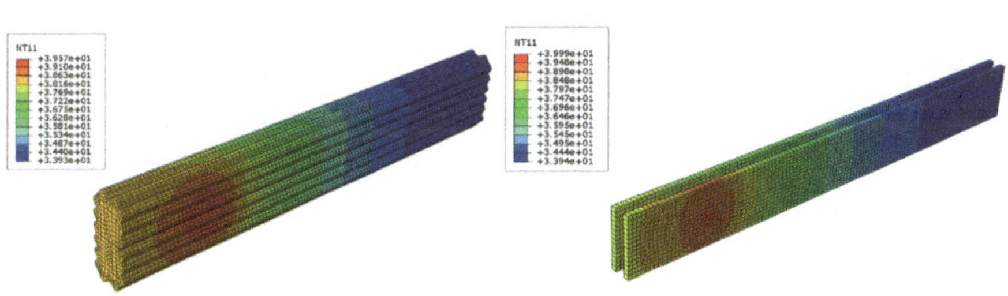

(a) Temperature responses on the outer surface (b) Temperature responses in the inner structure

**Fig. 13.7** FEM simulation of the AR glasses controller [18]

$$\begin{cases} T_A(\mathbf{X},\mathbf{U}) = 100.47 - 124.32X_1 - 14.39X_2 + 10.74X_3 + 1.50X_1X_2 + 10.10X_1X_3 \\ \quad +0.29X_2X_3 + 38.32X_1^2 + 4.28X_2^2 - 8.9X_3^2 + 13.65U_3 - 42.59U_4 + 17.51U_3U_4 \\ \quad +4.85U_3^2 + 145.24U_4^2 + 0.90U_1 - 34.41U_2 - 0.18U_1U_2 + 28.41U_2^2 \\ T_B(\mathbf{X},\mathbf{U}) = 73.27 - 88.5X_1 - 11.92X_2 + 10.13X_3 + 0.54X_1X_2 + 6.67X_1X_3 + 1.15X_2X_3 \\ \quad +27.81X_1^2 + 3.45X_2^2 - 7.51X_3^2 + 0.92U_1 - 34.73U_2 - 0.09U_1U_2 + 32.49U_2^2 \\ \quad +6.60U_3 - 28.69U_4 + 26.31U_3U_4 + 3.63U_3^2 + 81.92U_4^2 \\ T_C(\mathbf{X},\mathbf{U}) = 17.4 - 9.95X_1 - 7.82X_2 + 4.26X_3 - 1.96X_1X_2 + 0.47X_1X_3 + 1.82X_2X_3 \\ \quad +3.94X_1^2 + 2.66X_2^2 - 2.78X_3^2 + 1.02U_1 - 38.18U_2 - 0.06U_1U_2 \\ \quad +37.33U_2^2 + 63.86U_3 - 2.64U_4 + 3.03U_3U_4 - 22.3U_3^2 + 20.36U_4^2 \end{cases} \tag{13.18}$$

The optimization results with a $\lambda$ level of 1.0 for all the constraints are shown in Table 13.5. As seen in the table, the optimal design variables are 1.26, 0.92, and 2.40 mm, and the corresponding weight of the controller housing is only 4.42 g. For this optimal design, the response intervals of the three constraints are $T_A^I = [9.40\ °C, 30.00\ °C]$, $T_B^I = [10.57\ °C, 30.40\ °C]$, and $T_C^I = [25.62\ °C, 54.46\ °C]$, respectively, whose upper bounds are all less than the corresponding lower bounds of the given allowable intervals, 32 °C, 35 °C, and 55 °C. In addition, all possibility degree values of the three constraints are 1.0, which means the given possibility degree level requirements are fully satisfied. The convergence history of the optimization shown in Fig. 13.8 indicates again that the proposed IDE algorithm has a rapid convergence speed for a practical engineering problem.

**Table 13.5** Optimization results of the AR glasses controller $\lambda_i$ [2]

| $\lambda$ | Optimal objective function | Interval of constraint 1 | Interval of constraint 2 | Interval of constraint 3 | Possibility degrees of constraints | Optimum of **X** |
|---|---|---|---|---|---|---|
| 1.0 | 4.42 g | [9.40 °C, 30.00 °C] | [10.57 °C, 30.40 °C] | [25.62 °C, 54.46 °C] | 1.0, 1.0, 1.0 | (1.26 mm, 0.92 mm, 2.40 mm) |

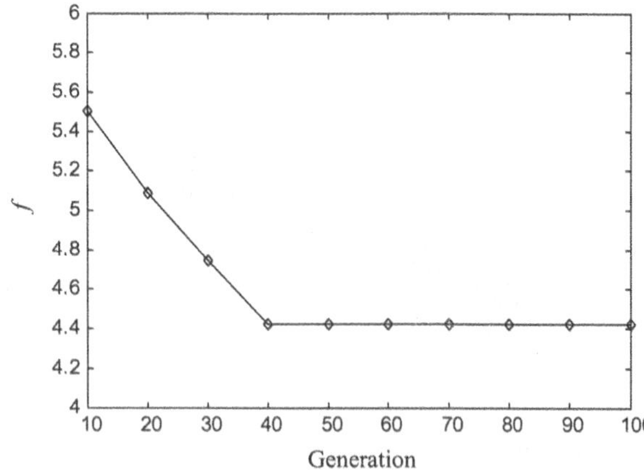


**Fig. 13.8** Convergence curve of the AR glasses controller design [2]

## 13.4 Summary

This chapter develops an IDE algorithm for nonlinear interval optimization. Compared to conventional interval optimization methods, the IDE algorithm directly solves the original interval optimization problem rather than transforming it into a deterministic optimization, which simplifies the solution process. The core of the IDE algorithm is the satisfaction value of interval possibility degree model used to deal with the uncertain constraints. Besides, an interval preferential rule is proposed for constructing the individual selection strategy. In the future, more efficient nested solution methods for IDE are expected to be developed to promote its computational efficiency.

## Appendix: Numerical Examples

(1) Function $f_1$

$$\begin{cases} \min\limits_{\mathbf{X}} f_1(\mathbf{X}, \mathbf{U}) = 5U_1 \sum\limits_{i=1}^{4} X_i - 5\sum\limits_{i=1}^{4} X_i^2 - \sum\limits_{i=5}^{13} X_i \\ \text{s.t. } g_1(\mathbf{X}, \mathbf{U}) = U_3X_1 + 2X_2 + U_1X_{10} + U_2X_{11} \leq [9, 10] \\ \quad g_2(\mathbf{X}, \mathbf{U}) = 2X_1 + 2X_3 + U_1X_{10} + U_2X_{12} \leq [9, 10] \\ \quad g_3(\mathbf{X}, \mathbf{U}) = 2X_2 + U_3X_3 + U_2X_{11} + X_{12} \leq [9, 10] \\ \quad g_4(\mathbf{X}, \mathbf{U}) = -8X_1 + U_2X_{10} \leq [0, 1] \\ \quad g_5(\mathbf{X}, \mathbf{U}) = -8X_2 + U_2X_{11} \leq [0, 1] \\ \quad g_6(\mathbf{X}, \mathbf{U}) = -8X_3 + U_1X_{12} \leq [0, 1] \\ \quad g_7(\mathbf{X}, \mathbf{U}) = -2X_4 - U_1X_5 + U_3X_{10} \leq [0, 1] \\ \quad g_8(\mathbf{X}, \mathbf{U}) = -U_3X_6 - U_1X_7 + U_2X_{11} \leq [0, 1] \\ \quad g_9(\mathbf{X}, \mathbf{U}) = -2X_8 - U_1X_9 + U_3X_{12} \leq [0, 1] \\ \quad 0 \leq X_i \leq 1, \ \ i = 1, 2, \ldots, 9, \ \ 0 \leq X_{10}, X_{11}, X_{12} \leq 100, \ \ 0 \leq X_{13} \leq 1 \\ \quad U_1, \ U_2 \in [0.9, 1.1], \ \ U_3 \in [1.8, 2.2] \end{cases} \tag{13.19}$$

(2) Function $f_2$

$$\begin{cases} \min\limits_{\mathbf{X}} f_2(\mathbf{X}, \mathbf{U}) = -U_1U_2\left(\sqrt{n}\right)^n \prod\limits_{i=1}^{n} X_i \\ \text{s.t. } g_1(\mathbf{X}, \mathbf{U}) = U_1U_2 \sum\limits_{i=1}^{n} X_i^2 \leq [0.9, 1.1] \\ \quad 0 \leq X_i \leq 1, \ \ i = 1, 2, \ldots, n \\ \quad U_1, U_2 \in [0.9, 1.1] \end{cases} \tag{13.20}$$

(3) Function $f_3$

$$\begin{cases} \min\limits_{\mathbf{X}} f_3(\mathbf{X}, \mathbf{U}) = 5.3578547U_1X_3^2 + 0.8356891X_1X_5 + 37.293239U_2X_1 - 40792.141 \\ \text{s.t. } g_1(\mathbf{X}, \mathbf{U}) = 85.334407 + 0.0056858U_1X_2X_5 + 0.0006262X_1X_4 - 0.0022053U_2X_3X_5 \le [92, 94] \\ \quad g_2(\mathbf{X}, \mathbf{U}) = -85.334407 - 0.0056858U_1X_2X_5 - 0.0006262U_2X_1X_4 + 0.0022053X_3X_5 \le [0, 1] \\ \quad g_3(\mathbf{X}, \mathbf{U}) = -80.51249 + 0.0071317U_1X_2X_5 + 0.0029955U_2X_1X_2 + 0.0021813X_3^2 \le [105, 120] \\ \quad g_4(\mathbf{X}, \mathbf{U}) = -80.51249 - 0.0071317U_1X_2X_5 - 0.0029955X_1X_2 - 0.0021813U_2X_3^2 \le [-95, -85] \\ \quad g_5(\mathbf{X}, \mathbf{U}) = 9.300961 + 0.0047026U_1X_3X_5 + 0.0012547X_1X_3 + 0.0019085U_2X_3X_4 \le [24, 26] \\ \quad g_6(\mathbf{X}, \mathbf{U}) = -9.300961 - 0.0047026U_1X_3X_5 - 0.0012547X_1X_3 - 0.0019085U_3X_3X_4 \le [-22, -20] \\ \quad 78 \le X_1 \le 102, \;\; 33 \le X_2 \le 45, \;\; 27 \le X_3, X_4, X_5 \le 45, \\ \quad U_1, \; U_2, U_3 \in [0.9, 1.1] \end{cases} \tag{13.21}$$

(4) Function $f_4$

$$\begin{cases} \min\limits_{\mathbf{X}} f_4(\mathbf{X}, \mathbf{U}) = U_1(X_1 - 10)^3 + U_2(X_1 - 20)^3 \\ \text{s.t. } g_1(\mathbf{X}, \mathbf{U}) = -U_1(X_1 - 5)^2 - U_2(X_2 - 5)^2 \le [-105, -95] \\ \quad g_2(\mathbf{X}, \mathbf{U}) = U_1(X_1 - 6)^2 + U_2(X_2 - 5)^2 \le [81, 84] \\ \quad 13 \le X_1 \le 100, \;\; 0 \le X_2 \le 100 \\ \quad U_1, \; U_2 \in [0.9, 1.1] \end{cases} \tag{13.22}$$

(5) Function $f_5$

$$\begin{cases} \min\limits_{\mathbf{X}} f_5(\mathbf{X}, \mathbf{U}) = X_1^2 + X_2^2 + U_1X_1X_2 - 14U_2X_1 - 16X_2 + (X_3 - 10)^2 + 4(X_4 - 5)^2 + (X_5 - 3)^2 \\ \qquad +2(X_6 - 1)^2 + 5X_7^2 + 7(X_8 - 11)^2 + U_3(X_9 - 10)^2 + (X_{10} - 7)^2 + 45 \\ \text{s.t. } g_1(\mathbf{X}, \mathbf{U}) = -105 + 4U_1X_1 + 5X_2 - 3X_7 + 9X_8 \le [0, 1] \\ \quad g_2(\mathbf{X}, \mathbf{U}) = 10X_1 - U_4X_2 - 17X_7 + U_3X_8 \le [0, 1] \\ \quad g_3(\mathbf{X}, \mathbf{U}) = -U_4X_1 + 2X_2 - 17X_7 + U_3X_{10} \le [12, 13] \\ \quad g_4(\mathbf{X}, \mathbf{U}) = 3(X_1 - 2)^2 + 4(X_2 - 3)^2 + U_3X_3^2 - 7X_4 \le [110, 120] \\ \quad g_5(\mathbf{X}, \mathbf{U}) = 5X_1^2 + U_4X_2 + (X_3 - 6)^2 - U_3X_4 \le [35, 45] \\ \quad g_6(\mathbf{X}, \mathbf{U}) = X_1^2 + 2(X_2 - 2)^2 - U_3X_1X_2 + 14X_5 - 6X_6 \le [0, 1] \\ \quad g_7(\mathbf{X}, \mathbf{U}) = 0.5(X_1 - U_4)^2 + U_3(X_4 - 4)^2 + 3X_5^2 - X_6 \le [25, 35] \\ \quad g_8(\mathbf{X}, \mathbf{U}) = -3X_1 + 6U_2X_2 + 12(X_9 - U_4)^2 - 7X_{10} \le [0, 1] \\ \quad -10 \le X_i \le 10, \;\; i = 1, 2, \ldots, 10 \\ \quad U_1, \; U_2 \in [0.9, 1.1], U_3 \in [1.9, 2.1], U_4 \in [7.5, 8.5] \end{cases} \tag{13.23}$$

(6) Function $f_6$

$$\begin{cases} \min\limits_{\mathbf{X}} f_6(\mathbf{X}, \mathbf{U}) = -\frac{U_1U_2\sin^3(2\pi X_1)\sin(2\pi X_2)}{X_1^3(X_1+X_2)} \\ \text{s.t. } g_1(\mathbf{X}, \mathbf{U}) = U_1X_1^2 - U_2X_2 \le [-1, 0] \\ \quad g_2(\mathbf{X}, \mathbf{U}) = -U_1X_1 + (U_2X_2 - 4)^2 \le [-1, 0] \\ \quad 0 \le X_1, \;\; X_2 \le 10 \\ \quad U_1, \; U_2 \in [0.9, 1.1] \end{cases} \tag{13.24}$$

(7) Function $f_7$

$$\begin{cases} \min\limits_{\mathbf{X}} f_7(\mathbf{X},\mathbf{U}) = (X_1-10)^2 + 5U_1(X_2-12)^2 + X_3^4 + U_3(X_4-11)^2 + 10X_5^6 \\ \qquad\qquad +7X_6^2 + X_7^4 - 4U_1X_6X_7 - 10X_6 - 8X_7 \\ \text{s.t. } g_1(\mathbf{X},\mathbf{U}) = 2U_1X_1^2 + U_3X_2^4 + X_3 + 4X_4^2 + 5X_5 \le [126, 127] \\ \quad g_2(\mathbf{X},\mathbf{U}) = 7X_1 + U_3X_2 + 10U_1X_3^2 + X_4 - X_5 \le [270, 282] \\ \quad g_3(\mathbf{X},\mathbf{U}) = 23X_1 + X_2^2 + 6U_2X_6^2 - 8X_7 \le [195, 196] \\ \quad g_4(\mathbf{X},\mathbf{U}) = 4U_2X_1^2 + X_2^2 - U_3X_1X_2 + 2X_3^2 + 5X_6 - 11X_7 \le [-1, 0] \\ \quad -10 \le X_i \le 10,\ i = 1, 2, \ldots, 7 \\ \quad U_1,\ U_2 \in [0.9, 1.1], U_3 \in [2.8, 3.2] \end{cases} \tag{13.25}$$

(8) Function $f_8$

$$\begin{cases} \min\limits_{\mathbf{X}} f_8(\mathbf{X},\mathbf{U}) = U_1X_1^2 + U_2(X_2-1)^2 \\ \text{s.t. } g_1(\mathbf{X},\mathbf{U}) = U_1X_2 - U_2X_1^2 \le [0, 0.5] \\ \quad -1 \le X_1,\ X_2 \le 1 \\ \quad U_1,\ U_2 \in [0.9, 1.1] \end{cases} \tag{13.26}$$

(9) Function $f_9$

$$\begin{cases} \min\limits_{\mathbf{X}} f_9(\mathbf{X},\mathbf{U}) = U_1 e^{X_1X_2X_3X_4X_5} \\ \text{s.t. } g_1(\mathbf{X},\mathbf{U}) = U_1X_1^2 + X_2^2 + X_3^2 + U_2X_4^2 + X_5^2 \le [10, 12] \\ \quad g_2(\mathbf{X},\mathbf{U}) = U_1X_2X_3 - 5U_2X_4X_5 \le [0, 0.5] \\ \quad g_3(\mathbf{X},\mathbf{U}) = U_3X_1^3 + X_2^3 \le [-1, 0] \\ \quad -2.3 \le X_1, X_2 \le 2.3, \quad -3.2 \le X_3, X_4, X_5 \le 3.2 \\ \quad U_1,\ U_2, U_3 \in [0.9, 1.1] \end{cases} \tag{13.27}$$

(10) Function $f_{10}$

$$\begin{cases} \min\limits_{\mathbf{X}} f_{10}(\mathbf{X},\mathbf{U}) = U_1(X_1-2)^2 + U_2(X_2-1)^2 + U_3X_3 \\ \text{s.t. } g_1(\mathbf{X},\mathbf{U}) = U_1X_1^2 - U_2^2X_2 + U_3X_3 \ge [6.5, 7] \\ \quad g_2(\mathbf{X},\mathbf{U}) = U_1X_1 - U_2X_2 + U_3^2X_3^2 + 1.0 \ge [10, 15] \\ \quad -2 \le X_1 \le 6,\ -4 \le X_2 \le 7, -3 \le X_2 \le 8 \\ \quad U_1 \in [0.6, 1.8], U_2 \in [0.5, 1.5], U_3 \in [0.6, 2] \end{cases} \tag{13.28}$$

## References

1. Cheng J, Liu Z, Wu ZY, Tang MY, Tan JR (2016) Direct optimization of uncertain structures based on degree of interval constraint violation. Comput Struct 164:83–94
2. Fu CM, Jiang C, Tang JC (2017) A novel interval differential evolution algorithm for uncertain optimization problems. IEEE Trans Evol Comput, submitted
3. Storn R, Price K (1997) Differential evolution–a simple and efficient heuristic for global optimization over continuous spaces. J Global Optim 11(4):341–359
4. Das S, Suganthan PN (2011) Differential evolution: a survey of the state-of-the-art. IEEE Trans Evol Comput 15(1):4–31
5. Das S, Mullick SS, Suganthan PN (2016) Recent advances in differential evolution–an updated survey. Swarm Evol Comput 27:1–30
6. Vasile M, Minisci E, Locatelli M (2011) An inflationary differential evolution algorithm for space trajectory optimization. IEEE Trans Evol Comput 15(2):267–281
7. Zhong JH, Shen M, Zhang J, Chung HS-H, Shi Y-H, Li Y (2013) A differential evolution algorithm with dual populations for solving periodic railway timetable scheduling problem. IEEE Trans Evol Comput 17(4):512–527
8. Roque CMC, Martins PALS, Ferreira AJM, Jorge RMN (2016) Differential evolution for free vibration optimization of functionally graded nano beams. Compos Struct 156:29–34
9. Ho-Huu V, Nguyen-Thoi T, Vo-Duy T, Nguyen-Trang T (2016) An adaptive elitist differential evolution for optimization of truss structures with discrete design variables. Comput Struct 165:59–75
10. Wang Y, Xu B, Sun G, Yang S (2017) A two-phase differential evolution for uniform designs in constrained experimental domains. IEEE Trans Evol Comput 21(5):665–680
11. Qiu X, Tan KC, Xu JX (2017) Multiple Exponential Recombination for Differential Evolution. IEEE Trans Cybern 47(4):995–1006
12. Wang Y, Cai Z, Zhang Q (2012) Enhancing the search ability of differential evolution through orthogonal crossover. Inf Sci 185(1):153–177
13. Liu J, Teo KL, Wang X, Wu C (2016) An exact penalty function-based differential search algorithm for constrained global optimization. Soft Comput 20(4):1305–1313
14. Tessema B, Yen GG (2009) An adaptive penalty formulation for constrained evolutionary optimization. IEEE Trans Syst, Man, Cybern-Part A: Syst Hum 39(3):565–578
15. Nocedal J, Wright SJ (1999) Numerical optimization. Springer, New York
16. Brest J, Greiner S, Boskovic B, Mernik M, Zumer V (2006) Self-adapting control parameters in differential evolution: a comparative study on numerical benchmark problems. IEEE Trans Evol Comput 10(6):646–657
17. Liang J, Runarsson TP, Mezura-Montes E, Clerc M, Suganthan PN, Coello CC, Deb K (2006) Problem definitions and evaluation criteria for the CEC 2006 special session on constrained real-parameter optimization. J Appl Mech 41(8)
18. Huang ZL, Jiang C, Zhang Z, Fang T, Han X (2017) A decoupling approach for evidence-theory-based reliability design optimization. Struct Multidiscip Optim 56(3):647–661

The manufacturer's authorised representative in the EU is Springer Nature Customer Service Centre GmbH, Europaplatz 3, 69115 Heidelberg, Germany. If you have any concerns regarding our products, please contact ProductSafety@springernature.com

Printed and bound by CPI Group (UK) Ltd, Croydon, CR0 4YY
30/09/2026
02225289-0006